VOLUME FIFTY SEVEN

Advances in
ECOLOGICAL RESEARCH

Networks of Invasion: Empirical
Evidence and Case Studies

ADVANCES IN ECOLOGICAL RESEARCH

Series Editors

DAVID A. BOHAN
Directeur de Recherche
UMR 1347 Agroécologie
AgroSup/UB/INRA
Pôle GESTAD, Dijon, France

ALEX J. DUMBRELL
School of Biological Sciences
University of Essex
Wivenhoe Park, Colchester
Essex, United Kingdom

VOLUME FIFTY SEVEN

ADVANCES IN

ECOLOGICAL RESEARCH

Networks of Invasion: Empirical
Evidence and Case Studies

Edited by

DAVID A. BOHAN
Directeur de Recherche
UMR 1347 Agroécologie
AgroSup/UB/INRA
Pôle GESTAD, Dijon, France

ALEX J. DUMBRELL
School of Biological Sciences
University of Essex
Wivenhoe Park, Colchester, Essex,
United Kingdom

FRANÇOIS MASSOL
UMR 8198 Evo-Eco-Paleo
CNRS, Université de Lille
Lille, France

ACADEMIC PRESS
An imprint of Elsevier

Academic Press is an imprint of Elsevier
The Boulevard, Langford Lane, Kidlington, Oxford OX5 1GB, United Kingdom
125 London Wall, London EC2Y 5AS, United Kingdom
50 Hampshire Street, 5th Floor, Cambridge, MA 02139, United States
525 B Street, Suite 1800, San Diego, CA 92101-4495, United States

First edition 2017

Notices
Knowledge and best practice in this field are constantly changing. As new research and experience broaden our understanding, changes in research methods, professional practices, or medical treatment may become necessary.

Practitioners and researchers must always rely on their own experience and knowledge in evaluating and using any information, methods, compounds, or experiments described herein. In using such information or methods they should be mindful of their own safety and the safety of others, including parties for whom they have a professional responsibility.

To the fullest extent of the law, neither the Publisher nor the authors, contributors, or editors, assume any liability for any injury and/or damage to persons or property as a matter of products liability, negligence or otherwise, or from any use or operation of any methods, products, instructions, or ideas contained in the material herein.

ISBN: 978-0-12-813328-6
ISSN: 0065-2504

For information on all Academic Press publications
visit our website at https://www.elsevier.com/books-and-journals

Working together
to grow libraries in
developing countries

www.elsevier.com • www.bookaid.org

Publisher: Zoe Kruze
Acquisition Editor: Alex White
Editorial Project Manager: Helene Kabes
Production Project Manager: Magesh Kumar Mahalingam
Cover Designer: Vicky Pearson Esser

Typeset by SPi Global, India

CONTENTS

CONTRIBUTORS

J.L. Abbate
Laboratoire MIVEGEC (UMR CNRS 5290, UR IRD 224, UM), Montpellier; UMI UMMISCO (UMI 209 IRD, UPMC), Bondy, France

M.E. Alexander
Institute for Biomedical and Environmental Health Research, School of Science and Sport, University of the West of Scotland, Paisley, United Kingdom

S. Alizon
Laboratoire MIVEGEC (UMR CNRS 5290, UR IRD 224, UM), Montpellier, France

E. Allen-Vercoe
University of Guelph, Guelph, Canada

L. Amsellem
Unité "Evolution, Ecologie, Paléontologie", UMR CNRS 8198, Université de Lille, Villeneuve d'Ascq, France

D. Andreou
Faculty of Science and Technology, Bournemouth University, Poole, United Kingdom

M. Baude
Laboratoire Biologie des Ligneux et des Grandes Cultures (EA 1207), Equipe entomologie et biologie intégrée, Université d'Orléans, Orléans, France

J.R. Britton
Faculty of Science and Technology, Bournemouth University, Poole, United Kingdom

C. Brouat
IRD, CBGP (UMR INRA/IRD/CIRAD/Montpellier SupAgro), Campus International de Baillarguet, Montferrier-sur-Lez, France

I. Dajoz
Institut d'écologie et des sciences de l'environnement de Paris (iEES-Paris UMR CNRS 7618), Equipe Ecologie et évolution des réseaux d'interactions, Université Paris-Diderot–CNRS—UPMC, Paris, France

J.T.A. Dick
Institute for Global Food Security, School of Biological Sciences, Queen's University Belfast, MBC, Belfast, United Kingdom

O. Duron
Laboratoire MIVEGEC, CNRS (UMR 5290), Université de Montpellier, IRD (UR 224), Montpellier, France

B. Facon
INRA, CBGP (UMR INRA/IRD/CIRAD/Montpellier SupAgro), Campus International de Baillarguet, Montferrier-sur-Lez Cedex; INRA, UMR PVBMT, Saint-Pierre, France

C. Firmat
INRA, UMR 1202, BIOGECO, Cestas, France

C. Fontaine
Centre d'Ecologie et des Sciences de la Conservation (CESCO UMR CNRS 7204), Equipe
Socio-écosystèmes, CNRS-Muséum national d'histoire naturelle, Paris, France

R. Froissart
Laboratoire MIVEGEC (UMR CNRS 5290, UR IRD 224, UM), Montpellier, France

B. Gauzens
German Centre for Integrative Biodiversity Research (iDiv) Halle-Jena-Leipzig, Leipzig,
Germany

B. Geslin
Institut Méditerranéen de Biodiversité et d'Ecologie marine et continentale (IMBE-UMR-
CNRS-IRD 7263), Equipe Ecologie de la Conservation et Interactions Biotiques, Aix
Marseille Univ, Univ Avignon, CNRS, IRD, IMBE, Marseille, France

J. Grey
Lancaster Environment Centre, Lancaster University, Lancaster, United Kingdom

M. Henry
INRA, UR406 Abeilles et Environnement, Avignon, France

M.C. Jackson
Centre for Invasion Biology, University of Pretoria, Hatfield, South Africa; Life Sciences,
Imperial College London, Ascot, Berkshire, United Kingdom

K. McCann
University of Guelph, Guelph, Canada

V. Médoc
Sorbonne Universités, UPMC Univ Paris 06, CNRS UMR 7618, Institute of Ecology and
Environmental Sciences—Paris, Paris, France

C.L. Murall
Laboratoire MIVEGEC (UMR CNRS 5290, UR IRD 224, UM), Montpellier, France

J. Pegg
Faculty of Science and Technology, Bournemouth University, Poole, United Kingdom

S.S. Porter
School of Biological Sciences, Washington State University, Vancouver, WA, United States

A. Ricciardi
Redpath Museum, McGill University, Montreal, QC, Canada

O. Rollin
ITSAP-Institut de l'Abeille; UMT PrADE, Avignon, France

L. Ropars
Institut Méditerranéen de Biodiversité et d'Ecologie marine et continentale (IMBE-UMR-CNRS-IRD 7263), Equipe Ecologie de la Conservation et Interactions Biotiques, Aix Marseille Univ, Univ Avignon, CNRS, IRD, IMBE, Marseille; Institut d'écologie et des sciences de l'environnement de Paris (iEES-Paris UMR CNRS 7618), Equipe Ecologie et évolution des réseaux d'interactions, Université Paris-Diderot–CNRS—UPMC, Paris, France

D.J. Sheath
Faculty of Science and Technology, Bournemouth University, Poole, United Kingdom

E. Thébault
Institut d'écologie et des sciences de l'environnement de Paris (iEES-Paris UMR CNRS 7618), Equipe Ecologie et évolution des réseaux d'interactions, Université Paris-Diderot–CNRS—UPMC, Paris, France

M. Puelma Touzel
Laboratoire de physique théorique (UMR 8549) CNRS and École Normale Supérieure, Paris, France

N.J. Vereecken
Agroecology & Pollination Group, Landscape Ecology & Plant Production Systems (LEPPS/EIB), Boulevard du Triomphe CP 264/2, Université Libre de Bruxelles (ULB), Brussels, Belgium

A. Vilcinskas
Fraunhofer Institute for Molecular Biology and Applied Ecology; Institute for Insect Biotechnology, Justus-Liebig-University of Giessen, Giessen, Germany

R.J. Wasserman
South African Institute for Aquatic Biodiversity, Grahamstown, South Africa; School of Science, Monash University, Jalan Lagoon Selatan, Selangor, Malaysia

PREFACE

Biological invasions are considered one of the preeminent sources of disturbance of ecosystems, causing marked loss of biodiversity at large spatial scales and concomitant changes in function and ecosystem service provision (Ehrenfeld, 2010; Murphy and Romanuk, 2014; Vilà et al., 2006). The past half century has seen increased concern about the impacts of biological invasions (Gurevitch and Padilla, 2004), from local species extinction to changes in ecosystem functioning, as well as effects on genetic diversity or facilitation of further invasions through invasion meltdown (Murphy and Romanuk, 2014; Paolucci et al., 2013; Simberloff et al., 2013; Strayer, 2012; Vilà et al., 2011). In order to prevent invasions, much effort has also been devoted towards understanding conditions favouring the success of invasions (Beisner et al., 2006; Pyšek et al., 2010; Romanuk et al., 2009). In spite of numerous empirical and theoretical developments, understanding, predicting and managing species invasions remain one of the main issues of ecology today, with potentially massive costs now and in the years to come (Bradshaw et al., 2016).

Invasion biology, born as a new discipline more than half a century ago (Elton, 1958), is only now coming of age. Despite an accumulation of detailed case studies, a growing theoretical corpus and connections made to many areas of ecology, evolutionary biology and economy, there remain some key gaps in the study of invasions. One is the need for a more integrative, systems-based view of an invasive species encountering an established, recipient community, with an emphasis placed on the complexity of reciprocal relationships, including the realization that the latter is itself a complex entity. The research that results will likely be neither a simple, linear addition of a species to a community nor the summation of all pairwise interactions between the constituent species.

Until recently, relatively little investigation of species invasions has been done using systems- or network-based approaches (but see Romanuk et al., 2009), despite growing evidence that consideration of the invader alone, or the species it appears to displace, is not enough. In biocontrol, whereby species have been deliberately introduced to counteract the effects of particular pest species, there is ample evidence that the insertion of a novel species into an established community can markedly modify community structure and functioning in ways that are often unexpected (Lombaert et al., 2010; Roy et al., 2011). The results of such introductions range from mere persistence of the biocontrol agent, to successful introduction and control effect, through to wholesale switching to other unintended prey species and the loss of these from the system instead (Clarke et al., 1984; Phillips et al., 2006). These findings in biocontrol, when placed alongside others from 'natural' invasion events, argue for the effects of invading species ramifying across whole ecosystems and that systems- or network-based approaches would have great power to resolve mechanisms and understand invasion.

'Networks of invasion', understood both through biological invasions into ecological interaction networks and through the web-like nature of invasion patterns, spatially and in terms of sequential invasions, are the subject of two thematic issues of *Advances in Ecological Research*. These deal, in turn, with papers synthesizing concepts in invasion ecology and the empirical evidence and case studies of invasions viewed with a network perspective. This first volume opened with a set of papers, both empirical and theoretical, which set out the general Perspectives in Invasion Ecology, when viewed from the network standpoint. These papers first synthesized known impacts of invaders on food webs, with both a general review focusing on the 'insertion point' of invaders within food webs (David et al., Chapter 1) and a meta-analysis (Mollot et al., Chapter 2) which detailed how often and under which conditions species invasions lead to species losses. Chapter 3 (Kamenova et al.) introduced current tools—genetic, isotopic, participatory, model based, etc.—available to monitor, detect and predict invasions in ecological networks. The following two chapters (Massol et al., Chapter 4; Romanuk et al., Chapter 5) presented two different models aimed at elucidating the effects of food web topology on invasion probability and impact. While Chapter 4 developed a spatial food web model based on MacArthur and Wilson's (1963) theory of island biogeography to understand the effects of food web properties on sequences of invasions, Chapter 5 focuses on what makes a food web robust to species invasions, elaborating on the niche model of Williams and Martinez (2000) to explore whether quantitative and qualitative robustness are traded off along the food web connectance gradient. The first volume closed with a general perspective paper (Pantel et al., Chapter 6) that focuses on 14 different questions for future studies on invasions in ecological networks.

In this second volume, the series of papers details invasions of particular types of invaders (e.g. parasites) or into particular types of ecosystems (e.g. freshwater ecosystems). It begins with Médoc et al., which synthesizes predictions of impacts of parasite invasions in food webs. The following chapter (Jackson et al., Chapter 2) focuses on shifts in freshwater food web topologies due to species invasions and emphasizes the possibility of trophic link disruption in such situations. Chapter 3 (Amsellem et al.) then synthesizes current evidence and hypotheses regarding the effects of microorganisms on species invasions, from the classic enemy release hypothesis to spillback and spillover effects linked to the coinvasion of organisms with their symbionts. Plant–pollinator interactions and the topology of the associated networks under a regime of massive species introductions are the focus of Chapter 4 (Geslin et al.), which detail the potential impacts of such massive management of both plants and insects on nearby, nonmanaged ecosystems. The volume then ends with Chapter 5, by Murall et al., with a change of scale to describe invasions in microbial networks within hosts, emphasizing what is already known regarding mammal microbiota and suggesting ways in which interdisciplinary dialogue between ecologists, microbiologists and physiologists might pave the way for a better understanding of problems such as the emergence of antibiotic resistance.

These two volumes present a snapshot of the current state of the art in the study of invasion ecology, emphasizing species interactions and using network-based approaches. The 11 chapters reveal that there is great value in using networks to study invasion, both from a conceptual viewpoint and in practical terms to measure, predict and monitor the impacts of invasions on ecosystems. When taken together, the work presented suggests that building a richer understanding of invasion, which networks afford, could be an important step forward in developing predictive approaches to managing or preventing invasions. We hope that the variety of approaches and questions to networks of invasion contained within these pages provides a fruitful and stimulating framework for future studies and research programmes aimed at tackling invasions.

F. MASSOL

P. DAVID

D.A. BOHAN

ACKNOWLEDGEMENTS

This series of papers came, in part, out of the COREIDS project, cofinanced by TOTAL and the Fondation pour la Recherche sur la Biodiversité (FRB), at the Centre for the Synthesis and Analysis of Biodiversity (CESAB) in Aix-en-Provence, France. The COREIDS project has the goals of unifying and analyzing databases on species invasions into sensitive communities and zones using network ecological approaches, with the aim of identifying the generic processes and mechanisms of invasion that might be used in invasion prediction and management.

REFERENCES

Beisner, B.E., et al., 2006. Environmental productivity and biodiversity effects on invertebrate community invasibility. Biol. Invasions 8, 655–664.

Bradshaw, C.J.A., et al., 2016. Massive yet grossly underestimated global costs of invasive insects. Nat. Commun. 7, 12986.

Clarke, B., et al., 1984. The extinction of endemic species by a program of biological control. Pac. Sci. 38, 97–104.

Ehrenfeld, J.G., 2010. Ecosystem consequences of biological invasions. Annu. Rev. Ecol. Evol. Syst. 41, 59–80.

Elton, C.S., 1958. The Ecology of Invasions by Animals and Plants. Methuen & Co Ltd., London, UK.

Gurevitch, J., Padilla, D.K., 2004. Are invasive species a major cause of extinctions? Trends Ecol. Evol. 19, 470–474.

Lombaert, E., et al., 2010. Bridgehead effect in the worldwide invasion of the biocontrol Harlequin Ladybird. PLoS One 5, e9743.

MacArthur, R.H., Wilson, E.O., 1963. An equilibrium theory of insular zoogeography. Evolution 17, 373–387.

Murphy, G.E.P., Romanuk, T.N., 2014. A meta-analysis of declines in local species richness from human disturbances. Ecol. Evol. 4, 91–103.

Paolucci, E.M., et al., 2013. Origin matters: alien consumers inflict greater damage on prey populations than do native consumers. Divers. Distrib. 19, 988–995.

Phillips, B.L., et al., 2006. Invasion and the evolution of speed in toads. Nature 439, 803.

Pyšek, P., et al., 2010. Disentangling the role of environmental and human pressures on biological invasions across Europe. Proc. Natl. Acad. Sci. U.S.A. 107, 12157–12162.

Romanuk, T.N., et al., 2009. Predicting invasion success in complex ecological networks. Philos. Trans. R. Soc. B 364, 1743–1754.

Roy, H.E., et al., 2011. Living with the enemy: parasites and pathogens of the ladybird Harmonia axyridis. Biocontrol 56, 663–679.

Simberloff, D., et al., 2013. Impacts of biological invasions: what's what and the way forward. Trends Ecol. Evol. 28, 58–66.

Strayer, D.L., 2012. Eight questions about invasions and ecosystem functioning. Ecol. Lett. 15, 1199–1210.

Vilà, M., et al., 2006. Local and regional assessments of the impacts of plant invaders on vegetation structure and soil properties of Mediterranean islands. J. Biogeogr. 33, 853–861.

Vilà, M., et al., 2011. Ecological impacts of invasive alien plants: a meta-analysis of their effects on species, communities and ecosystems. Ecol. Lett. 14, 702–708.

Williams, R.J., Martinez, N.D., 2000. Simple rules yield complex food webs. Nature 404, 180–183.

CHAPTER ONE

Parasites and Biological Invasions: Predicting Ecological Alterations at Levels From Individual Hosts to Whole Networks

V. Médoc*,1, C. Firmat†, D.J. Sheath‡, J. Pegg‡, D. Andreou‡, J.R. Britton‡

*Sorbonne Universités, UPMC Univ Paris 06, CNRS UMR 7618, Institute of Ecology and Environmental Sciences—Paris, Paris, France
†INRA, UMR 1202, BIOGECO, Cestas, France
‡Faculty of Science and Technology, Bournemouth University, Poole, United Kingdom
1Corresponding author: e-mail address: vincent.medoc@upmc.fr

Contents

Advances in Ecological Research, Volume 57
ISSN 0065-2504
http://dx.doi.org/10.1016/bs.aecr.2016.10.003

Abstract

The network approach is increasingly used by food-web ecologists and ecological parasitologists and has shed light on how parasite–host assemblages are organized, as well as on the role of parasites on the structure and stability of food webs. With accelerating rates of nonnative parasites being introduced around the world, there is an increasing need to predict their ecological impacts and the network approach can be helpful in this regard. There is inherent complexity in parasite invasions as parasites are highly diverse in terms of taxa and life strategies. Furthermore, they may depend on their cointroduced host to successfully overcome some crucial steps in the invasion process. Free-living introduced species often experience enemy release during invasion, which reduces the number of introduced parasites. However, introduced parasites that successfully establish may alter the structure of the recipient network through various mechanisms including parasite spill-over and spill-back, and manipulative and nonmanipulative phenotypic alterations. Despite limited literature on biological invasions in infectious food webs, some outstanding methodological issues and the considerable knowledge gaps that remain, the network approach provides valuable insights on some challenging questions, such as the link between structure and invasibility by parasites. Additional empirical data and theoretical investigations are needed to go further and the predictive power of the network approach will be improved by incorporating weighted methods that are based on trophic data collected using quantitative methods, such as stable isotope analyses.

1. INTRODUCTION

Characteristics of food-web structure can be expressed through features including species richness, functional diversity and topology, with the latter enabling calculation of network properties such as connectivity and chain length (Rooney and McCann, 2012). These characteristics and properties all contribute to building understandings of food-web stability and complexity (Lafferty and Kuris, 2009). Predictive food-web networks are often based on probabilistic rules that dictate that the diet choices of species are largely restricted to items in lower trophic positions as bound by predator–prey body sizes (Svanbäck et al., 2015).

Despite having body sizes much smaller than their hosts, parasite infections often have significant consequences for host biology; these are often nutritional due to the parasite feeding on or within the host (Barber et al., 2000). Parasite behaviour and development may also be associated with host pathological, physiological and/or behavioural changes, with likely adverse consequences for their growth, survival and fitness (Britton et al., 2011; Johnson and Hoverman, 2012; Johnson et al., 2008).

Consequently, parasites can profoundly shape the dynamics of their host populations and communities, alter intra- and interspecific competition, influence trophic relationships and are important drivers of biodiversity through their contribution of large numbers of species and causal links between density-dependent transmission and host specificity that increases biodiversity (Dunn et al. 2012; Hatcher and Dunn, 2011; Hatcher et al., 2006, 2012a, 2014; Lafferty et al., 2006; Luque and Poulin, 2007).

Trophic networks that have been developed using the general rules of predator–prey body sizes (Emmerson and Raffaelli, 2004; Warren and Lawton, 1987) have traditionally tended to overlook the inclusion of parasites (Beckerman and Petchey, 2009). In doing so, they may underestimate food-web connectivity and complexity (Marcogliese and Cone, 1997; Thompson et al., 2005). This is because insights provided by 'infectious food webs' (i.e. food webs with both predators and parasites) in aquatic ecosystems have revealed that parasites uniquely alter food-web structure and stability through, for example, substantially increasing connectivity, nestedness and linkage density (Amundsen et al., 2009; Byers, 2008; Lafferty et al., 2006, 2008). The inclusion of parasites in trophic networks has enabled them to become considered as crucial components of ecosystems. They often dominate food–web links (Lafferty et al., 2006), comprise substantial amounts of biomass (Kuris et al., 2008) and influence major aspects of ecological community structure (Wood et al., 2007).

The continuing acceleration of rates of international trade, travel and transport has resulted in a progressive mixing of biota from across the world via introductions of nonnative species into many new regions (Hulme et al., 2009; Pyšek et al., 2010). Whilst the introduction process can filter out many parasites that would otherwise move with their nonnative hosts, it enables others to be released, providing the opportunity for their 'host-switching' to native species (Blakeslee et al., 2012; Sheath et al., 2015; Torchin et al., 2003). The lack of coevolution between the introduced parasite and their potential new hosts suggests that transmission may occur due to low resistance arising from poor immune responses and antipredator behaviours (Taraschewski, 2006), although resistance will also depend on a wide range of environmental and biological factors (Blanchet et al., 2010; Penczykowski et al., 2011). Host responses to infection will vary according to factors including the complexity of the parasite's life cycle and host resilience (e.g. their ability of hosts to adapt to infection via alterations in life-history traits and their immune response that influences the severity of the infection) (Blanchet et al., 2010; Hawley et al., 2010) and high mortality rates might be incurred (Frick et al., 2010). Sublethal consequences for hosts include shifts in energetics and behaviour that may result in reduced host activity, growth, condition and fitness (Barber et al., 2000; Johnson and Hoverman, 2012).

From the perspective of the trophic network, the inclusion of an introduced parasite increases the number of connections (Britton, 2013). However, the extent of the network alteration is likely to be influenced by issues including the complexity of the parasite life cycle. These life cycles can be relatively simple, where transmission is host to host. They can also be complex, especially when the final host is a fish or bird at a high trophic position and the parasite is of low trophic position (Jansen and Bakke, 1993; Macnab and Barber, 2012). To overcome this discrepancy in their respective trophic positions, the parasite must navigate through a series of intermediate hosts before reaching their final host in which they sexually mature (Britton et al., 2009; Macnab and Barber, 2012). The parasite thus overcomes the 'trophic vacuum' between their low trophic position and the high trophic position of their final hosts (Parker et al., 2015), creating a series of new connections in the food web that connects species in low trophic positions with those at the top.

Given these different perspectives of trophic networks, parasites and introduced species and their parasites, the aim of this review is to outline how alterations in the properties of trophic networks from introductions of free-living species and their parasites can be predicted. It initially provides an overview of parasite strategies and network analysis (Section 2), before discussing the effect of including parasites in trophic networks (Section 3), the establishment process of introduced parasites with a case study of enemy release (Section 4) and the processes that determine the extent of network alterations by established parasites (Section 5). These aspects are then all integrated in Section 6 that reviews how the consequences of biological invasions can be predicted using network analyses, and how food-web structure can determine invasibility by parasites. Section 7 provides perspectives on analysing the influence of parasites on food-web structure through comparing more quantitative approaches with the qualitative approaches outlined in previous sections. Final conclusions are then drawn in Section 8. Throughout the review, when the term 'parasite' is used, it refers primarily to 'macroparasites', an artificial group of metazoan parasites, composed mainly of members of the Platyhelminthes (flatworms, including monogenean and digenean trematodes and cestodes), Nemathelminthes (roundworms and allies, including nematodes and acanthocephalans) and Arthropods (true lice and parasitic copepods) (Barber et al., 2000). The terms 'infectious food web' or 'whole food web' are used to describe a trophic network that includes both free-living species and their parasites. Although all parasite taxa and strategies are concerned with biological invasions, many examples are from aquatic food webs, a reflection of the current literature.

2. OVERVIEW OF PARASITES AND NETWORK ANALYSES

2.1 The Diversity of Parasite Strategies

Whilst micro-predation suggests that the boundary between predation and parasitism can be blurred, it is generally accepted that parasitism differs from predation in the durability of the attack (brief for predators and durable for parasites) and the number of victims per life stage (one host for parasites and more than one prey for predators). Within parasitism, life-history dichotomies have been proposed to categorize different parasite strategies, such as castrators, trophic transmission and intensity-dependent pathology (e.g. Anderson and May, 1979; Kuris and Lafferty, 2000; Lafferty and Kuris, 2002; May and Anderson, 1979; Poulin, 2011; Table 1). These strategies are life stage specific rather than species specific. They can thus differ across the life cycle of a parasite, such as between larval and adult stages (Poulin, 2011).

As parasite strategies do not have a single evolutionary origin then they have representative species from many phyla, especially in directly transmitted parasites (Poulin, 2011). However, some strategies are dominated by particular phyla, such as microparasites that tend to be mostly protozoans, bacteria or viruses, and trophically transmitted parasites that include many helminth species. Consequently, different parasite phyla potentially result in contrasting outcomes for food-web structure. Of the parasite strategies outlined in Table 1, the dichotomies that arguably will have greatest influences on trophic networks are whether the parasite exploits a single host/has multiple host species during its life cycle and whether the parasite exploits a free-living species or another parasite, especially in relation to free-living species and multiparasitism. Both of these will strongly influence the number of links in the network and its connectivity.

2.2 Network Analysis as a Tool for Ecological Parasitologists

A network is simply any collection of discrete entities potentially interacting as a system, which are usually represented by a set of nodes

Table 1 Parasite Life-History Dichotomies and Their Implications for Parasite Strategies

Life-History Dichotomy	Implication for Parasite Strategy	Examples
The parasite has an obligate association with the host/has a physical association with the host	Parasites that live in or on their host and are physiologically dependent on it (e.g. nutrient acquisition) differ from behavioural parasites, such as brood parasites exploiting the parental care of their hosts, whilst klepto-parasites steal their food items	Brood parasites: mainly birds (cuckoos, cowbirds), some fish and insects Klepto-parasites: vertebrates (birds, mammals), arthropods (bees, wasps, flies, ants, spiders)
The parasite exploits a single host/has multiple host species during its life cycle	Parasites with a simple life cycle exploit only a single individual from one host species; transmission can be vertical (parent to offspring) and/or horizontal (between unrelated individuals). Parasites with complex life cycles undergo larval development in one or up to three successive intermediate hosts before reproducing in the definitive host	Simple life cycle parasites: nematodes (hookworms), monogeneans, arthropods (crustaceans, lice and mites), protozoans, bacteria, viruses, fungi, plants Complex life cycle parasites: helminths (trematodes, cestodes, acanthocephalans), protozoans, bacteria, viruses
Parasite infection affects host fitness/does not affect host fitness	Parasitic castrators suppress host reproduction either directly, by feeding on gonads, or indirectly, by diverting energy away from gonad development or by secreting "castrating" hormones. Alternatively, fitness is affected by death before reproduction, such as where parasitoid larvae are laid in or on a single host, or sometimes penetrate it via direct contact or through ingestion, and consume host tissues to grow until emergence, which leads to host death	Parasitic castrators: crustaceans, some insects and also some helminths at a larval stage (trematodes, cestodes, acanthocephalans)

Infection is lethal/sublethal	In parasitoids, host death tends to be an inevitable outcome of their own development. By contrast, although parasitic castrators reduce host fitness to zero, they benefit from the long life of their host. Also, host death is required for trophically transmitted parasites whose transmission relies on the consumption of an infected intermediate host by a predatory host	Parasitoids: insects (braconid wasps), fungi, nematomorphs (hairworms) Trophically transmitted parasites: helminths (trematodes, cestodes, acanthocephalans)
The pathology of parasite infection is intensity-dependent/intensity-independent pathology	This separates microparasites from macroparasites. Microparasites multiply directly within the host whereas macroparasites do not and their impact is proportional to the number of separate individual parasites that infect the host	Microparasites: protozoans, bacteria, viruses, fungi Macroparasites: metazoans
The parasite exploits a free-living species/exploits another parasite	A hyperparasite is a parasite that uses another parasite as host. By contrast, super-parasitism is where a parasite infects an already infected host individual (e.g. in insect parasitoids), and multiparasitism is where an individual host is infected by more than one parasite species	Hyperparasites: arthropods (ticks, hymenopteran insects), fungi, bacteria, viruses
Parasitism is the only trophic strategy for the parasite/the parasite has multiple trophic strategies, of which parasitism is one	Obligate parasites are completely dependent on parasitism for their nutritional requirements. By contrast, facultative parasites only engage in parasitism under appropriate circumstances and are able to obtain resources through other means under different circumstances	Facultative parasites: fungi, bacteria, behavioural parasites

connected with directed or undirected edges (links) (Proulx et al., 2005). Network analysis consists in characterizing, through various mathematical and statistical tools, the structure of such relational data and their emerging properties. It originates from graph theory, a branch of applied mathematics, and was quickly adopted by social scientists and psychologists to study human social organization, and later by biologists to better understand metabolic pathways, gene and protein regulatory networks or neural networks, and by ecologists to investigate biotic interactions (Scott, 2000). For ecologists, nodes are usually distinct species (but it can be aggregates of species, life stages of species, nontaxonomic groups or individuals) and links represent antagonistic (predation or parasitism) or mutualistic interactions. Biotic interactions are the backbone of ecological communities and the underlying ideas of global interdependence between species behind the concept of food webs were already mentioned by Darwin (1859) and Elton (1927) (Pocock et al., 2016). Nevertheless, a big step was taken when food-web theory borrowed concepts and tools from network theory.

In its basic version, an ecological network can take the form of a species-by-species matrix of interactions, where columns represent consumers (predators, parasites, pollinators) and rows represent resources (prey, hosts, plants). At the intersections of each row and column (i.e. the cells of the matrix), binary entries indicate whether a link exists between the two species (1: yes, 0: no) (Fig. 1A). Such a matrix leads to a bipartite interaction network (Fig. 1B), where links between members of the same group (i.e. between consumers or between resources) are not allowed. Bipartite networks can be particularly relevant for mutualistic and parasite–host interactions. When there is no group distinction between members (i.e. all links allowed), they all appear in both columns and rows of the matrix (Fig. 1C). This results in a full interaction network (Fig. 1D), which is particularly relevant for predator–prey and parasite–host interactions, and especially when it comes to mixing both types of interactions in the same network (see later). The matrix is then used to calculate network statistics (metrics) like connectance, nestedness and modularity (Fig. 1E and F). They describe the structural attributes (topology) of the ecological network (Table 2). It is possible to consider species abundance and interaction frequency through quantitative approaches (Fig. 1G) with weighted metrics (Bersier et al., 2002). A range of computational approaches exists to visualize the network, the choice of a particular method depending on the structural attribute to be highlighted (Pocock et al., 2016).

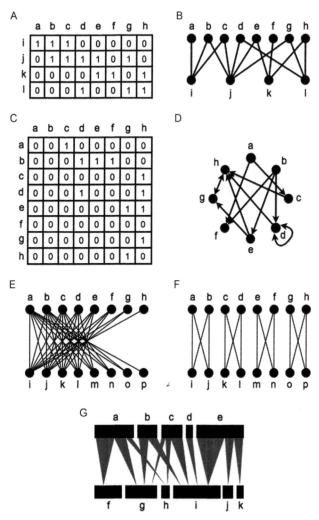

Fig. 1 Examples of interaction matrices and networks. Letters indicate entities that potentially interact (e.g. taxonomic species) and binary entries indicate the links (1: yes, 0: no). (A) Interaction matrix and (B) its visualization for a qualitative bipartite network where the entities form two groups (a–h and i–l) with no within-group interaction allowed. (C) Interaction matrix and (D) its visualization for a qualitative network where all interactions are allowed. *Arrows* indicate the direction of the interaction (e.g. from a resource to its consumer). (E) Nested bipartite network where specialist species interact with proper subsets of the species that interact with more generalist ones. (F) Modular bipartite network where species are organized in modules and interact more within than among modules. (G) Quantitative bipartite network where *bars* indicate species abundances and linkage width indicates interaction frequency.

Table 2 Summary of Some Metrics Commonly Used to Characterize the Structure of Antagonist Networks (Predator–Prey Webs, Parasite–Host Webs or Whole Food Webs Where Both Free-Living Species and Parasites Are Included)

Metric	Definition
Richness	Number of nodes
Connectance	Proportion of possible links that are realized
Link density	Mean number of links per node
Generality	Mean number of prey or host per predator of parasite species
Vulnerability	Mean number of predator or parasite per prey or host species
Omnivory	Proportion of taxa that feed on more than one trophic level
Chain length	Total number of trophic levels
Nestedness	Nonrandom pattern of link distribution where specialist taxa interact with a proper subset of the group of taxa with which generalists interact
Modularity	Nonrandom pattern of link distribution where taxa form groups of highly connected taxa (i.e. modules) with more links among themselves than with the taxa of other groups

For more details on the metrics, their quantitative versions and their calculation, see Bersier et al. (2002), Bascompte et al. (2003), Olesen et al. (2007) and Fortuna et al. (2010).

Linking the structural attributes of networks with ecosystem-level properties, such as functionality (efficiency of energy transfer), stability (persistence of species through time), resilience (time needed to recover following perturbation) or robustness (resistance to species removal) is one of the most challenging and exciting issues in the study of ecological networks. For instance, compartmentalization is supposed to limit the spread of disturbances though the whole network, whilst nestedness could limit the risk of secondary extinctions (see Tylianakis et al., 2010 and references therein). The role of connectance is more complex and falls into the old diversity–stability debate (see Rooney and McCann, 2012). On the one hand, high connectance means increased generalism. This should stabilize the rate of ecosystem processes through time and promote robustness by providing a buffer in the responses of predators to fluctuating prey abundances (Dunne et al., 2002; Tylianakis et al., 2010). On the other hand, species diversity is supposed to confer stability and decreases with connectance. It follows that highly connected networks should be less stable. In such networks, apparent competition (indirect interaction in which two victim species negatively affect each other by enhancing the equilibrium density of a shared natural enemy; Holt, 1977) should be greater and its negative impact could result in a lower stability in terms of total biomass (Thébault and Fontaine, 2010). More than species diversity, it is the way diversity is distributed that could be important (Rooney and McCann, 2012). A pattern of interaction strengths with a few strong links for a majority of weak interactions has been shown to be powerfully stabilizing (McCann et al., 1998). Such skewed distribution could confer stability to highly diverse and connected networks.

Network analysis has been applied extensively to predator–prey interactions before being used to study mutually beneficial interactions, such as those between plants and pollinators or seed dispersers, and is just beginning to be used by ecological parasitologists. Typical parasite–host networks are bipartite and the first relationship between metrics that emerged was a rapid decrease in connectance with increasing species richness (or network size) (Mouillot et al., 2008; Poulin, 2010). Such a relationship is expected for antagonistic interactions, where there is a greater level of specialization (coevolution) between partners. As a result of the constant arm race between parasites and hosts, parasites show some degree of specificity, which results in not all of the links being realized (Poulin, 2010). This decreases connectance, especially in species-rich communities. Two other main structural patterns emerge from parasite–host networks: nestedness and, most importantly, modularity, which are partly driven by the intimacy of interactions (Fontaine et al., 2011) and are both extremes of the same continuum (Fig. 1E and F). Nevertheless, nestedness and modularity are not mutually exclusive, as interactions within a module may be nested (Lewinsohn et al., 2006). Differences in nestedness and modularity among parasite–host networks are beginning to be understood and may be explained by constraints linked to phylogeny (Bellay et al., 2011, 2013, 2015; Braga et al., 2014, 2015; Brito et al., 2014; Krasnov et al., 2012; Lima et al., 2012), geography (Braga et al., 2014, 2015), species abundances (Vázquez et al., 2005), host traits such as body size (Campião et al., 2015), foraging strategy or habitat use (Brito et al., 2014), or parasite strategies (Bellay et al., 2013, 2015; Graham et al., 2009; Lima et al., 2012) (Table 3).

In addition to the detection and understanding of structural patterns in parasite–host assemblages, network analysis has been used by ecological parasitologists to investigate the effects of climate change (Maunsell et al., 2015;

Table 3 Factors That May Explain Differences in Nestedness and Modularity in Parasite–Host Networks

Factor	How It May Promote Nestedness or Modularity
Phylogeny	Phylogenetic continuity among hosts promotes parasite sharing and a nested pattern. Alternatively, phylogenetic gaps promote modularity with the detection of a phylogenetic signal within modules
Spatial overlap	At a large scale, overlapping geographical ranges promote parasite sharing and therefore nestedness. At a smaller scale, host species using the same microhabitats a more likely to belong to the same modules and to share the same parasites
Species abundances	Abundant species tend to have more links than rare species. Asymmetry in species abundance may thus promote a nested pattern where abundant hosts harbour many parasites, with a high proportion of specialists, whereas rare hosts tend to be parasitized by generalists
Host size	Large host species tend to harbour more parasites than small hosts. Asymmetry in host size may thus promote nestedness.
Foraging strategy	Host species having the same foraging strategy are more likely to belong to the same modules and to share the same parasites. Host species with a more diversified diet are more exposed to parasites than specialists. Asymmetry in diet diversity may thus promote nestedness
Host–parasite specificity	A high degree of specialization promotes modularity, whereas a lower specificity creates links between modules and therefore decreases modularity
Life cycle complexity	Parasites with a complex life cycle and experiencing strong ontogenetic shifts may connect distinct modules and decrease modularity

Morris et al., 2015), land use (Albrecht et al., 2007; Tylianakis et al., 2007), habitat fragmentation (Gagic et al., 2011; Kaartinen and Roslin, 2011; Murakami et al., 2008; Valladares et al., 2012), nutrient enrichment (Fonseca et al., 2005) and biological invasions (Amundsen et al., 2013; Timms et al., 2012). Epidemiological networks, i.e. unipartite networks that connect hosts sharing at least one parasite species, can be derived from bipartite parasite–host networks to explore dynamics of parasite transmission among host populations (Pilosof et al., 2015).

3. INFECTIOUS FOOD WEBS: INCLUDING PARASITES IN TROPHIC NETWORKS

3.1 From Bipartite Interaction Networks to Whole Food Webs

Members of bipartite networks may be under control of other species that are not depicted in the network (Fontaine et al., 2011). For instance, flower visitors in a mutualistic network experience a wide range of predators that are likely to influence visitation patterns. Similarly, parasitoids can be the victims of secondary parasitoids (hyperparasitoids). Therefore, it can be difficult to fully understand the dynamics of interacting species. To account for this, a third group of species with no within-group links can be added to the classical bipartite network so as to generate a tripartite interaction network, for instance a plant–pollinator–predator network (Marrero et al., 2013), a plant–herbivore–parasitoid network or a herbivore–parasitoid–hyperparasitoid network (Lohaus et al., 2013). However, network analysis in tripartite networks is rather limited and comes down to analysing the two stacked, bipartite networks separately.

Given the ubiquity of parasitism as a consumer strategy, food-web ecologists, who traditionally focus their attention on predator–prey interactions, were called to accommodate parasites into their network analyses (Cohen et al., 1993; Fontaine et al., 2011; Marcogliese, 2003; Marcogliese and Cone, 1997). This can be achieved by adding parasites to the list of entities that appear in both columns (consumers) and rows (resources) of the predation matrix (Fig. 2). The resulting 'whole food web' or 'infectious food web' is then a powerful tool to provide new insights on the ecological significance of parasites through their effect on food-web structure and stability (see later), and on the fundamental principles of trophic organization in ecosystems.

In addition to the predator–prey links of traditional food webs, parasite inclusion introduces new types of links (Fig. 2A). Parasite–host links are those involved in almost all parasite strategies, whereby parasites feed on the free-living species they exploit as hosts. Predator–parasite links occur when free-living species prey upon parasites either directly when parasites are at a free-living, nonfeeding life stage or indirectly when parasites are ingested along with the prey they exploit (concomitant predation) (Johnson et al., 2010; Thieltges et al., 2008, 2013). Trophically transmitted parasites use concomitant predation to reach the next host. Finally, parasite–parasite

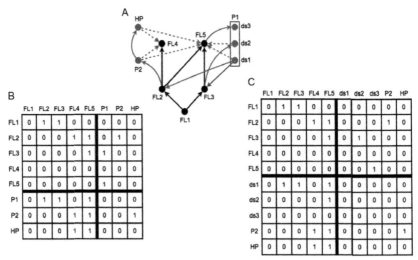

B	FL1	FL2	FL3	FL4	FL5	P1	P2	HP
FL1	0	1	1	0	0	0	0	0
FL2	0	0	0	1	1	0	1	0
FL3	0	0	0	0	1	1	0	0
FL4	0	0	0	0	0	0	0	0
FL5	0	0	0	0	0	1	0	0
P1	0	1	1	0	1	0	0	0
P2	0	0	0	1	1	0	0	1
HP	0	0	0	1	1	0	0	0

C	FL1	FL2	FL3	FL4	FL5	ds1	ds2	ds3	P2	HP
FL1	0	1	1	0	0	0	0	0	0	0
FL2	0	0	0	1	1	0	0	0	1	0
FL3	0	0	0	0	1	0	1	0	0	0
FL4	0	0	0	0	0	0	0	0	0	0
FL5	0	0	0	0	0	0	0	1	0	0
ds1	0	1	1	0	1	0	0	0	0	0
ds2	0	0	0	0	1	0	0	0	0	0
ds3	0	0	0	0	0	0	0	0	0	0
P2	0	0	0	1	1	0	0	0	0	1
HP	0	0	0	1	1	0	0	0	0	0

Fig. 2 Example of an ecological network (A) where black nodes are free-living species (FL1 to FL5) and grey nodes are parasites (P1, P2 and HP). *Black arrows* indicate predator–prey links and *grey arrows* those that involve parasites: parasite–host links (*curved arrows*), predator–parasite links (*solid arrows*) and parasite–parasite links (*curved arrow* between HP and P2). P1 is a complex life cycle with three developmental stages: (ds1) the free-living stage infective to the intermediate host FL3, (ds2) the larval stage found in FL3 and (ds3) the adult reproducing in the definitive host FL5, a predator of FL3. P2 is a simple life cycle parasite exploiting FL2, and itself exploited by HP, a hyperparasite. Predator–parasite links include direct predation on parasites at a free-living stage (e.g. between FL2 and ds1) and concomitant predation, when parasites are consumed along with they host (e.g. between FL5 and ds2, FL4 and P2 with HP). The corresponding interaction matrix (B and C) include a predator–prey subweb (*upper left quadrant*), a parasite–host subweb (*upper right quadrant*), a predator–parasite subweb (*lower left quadrant*) and a parasite–parasite subweb (*lower right quadrant*). Nodes can be defined as taxonomic species (B) or developmental stages (C), here to account for the strong ontogenetic shifts experienced by P1.

links describe hyperparasitism (frequent among parasitoids) or predation among parasites (larval trematodes commonly engage in intraguild predation within their snail intermediate host; Sousa, 1992). It follows that the whole interaction matrix can be divided into four submatrices corresponding to the four types of links (Lafferty et al., 2006; Fig. 2B and C). Network statistics can be calculated either from the whole matrix or from a particular submatrix.

Including parasites in food webs raises some methodological issues. For instance, differences in how the links are described can substantially alter the way connectance is calculated because, of all the possible links that are added with parasites, some might appear illogical. Concerning the nodes, the complexity of the life cycle of some parasites raises the question of node resolution (Fig. 2B and C). Indeed, many parasites undergo extreme ontogenetic niche shifts during development to such an extreme degree that nodes could be viewed as life stages and not as trophic species, as it is often the case in traditional food webs (Preston et al., 2014).

Ontogenetic niche shifts may also occur among free-living species and those that experience shifts in predators and/or diet throughout their development should be disaggregated into life stages, as well as parasites, to minimize the asymmetry in node resolution between the species interaction types of an infectious network. One possible drawback of species disaggregation is that the different life stages are treated as separate populations in the network approach even though their success is tightly coupled (Preston et al. 2014).

3.2 Impact of Parasite Inclusion on Network Metrics and Properties

We examined one terrestrial and ten aquatic food webs including both predation and parasitism (Table 4; aquatics webs are reviewed in Jephcott et al., 2016). Of the ten aquatic webs, eight are estuarine and only one is from the Southern Hemisphere, which limits extrapolation of results. The most widely reported finding is that parasite inclusion alters several key statistics calculated from the interaction matrix, with an increase in species richness, chain length, linkage density, nestedness and connectance (Amundsen et al., 2009, 2013; Hernandez and Sukhdeo, 2008; Huxham et al., 1995; Kuang and Zhang, 2011; Lafferty et al., 2006; Memmott et al., 2000; Thompson et al., 2005). However, the effect of parasite inclusion on nestedness and connectance, which should influence stability, has proved to be sensitive to the way nodes are defined (Fig. 2B and C). Preston et al. (2014) investigated three versions of a freshwater pond food web (Quick Pond in the San Francisco Bay Area of California; Preston et al., 2012) with varying levels of node resolution. Consistent with previous findings, nestedness and connectance increased after parasite inclusion when parasites were included as taxonomic species (a single node per parasite species), whereas the opposite occurred when nodes were disaggregated into parasite life stages. This is because in the highly resolved food web, the number of observed links did not increase in proportion with the number of possible links added by the parasite life stages (Preston et al., 2014). Clearly, this question over node resolution has to be further explored, for instance using the

Table 4 List of Eleven Model Food Webs That Include Both Predators and Parasites

Name	Location	References
Loch Leven	United Kingdom	Huxham et al. (1995)
Ythan Estuary	United Kingdom	Huxham et al. (1995)
Broom fauna at Silwood Park	United Kingdom	Memmott et al. (2000)
Otago Harbour	New Zealand	Thompson et al. (2005) and Mouritsen et al. (2011)
Carpinteria Salt Marsh	USA, California	Lafferty et al. (2006), Kuris et al. (2008) and Hechinger et al. (2011)
Estero de Punta Banda	Mexico	Kuris et al. (2008) and Hechinger et al. (2011)
Bahía Falsa	Mexico	Kuris et al. (2008) and Hechinger et al. (2011)
Muskingum Brook	USA, New Jersey	Hernandez and Sukhdeo (2008)
Lake Takvatn	Norway	Amundsen et al. (2009, 2013)
Flensburg Fjord	Germany/Denmark	Zander et al. (2011)
Sylt Tidal Basin	Germany/Denmark	Thieltges et al. (2011)

methodology of Preston et al. (2014) with other well-described communities to test whether the structural attributes of networks and their associated ecological properties are robust to node disaggregation.

The ecological significance of parasite-induced changes in food-web metrics generates a lot of interest but is far from being resolved. The relationship between structure and stability being complex, the role of increased connectance following parasite inclusion is not clear. It might be that parasites stabilize food webs, because they are engaged in weak interactions with their hosts. Their inclusion in food webs therefore adds weak interactions and increases the heterogeneity of interaction strengths (few strong links for a majority of weak links), a pattern assumed to promote stability (McCann et al., 1998). Conversely, parasites make food webs more susceptible to secondary extinctions, especially where the parasites have complex life cycles, as the removal of a single free-living would induce the loss of its parasites (Chen et al., 2011; Lafferty and Kuris, 2009). Several studies have highlighted that parasite-induced changes in food-web metrics could be the generic effects of adding complexity and diversity to the matrix through new nodes and links (Lafferty et al., 2008; Marcogliese, 2003; Sukhdeo, 2010, 2012). This raises the argument as to what extent are the topological roles of parasites unique compared to those of free-living species? Of the links that involve parasites (Fig. 2A), it has been shown that concomitant links (between predators and the parasites of their prey) are those that have the greatest effect on network structure (Cirtwill and Stouffer, 2015; Dunne et al., 2013; see also Thieltges et al., 2013). Concomitant predation might thus be what makes parasites unique (but see Jacobs et al., 2015). Parasites also differ from predators in that their trophic niches tend to be broader and have more gaps (Dunne et al., 2013). This may result from the complex life cycles of many parasites, with the successive exploitation of very different hosts in terms of body size or phylogeny. Generative network models, such as the probabilistic niche model, provide powerful and new tools to better understand the roles of parasites in food webs and the relative importance of predation, parasitism and concomitant predation (see Jacobs et al., 2015).

Besides changing network metrics, including parasites also changes the frequency distribution of network motifs among food webs (Cirtwill and Stouffer, 2015; Dunne et al., 2013). Again, the role of concomitant predation was highlighted but the ecological significance of the over- or under-representation of network motifs following parasite inclusion remains to be explored.

Network theory also gives insights on how parasites, especially those with complex life cycle and trophic transmission, establish in host communities. For instance, it has been shown that parasites are more likely to colonize host species that are central in the food web, i.e. those that are highly connected and contained within modules of tightly interacting species (Anderson and Sukhdeo, 2011; Chen et al., 2008; David et al., 2017). This is because predators with wide diet ranges are more likely to eat prey serving as intermediate hosts, whilst prey with many predators may represent efficient transmission pathways. Consistently, Locke et al. (2014), who searched for the factors that best predict the diversity of both larval and adult stages of the trophically transmitted parasites of 25 fish species from the Bothnian Bay, Finland, found that adult parasite diversity increased with prey diversity (i.e. diet range, generality) and larval diversity increased with predator diversity (i.e. vulnerability). Finally, highly connected species within core modules are less susceptible to extinction compared to less connected species, and thus may represent stable units for the persistence and evolution of complex life cycle parasites (Anderson and Sukhdeo, 2011).

In summary, the analysis of how parasites alter network properties has emerged as a major research theme in the last decade and has provided insights into how food-web connectivity, stability and robustness alter when parasites are included. Although there remains a series of questions that have yet to be fully resolved in these networks, the studies completed to date suggest that network analysis will provide a strong tool for predicting how introductions of free-living species and their parasites will alter food-web structure. However, before this is discussed, the processes that influence the establishment and trophic consequences of introduced nonnative parasites are reviewed. This is because these processes have important implications for the number of introduced parasites that establish and thus the extent of the network alterations.

4. PARASITES AS ALIEN SPECIES

4.1 The Establishment of Introduced Parasites

Similar to introduced free-living species, for a nonnative parasite to become invasive in a new range it has to overcome the barriers of transport, establishment and dispersal (Fig. 3; Lymbery et al., 2014). Cointroduced parasites are those introduced with their free-living host (Fig. 3A). They are described as established when they maintain self-sustaining populations in the new area through survival, reproduction and dispersal in their cointroduced hosts (Fig. 3B). Invasion includes a switch from exotic to native hosts (Fig. 3C). For parasites that have been introduced without their free-living host (Fig. 3D), such as free-living infective stages within ballast water (Goedknegt et al., 2015), establishment and invasion are confounded as they have to establish self-sustaining populations on or in native hosts to persist (Fig. 3E). In reality, there tends to be little distinction made between established, cointroduced and introduced parasites, given that parasites which are introduced but do not establish are unlikely to ever be recorded (Lymbery et al., 2014).

Of 98 cointroduced parasites recorded by Lymbery et al. (2014) at a global scale, 49% were helminths, 17% were arthropods and 14% were protozoans. Fish were by far the most common nonnative hosts, with many freshwater species, which reflects the vulnerability of aquatic ecosystems to species introductions due to their high connectivity and the increase in

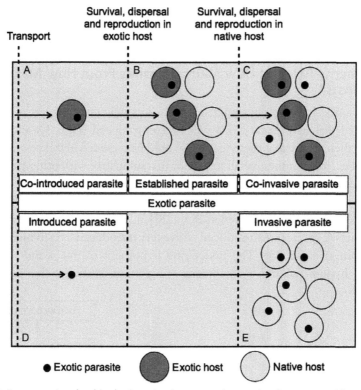

Fig. 3 Processes involved in the invasion by an exotic parasite that comes either along with its exotic host as a cointroduced parasite or alone as an introduced parasite. For a cointroduced parasite (A), establishment in the recipient ecosystems occurs in case of survival, dispersal and reproduction in the exotic host (B), and invasion occurs when it spreads among native hosts (C). For an introduced parasite (D), establishment and invasion are not distinguishable as its persistence in the recipient ecosystem depends on its ability to infect and spread among native hosts (E).

global shipping and aquaculture activities (Goedknegt et al., 2015). For instance, many organisms, including bacteria, viruses and other free-living infective stages of metazoan parasites, are moved out of their natural range with ballast water. Given the high rate of introductions of fish hosting non-native parasites then, intuitively, structural alterations to freshwater food webs might be most likely to occur from introduced fish parasites, although this will be dependent upon the life cycles of those parasites. For example, where these parasites have complex life cycles, they potentially form multiple new connections in the food web via filling the trophic vacuum, especially where they switch hosts to native species, altering trophic links within the food web. Where the parasites have direct life cycles, however, their effects on food–web structure might be limited.

4.2 Enemy Release: How Many Parasites From How Many Hosts?

When a free-living species is introduced into a new range then it might be expected that its parasite fauna will be cointroduced (Figs. 3A and 4). It would then follow that these parasites would be included in the new trophic network, either in the overall matrix or in a submatrix with other parasites, depending on the approach being used (Section 3.1). However, the introduction process often filters out many of the parasites that would otherwise have been introduced (Blakeslee et al., 2012). Estimates are that only two new parasite species are introduced with every introduced free-living species (Torchin et al., 2003). The underlying hypothesis of this is the 'Enemy release hypothesis' and is an integral component of studies that consider

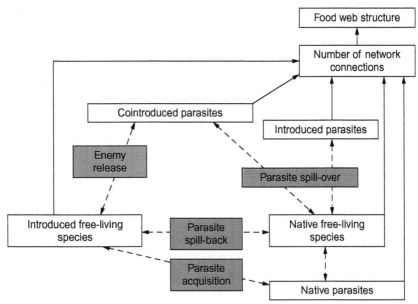

Fig. 4 Influence of enemy release, parasite spill-over, parasite spill-back and parasite acquisition on the number of network connections in a food web invaded by introduced free-living species, cointroduced parasites and introduced parasites. *Shaded boxes* denote processes, *dashed lines* denote interactions between the species via the different processes and *solid lines* denote the contributions of each category of parasite and free-living species to the number of network connections and, ultimately, food web structure.

how nonnative parasites might influence native species, trophic networks and ecosystems. It predicts that the parasite loss experienced by introduced free-living species enhances their ability to establish and invade (Hatcher and Dunn, 2011; Keane and Crawley, 2002; Mitchell and Power, 2003). Torchin and Mitchell (2004) suggested that introduced species escape at least 75% of their parasites from their native range and thus will gain substantial benefits regarding their fitness and survival in the invasive range (Torchin et al., 2003). It has been used as the basis to explain the invasion success of a diverse range of species, including nonnative slugs (Ross et al., 2010), mosquitoes (Aliabadi and Juliano, 2002) and frogs (Marr et al., 2008).

Specific examples of enemy release include the invasive European green crab *Carcinus maenas* that has significantly reduced parasite diversity and prevalence in its invasive range compared with its natural range, with their greater population biomasses in the invasive range attributed to this (Torchin et al., 2001). Several amphipod species that have invaded British waters host a lower diversity, prevalence and burden of parasites than the native amphipod *Gammarus duebeni celticus* (MacNeil et al., 2003; Prenter et al., 2004). Of the five parasite species that have been detected, three are shared by both the native and invasive amphipod species, but two are restricted to *G. duebeni celticus* (Dunn and Dick, 1998; MacNeil et al., 2003). Torchin et al. (2005) found a similar pattern in mud-snail communities in North America; whilst the native snail *Cerithidea californica* was host to 10 trematode species, the invader *Batillaria cumingi* was host to only one. These specific examples are supported by meta-analyses of native and invasive animals and plants which have revealed a higher-than-average parasite diversity in native populations; for example 473 plant species naturalized to the United States from Europe had, on average, 84% fewer fungal pathogens and 24% fewer virus species than native fauna (Mitchell and Power, 2003), whilst introduced fishes in the England and Wales had, on average, less than 9% of the number of macroparasites they had in their native range (Sheath et al., 2015).

The processes underlying 'enemy release' are therefore important when building trophic networks that incorporate introduced species. Enemy release suggests that the probability of the free-living species developing an established population is enhanced by the loss of substantial proportions of their natural parasite fauna, thus providing a new node. However, it also suggests that only a small proportion of nonnative parasites will be cointroduced with free-living hosts, limiting the number of new parasite nodes and thus limiting the potential for the introduction to substantially alter the network properties. The potential for connectivity to then increase is also diminished when the introduced parasites that do establish are those with direct life cycles that involve a single host species (Sheath et al., 2015).

4.3 Enemy Release: A Case Study With Invasive Cichlid Fish

Whilst enemy release in introduced free-living species is a well-established process (Section 4.2), the actual pattern of parasite loss in invasive species tends to be highly variable, even among closely related host species or populations (e.g. Benejam et al., 2009; Blakeslee et al., 2012). This variation is likely driven by numerous factors related to invasion history, or to the biological characteristics of the host and/or parasite (Lafferty et al., 2010; MacLeod et al., 2010). As regional parasite species richness is strongly predicted by the richness of the available host species (Kamiya et al., 2014), then this pattern might also occur in a biological invasion context, i.e. the greater the number of introduced free-living species, the greater the number of parasites that can be both introduced and acquired by a given free-living species. The rapid formation of a community of nonnative host species within an invaded territory is therefore expected to be associated with a simultaneous rapid formation of a parasite community. The transfer rate of parasites is thus expected to depend on the transfer rate of their hosts and so, therefore, the richest nonnative host communities should be associated to the richest parasite communities. This prediction could have a central influence on the contemporary emergence and the evolution of host-parasite networks.

This suggests that the regional species richness of introduced parasites can be predicted from the richness of their potential introduced host species and, conversely, that the number of parasite species found in a region can be indicative of the pattern of free-living species introduction at the community scale in this region (in the case of a community of free-living species sharing the same pool of parasites). This was tested here using the taxonomic data of Firmat et al. (2016) concerning invasive cichlid fish (commonly and thereafter called 'tilapia') that, whilst native to Africa, have been moved around the world for aquaculture purposes. They are now harmful invaders in several tropical and subtropical regions of the World (Canonico et al., 2005), providing replicated 'natural experiments' of species range expansion. The parasites used were Ancyrocephalidae, the super diverse species group of monogenean gill parasites harbouring a low level of host specificity (Mendlova and Simkova, 2014; Van Steenberge et al., 2015). These gill parasites have simple life cycle, a low level of host specificity (e.g. Jimenez-Garcia et al., 2001) and their taxonomy, with more than 120 described species, is well established (Pariselle and Euzet, 2009; Vanhove et al., 2016). These traits thus enable evaluation of the variation in parasite species richness to be tested against tilapia invasions.

For investigating the relationship between free-living vs parasite species richness in a biological invasion context, the data used here were based on literature and epidemiology survey data of Firmat et al. (2016) that provided a total of 14 case studies. Each case study represented a locality outside of Africa where data on parasite species richness and the number of host species was available (cf. Appendix reproduced from Firmat et al., 2016). The number of introduced host species per regional case study was extracted from Fishbase (Froese and Pauly, 2011) and complemented with the comprehensive register on tilapias from the Food and Agriculture Organization (2004).

Before these data were analysed, the species richness of hosts (R_h) and parasites (R_p) were both $\log(R + 1)$ transformed, with an ordinary least-squares regression then performed on $\log(R_p + 1)$ vs $\log(R_h + 1)$. Although log transformation for count data is generally not recommended (O'Hara and Kotze, 2010), here it allowed a biologically meaningful and simple modelling of the influence of host species richness on parasite species richness. This was because: (i) the slope of this log–log regression provided a direct estimate of the relative (i.e. proportional) increase in parasite richness induced by an increase in host richness in the system; and (ii) the procedure allowed fixing of the intercept to zero, a biologically meaningful value as no parasites are expected to be detected in the absence of host (i.e. when $R_h = 0$, $\log(R_h + 1) = 0$ and R_p should be null, then a zero intercept corresponds to $\log(R_p + 1) = 0$).

Variation in the reported parasite richness among the case studies ranged from 0 to 12 species (mean $\pm$ SD: 3.86 ± 2.82). Fig. 5 revealed that the number of parasites increased markedly with the number of hosts established during the twentieth century. The variance in host species richness explains 91% of the variance in parasite richness across invaded regions. The regression slope (*b*) was 0.86 (95% confidence interval: 0.70–1.03), indicating that for a 10% increase in host species richness, parasite species richness increases by 8.6%. Thus, each new introduced host species is expected to add almost one new parasite species to the food web. In other words, this suggests that the number of host species introductions in a new territory drives the number of introduced parasite species and then, potentially, the level of the enemy release phenomenon. Due to the low level of host specificity of

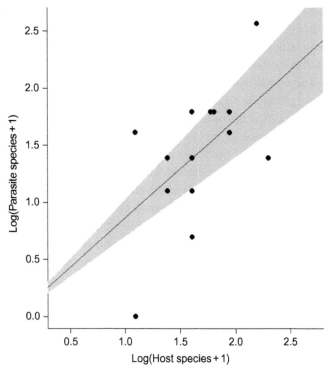

Fig. 5 Effect of the number of cichlid hosts introduced in a world region on the observed regional species richness of parasites ($n = 14$). Each point represents one region of the World out of hosts' native ranges (Africa) where Ancyrocephalid parasites were sampled. The *grey line* represents the predictions of a log–log linear regression and the grey area its 95% confidence interval ($R^2 = 0.91$).

monogenean parasites in this system, the potential number of 'enemies' of a given introduced free-living species will be increased by the number of other introduced free-living species. At the host species level, the potential number of parasites is broadly determined by the number of introduced hosts, i.e. the nonnative host community richness, a driver of the number of cointroduced parasites.

Taking the extreme case of the Grande Terre Island (New Caledonia, South West Pacific), only one introduction event was reported and no gill parasites were detected (Firmat et al., 2016). This exemplifies that where free-living species introduction are rare, this is then reflected in the parasite fauna. Conversely, the regions of highest parasite richness were those that had been highly exposed to the trading of live tilapias during recent decades (e.g. Madagascar, Mexico, China), indicating the initial parasite loss of enemy release can be partly compensated by the repeated introductions of new host species and therefore of new parasites. This raises questions over the effects of multiple parasitic infections on host fitness. In these examples, an increased cost of multispecies infection might be expected given that Ancyrocephalid parasites species appear to occupy different microhabitats on the tilapia gills (Madanire-Moyo et al., 2011), potentially leading to higher energetic costs of infection.

5. HOW EXOTIC PARASITES ALTER NETWORKS: PARASITE–HOST INTERACTIONS AND CONSEQUENCES

5.1 Parasite Spill-Over

Where nonnative parasites are cointroduced into a new range, despite enemy release, then their introduced free-living hosts can then act as a 'reservoir' of potential transmission to native species. This source of infection and subsequent transmission to native hosts is referred to as parasite 'spill-over' (Britton, 2013; Thompson, 2013) (Fig. 4). This spill-over can also occur when an introduced parasite is able to infect native hosts. An example from European freshwaters is the nematode parasite *Anguillicoloides crassus* (Kirk, 2003). Native to the Japanese eel *Anguilla japonica*, it was introduced into Europe via their movements in the global aquaculture trade. It has since spilled-over into the European eel *Anguilla anguilla* and is now widely distributed in their range (Kirk, 2003). Infections of *A. anguilla* are then caused by both consumption of both intermediate hosts (copepods) but also of paratenic hosts, i.e. via postcyclic transmission (Pegg et al., 2015a). Thus, from a network perspective, the spill-over of a cointroduced nonnative parasite into the native community increases the number of new links in the network from the same number of nodes, increasing connectance and, potentially, linkage density. Where an introduced parasite infects final hosts via two mechanisms, such as in *A. crassus*, then increased connectance should also result.

5.2 Parasite Acquisition and Spill-Back Processes

The process of parasite 'spill-back' can occur following the introduction of a free-living species. This is where the introduced species becomes infected with native parasites and then act as 'reservoirs' of infection for the subsequent spill-back of these parasites to their native hosts (Kelly et al., 2009; Fig. 4). For example, in Australia, the invasive Cane toad *Bufo marinus* played an important spill-back role in the emergence of two myxosporean parasites of native frogs, the Green and golden bell frog *Litoria aurea* and the Southern bell frog *Litoria raniformis*, facilitating parasite dispersal and transmission, and the consequent population declines of the frogs (Hartigan et al., 2011). The invasive crayfish *Pacifastacus leniusculus* displays both spill-over and spill-back. For spill-over, it is an asymptomatic host for the introduced fungus *Aphanomyces astaci* (crayfish plague) that is subsequently transmitted to white-clawed crayfish *Austropotamobius pallipes* (Kelly et al., 2009). For spill-back, it hosts the native microsporidian *Thelohania contejeani* where it acts as a reservoir of infection for *A. pallipes* which then tends to also cause mortality (Dunn et al., 2009).

Given that enemy release is a process common to many introduced free-living species, then their parasite fauna in their new range are often dominated by native parasites acquired in their new locality (Kelly et al., 2009; Sheath et al., 2015; Torchin et al., 2003; Fig. 4). This is the case for many exotic fish species that have been translocated globally as parasite-free eggs or juveniles, such as rainbow trout *Oncorhynchus mykiss* (Ortubay et al., 1994) and brown trout *Salmo trutta* (Hine et al., 2000). Thus, the threats of parasitism from nonnative species often include altered native host–parasite dynamics (Tompkins et al., 2011), which potentially results in either the spill-back (Daszak et al., 2000, Kelly et al., 2009), or dilution of, infections by native parasites (Telfer et al., 2005; Thieltges et al., 2009). Both can have considerable influences on food-web topology (Fig. 4).

The extent of parasite acquisition by introduced free-living species is variable according to a number of factors that affects their exposure and vulnerability to infection. Paterson et al. (2012) suggested five factors that strongly determined these outcomes for fish: total body length, time since introduction, phylogenetic relatedness to the native fish fauna, trophic level and native fish species richness. Kelly et al. (2009) suggested that parasite spill-back occurred when an introduced free-living species was a competent host for a native parasite, with the presence of this additional host then increasing disease impacts in native species as they act as a reservoir. In their review of animal–parasite literature, they revealed that native parasites accounted for 67% of the parasite fauna of introduced free-living species from across a range of taxonomic groups, indicating their competence for hosting native parasites and their potential for creating novel network links.

In considering spill-over and spill-back processes for infectious food webs, their key influence on network properties appears to be increasing connectance by providing a greater number of realized links for the same number of nodes. However, given that parasite infections can result in population declines and, in cases of disease emergence, population extirpations, the longer term consequences of these processes might differ with, for example, potentially deleterious effects on the nodes and links resulting from these transmission processes.

5.3 Parasite Life Cycles and the Trophic Vacuum

Some parasites have evolved complex life cycles in which they are trophically transferred up food chains in order for them to overcome the issues of their low trophic position vs the high trophic position of their final host. Recent studies have termed this difference in parasite and host trophic position as the 'trophic vacuum', given that most adult helminth parasites sexually reproduce in vertebrates that have high positions in food chains, with their free-living propagules unable to be transmitted directly to these hosts. This trophic vacuum is thus filled by one or more intermediate hosts (Benesh et al., 2014; Parker et al., 2015). This raises a number of questions over why the parasite then does not grow and develop further in the intermediate hosts, and instead shows suppressions of growth and reproduction until transmission to the final host, a process that can involve being transmitted through multiple intermediate hosts (Parker et al., 2015). It has been suggested that it relates to selection pressures associated with the increased longevity and higher growth that is possible by the parasite in the final host (due to their relatively large body size of these hosts) vs intermediate hosts (that are often copepods or gammarid species). The selection pressure is thus for larger parasite body size and higher fecundity at sexual maturity that is only possible in the relatively large final host (Parker et al., 2015).

Within an infectious food web, these complex life cycles can form chains that link species of very low trophic position with those much higher in the network, potentially leading to increased connectance and linkage density. For an introduced parasite with a complex life cycle, it also suggests that their likelihood of establishment is probably going to be diminished vs a parasite with a direct life cycle unless appropriate native species are present that can act as intermediate hosts. Accordingly, of the 98 examples of parasite cointroductions recorded by Lymbery et al. (2014), 64% had a direct life cycle and 36% had a complex life cycle. However, should these complex parasites establish then there is potential for a greater shift in network properties.

5.4 Parasite Impacts on Hosts: Manipulative and Nonmanipulative Alterations

Host manipulation is the alteration of the host phenotype in a way that promotes the reproductive success of the parasite. It is found among various parasite strategies including parasitoids, vector-transmitted parasites and trophically transmitted parasites (Poulin and Maure, 2015). It helps trophically transmitted parasites to fill the trophic vacuum and usually involves the behavioural manipulation of the intermediate host that increases the likelihood of that host being predated by the next host in the life cycle (Britton et al., 2009; Loot et al., 2001). Amphipods provide strong examples of intermediate hosts that are manipulated by their parasites to facilitate their predation by a fish or bird final host (Britton and Andreou, 2016). Infections by the trematode parasite *Microphallus papillorobustus* divides populations of *Gammarus insensibilis* into two groups: an infected group of individuals that inhabit the surface of salt marshes and an uninfected group of individuals that remains near the bottom (Ponton et al., 2005). This shift in habitat use promotes predation of *G. insensibilis* by bird final hosts (Britton and Andreou, 2016). Infected *Gammarus roeseli* with *Polymorphus minutus* exhibit reverse geotaxis, elevating their time spent at the water surface (Bauer et al., 2005; Médoc et al., 2006). In conjunction with reduced activity, this increases their predation risk by bird final hosts (Jacquin et al., 2014). This also decreases their predation risk by three-spined sticklebacks *Gasterosteus aculeatus*, a nonhost for *P. minutus* (Médoc et al., 2009), highlighting that the effects of parasite manipulation on network properties might be more complex than simply strengthening the links involved in trophic transmission (Médoc and Beisel, 2009, 2011).

Parasite manipulation has also been detected in fish intermediate hosts. The cestode parasite *Ligula intestinalis* is generally recognized as modifying the behaviour of its intermediate fish hosts (Britton and Andreou, 2016; Loot et al., 2001). Where fish are infected, they are increasingly encountered in the littoral zone, increasing their predation risk to the final bird host (Britton et al., 2009; Loot et al., 2001). In doing so, the infected fish potentially exploit different food resources to their uninfected conspecifics (Adamek et al., 1996), thus directly and indirectly creating a series of novel links within the food web. Another, and spectacular, example of link creation by parasites is the host manipulation by nematomorphs, whereby infected crickets are manipulated to commit suicide by jumping into water, where the adult parasites reproduce (Thomas et al., 2002). This makes a new and valuable food source for fish and modifies the food web (Sato et al., 2012).

Thus, the manipulation of new hosts by an introduced parasite with a complex life cycle has the potential to alter the links within the network, especially where weighted approaches are being used. The example of the nematomorphs also highlights the potential for manipulation to connect food webs across ecosystem boundaries.

Parasite infections can also modify the host phenotype through consequences that are not associated with manipulation, such as impaired traits and altered behaviours that result from pathological or physiological impacts (Knudsen et al., 2004) which can then affect other behaviours, such as foraging and prey selectivity (Pegg et al., 2015b). As example is the common carp *Cyprinus carpio* when infected, as the final host, with *Bothriocephalus acheilognathi*, a nonnative intestinal cestode parasite. Infections impair the foraging ability of hosts through reducing, for example, their consumption rates (Britton et al., 2011, 2012). Infected individuals then increasingly specialize on feeding on less motile food sources that divides their population trophic niche into infected and uninfected subgroups (Britton and Andreou, 2016; Pegg et al., 2015b), increasing connectance. This also raises the question as to whether *C. carpio* should be included in the network as a single node or as two nodes split between uninfected and infected subgroups.

Infections by parasites can also be important through their effects on host foraging time budgets and the associated selectivity in prey items. For example, when *Schistocephalus solidus* infect *G. aculeatus*, the foraging time of the fish increases and they invest less in antipredator behaviours (Milinski, 1985). The fish also select smaller prey than their uninfected conspecifics (Cunningham et al., 1994; Milinski, 1984). This finding is, however, contrary to Ranta (1995), who suggested that larger items were taken by infected individuals compared with uninfected conspecifics, with this being a compensatory mechanism to overcome some of the energy costs caused by the parasite.

Consequently, irrespective of manipulation, parasite infections can markedly alter the host phenotype and this can have marked consequences for the trophic ecology of the host. These consequences can then modify the links within the food web, including through processes such as parasite-mediated competition.

5.5 Parasite-Mediated Competition and Coexistence

Parasite-mediated competition can have important implications for the prey selectivity of parasitized individuals and thus the connections within the trophic network (Hatcher et al., 2006, 2012a,b; Holt and Dobson, 2006; Holt and Pickering, 1985). In essence, the infection by a focal parasite on a free-living species potentially alters the competitive ability of the host, resulting in either increased or reduced access to a shared resource (Dunn et al., 2012; Hatcher et al., 2006, 2012a). If reduced, then it might lead to that host having to exploit other resources to maintain their energetic requirements and thus could be a driver of the formation of new network links.

At an intraspecific level, the competitive ability of infective hosts may be reduced compared to their uninfected conspecifics, potentially leading to intraspecific niche partitioning (Hatcher et al., 2012a). For interspecific competitive interactions, parasites can reverse competitive interactions where one host species outcompetes another in the absence of a parasite; however, due to lower resistance to the parasite, they develop infections and subsequently the competitive interaction becomes more symmetrical (Hatcher et al., 2006). Apparent competition occurs when two host species that do not normally compete are infected by the same parasite species that creates a link between them and creates an indirect competitive interaction (Dunn et al., 2012; Hatcher et al., 2006, 2012a). This competition is generally driven by one host species being more resistant to the parasite and acting as a reservoir that feedbacks greater parasite pressure on to the other host species (Hatcher et al., 2006).

Parasite-mediated coexistence is where the infection by a parasite suppresses the interspecific competitive ability of its host sufficiently to enable the host species to coexist with an otherwise inferior competitor species. The effect can even be as extreme as to allow a species to colonize an area when it would otherwise be competitively excluded by the host species. For example, two Caribbean *Alonis* lizard species are only able to coexist when the malarial parasite (*Plasmodium azurophilum*) reduces the competitive ability of *Alonis gingivinius* (Schall, 1992). In this way, parasite-mediated coexistence can be important in maintaining species richness and patterns of biodiversity. It will then, by extension, also be important in regulating trophic relationships and thus strongly influence the extent of the connections within the trophic network.

6. INTEGRATING BIOLOGICAL INVASIONS INTO INFECTIOUS FOOD WEBS

6.1 Linking Network Structure and Invasibility by Parasites

Invasibility, along with invasiveness, are two major concepts in invasion ecology (Richardson and Pyšek, 2006). Invasiveness is the propensity of an introduced species to invade a recipient ecosystem, with its expected determinants including introduction history, species traits and ecological and evolutionary processes (van Kleunen et al., 2010). It is beyond the scope of this section to discuss this further. Invasibility is the susceptibility of the recipient ecosystem to the establishment and spread of introduced species (Lonsdale, 1999). Whilst the roles of species diversity and the patterns of resource use and competition in resistance to invasion are important in determining the outcomes of introductions (Kennedy et al., 2002), network approaches potentially provide a new effective, integrative tool to search for structural features that determine invasibility. The relationship between food-web structure and invasibility has been investigated almost exclusively regarding introduced free-living species. However, their main results can be used to address the case of cointroduced parasites, given that their success depends on that of their cointroduced host. Again, the structural features that are of most interest are connectance, nestedness and modularity. Here, the effects of connectance are primarily discussed, since patterns of nestedness and modularity are often the consequences of varying connectance values (Lurgi et al., 2014; Riede et al., 2010).

Connectance has been found to either constrain or promote invasion. A negative relationship between connectance and invasibility was first reported in the theoretical work by Romanuk et al. (2009) and later in the empirical study by Wei et al. (2015), who tested experimentally how the resource competition networks of resident bacterial communities affected invasion resistance to the plant pathogen *Ralstonia solanacearum*. In highly connected webs, all members efficiently exploit most resources. This results in increased competition and a crowded niche space, constraining invasion. Nevertheless, the effect of connectance on invasibility might depend on the trophic level of the invader (Baiser et al., 2010). This is because connectance seems to be positively linked to the fraction of intermediate species (Vermaat et al., 2009), which may constitute prey or predators for the invader depending on its position in the food web. It follows that invasion success is predicted to decrease with connectance for exotic herbivores due to a high number of potential predators, and to increase with connectance for exotic top predators due to a high number of potential resources (Baiser et al., 2010). More recently, and contrary to previous findings, the theoretical works by Lurgi et al. (2014) and Hui et al. (2016) illustrated how highly connected networks might be more vulnerable to invasions. This is because of the link between connectance and stability; highly connected networks could encompass more reinforcing feedbacks between species, which makes them unstable and might create opportunity niches for invasion (Hui et al., 2016). Conversely, networks with low connectance are more stable in terms of total biomass even when individual species abundances fluctuate, and are thus more robust to biological invasions (Lurgi et al., 2014). Another explanation lies on species diversity. Higher levels of diversity are known to confer resistance to invasion and more connected food webs are often less diverse (Lurgi et al., 2014). This could explain the positive relationship between connectance and invasibility.

In addition to the successful establishment of their host, cointroduced parasites with a complex life cycle must overcome another critical step before establishing which is to find the next host(s) and the transmission pathway(s) needed to complete their cycle. Intuitively, highly diverse networks should provide more opportunities to find a suitable host, but they generally exhibit low connectance. As previously discussed, this might decrease invasibility by the cointroduced host. Similarly, it should be easier to find suitable transmission pathways in highly connected networks, which are also supposed to facilitate invasion by the cointroduced host, but where parasites might face the problem of finding a suitable host as these networks are often less diverse. Therefore, it is difficult to predict how the structure of the recipient food web may determine invasibility by cointroduced parasites as antagonistic effects can be confounded. A rough prediction could be that the invasion success of parasites with complex life cycles decreases with the level of host specificity and the number of successive hosts involved in the cycle.

Spill-over to native hosts may allow cointroduced parasites to establish in the recipient food web even when its structure constrains invasibility by their cointroduced host. Infecting a native host is the first step of the invasion process for nonnative parasites that are introduced alone at a free-living infective stage. Once in the native hosts, nonnative parasites (either introduced or cointroduced) have first to meet life cycle requirements in terms of successive hosts and transmission pathways. As discussed earlier, this might depend on host specificity, the complexity of the cycle and the compromise between diversity and connectance. Second, they have to cope with the recipient parasite–host network, as the native hosts they exploit are themselves likely to be already exploited by a range of native parasites. The role of the structure of recipient parasite–host networks on invasibility by parasites is a question that remains to be addressed. For instance, connectance in parasite–host networks decreases rapidly with increasing number of species (Poulin, 2010). It follows that it might be easier for exotic parasites to avoid competition with native parasites in large networks. Modularity might also play a role in invasibility by parasites in that highly compartmentalized networks are formed by clearly bounded modules, with few interactions between modules, corresponding to spatially or temporally partitioned niches and habitats that are potentially available. This suggests that invasive species experiencing strong ontogenetic niche shifts, such as some complex life cycle parasites, are more likely to invade highly modular networks (Hui et al., 2016).

6.2 Impact of Exotic Parasites on Food-Web Structure

As when it comes to infectious food webs, the introduction of new parasites first adds greater complexity via new nodes and links. Once past this generic effect and in the case of invasion, which implies spill-over to native hosts, long-term effects are more difficult to predict as they depend on parasite strategy, the levels of virulence, pathogenicity and specificity, the presence of natural enemies that confer biotic resistance (e.g. predators, other parasites and hyperparasites), and the evolutionary responses of both hosts and parasites (see table 1 in Britton, 2013).

A rough categorization can be made regarding the level of pathogenicity (i.e. the extent to which infection reduces host fitness). Highly pathogenic parasites might cause the decline or even the exclusion of native hosts, sometimes followed by secondary extinctions. There is, however, no empirical work illustrating the structural effect of such parasites, although it can be argued that in the extreme cases of species exclusion, this results to network simplification through a reduction in the number of nodes and links. These parasites are sometimes referred as 'emerging infectious diseases', which are diseases that have appeared in a population for the first time and have significant social, economic and ecological costs (Hatcher et al., 2012b), or as 'biological weapons', because they facilitate invasion by their cointroduced host through their negative impacts on native hosts, which constitute natural enemies for invaders (Strauss et al., 2012). They include parasitoids, and particularly those used in classic biocontrol, microparasites like fungi, protozoans, bacteria and viruses, or macroparasites (see examples in Dunn and Hatcher, 2015; Gendron et al., 2012, 2012a; Strauss et al., 2012). The fungus that causes crayfish plague and discussed in Section 5.2 is a good example. Another popular example is the parapoxvirus that spilled-over from grey squirrels to red squirrels in which it causes a deleterious disease (Tompkins et al., 2003). The releases of insect parasitoids to control agricultural pests provide good examples of the intentional introduction of pathogenic parasites. Spill-over to nontarget species frequently occurs and may cause local extinctions and profoundly reorganize the networks (Boettner et al., 2000; Hawkins and Marino, 1997; Henneman and Memmott, 2001; Parry, 2009).

In contrast, invasive parasites of low pathogenicity can make invaded networks more complex through the creation and persistence of new nodes and links. This might be the case with many macroparasites, like helminths with complex life cycles, with the assumption that the more complex the life cycle, the greater the effect on network complexity. To our knowledge, the empirical work by Amundsen et al. (2013) is currently the only available study that compares the structure of an infectious food web pre- and postinvasion. The study evaluated how the introduction of two fish species, arctic charr *Salvelinus alpinus* and *G. aculeatus*, altered the pelagic network of Lake Takvatn, a subarctic lake in northern Norway. Artic charr and stickleback facilitated the arrival of four new birds feeding almost exclusively on these two fish, and five new parasites, including one parasite that uses the fish as an obligate host. Several of these parasites also infected native species in the food web during the completion of their life cycles. In comparing the pre- and postinvasion infectious food webs, the postinvasion web had greater complexity with more nodes (39 vs 50 species), more links (282 vs 440), and an increase in linkage density, mean trophic level, omnivory, vulnerability and nestedness. Only connectance slightly decreased, with this a common pattern when adding consumers with some degree of specificity, such as parasites. Most interestingly, 79% of the 158 new-established links involved parasites with many predator–parasite links. Trophically transmitted parasites were highly connected, thus highlighting the important contribution of this parasite strategy to the complexity of networks.

Amundsen et al. (2013) did not, however, address the question of modularity. As discussed in Section 6.1, highly compartmentalized networks might more easily accommodate complex life cycle parasites than nonmodular networks because the distinct modules correspond to partitioned niches and habitats. In return, the introduction of parasites with complex life cycles should decrease modularity as they create links between modules. This effect should increase with the length of the cycle (i.e. the number of successive hosts) and the magnitude to the ontogenetic shifts. We are not aware of empirical works that could illustrate this effect.

Parasites of low pathogenicity are more likely to be cointroduced as the probability of their cointroduced hosts surviving the translocation process is expected to decrease with the pathogenicity of the parasites they carry

(Blackburn et al., 2011; Lymbery et al., 2014; Strauss et al., 2012). Consequently, it might be expected that cointroduced parasites are more likely to make invaded networks more complex than to simplify them. However, the effects of cointroduced parasites can be confounded with the concomitant effects of their cointroduced hosts, which can drastically disturb the structure of the invaded network and at the same time threaten native parasites.

A further consideration of how exotic parasites might alter food-web structure is context dependency that is often apparent in how parasites establish in new environments, and infect and impact new host species (Tompkins and Poulin, 2006). This is because differences in the prevalence, abundance and pathogenicity of parasites can occur over environmental and biological gradients, including climate (Lafferty, 2009), latitude (Rohde, 1999) and biodiversity (Lafferty, 2012). Thus, the ability of an introduced parasite to establish and form new nodes and links within a network will also be a function of the biotic and abiotic conditions of the recipient ecosystem. For example, Macnab and Barber (2012) working on *S. solidus* and Sheath et al. (2016) working on *Pomphorhynchus laevis* both revealed that higher parasite prevalence was apparent in fish hosts under conditions that represented climate warming, with Sheath et al. (2016) indicating that infections developed in warmer conditions from lower exposure to the intermediate hosts. Thus, the ability of the parasites to establish and complete their life cycles will vary over thermal gradients, with these associated with gradients including climate, latitude and altitude.

6.3 How Invader-Induced Changes in Network Structure Affect Parasites

Invasive species (which refers only to free-living species in this section) can modulate the dynamics of native parasites in many ways, either directly of indirectly (Table 5; see examples in Johnson et al., 2010; Poulin et al., 2011). Here, we focus on the structural changes induced by biological invasions to discuss their potential effects on native parasites.

Few studies used the network approach to investigate how invaders affect native parasites, and it has been done mostly on parasitoids. Heleno et al. (2009) investigated how alien plants integrated into the native plant–herbivore–parasitoid tripartite network of the laurel forest in the Azores archipelago. They found no significant effect on network structure but a significant decrease in network size as plant and insect (both herbivores and parasitoids) richness decreased with plant invasion. Although insect abundance was not significantly affected by alien plants, insect biomass

Table 5 How Invasive Species Can Modulate the Dynamic of Native Parasites

	Positive Effect on Transmission	Negative Effect on Transmission
The invasive species serve as host (spill-back)	Better transmission in the invasive species than in the natives	Better transmission in the native species than in the invasives
The invasive species do not serve as host	Decreased abundance of the natural enemies (predators, other parasites and hyperparasites) of the native parasites (e.g. through predation or competition) Increased abundance of the native hosts (e.g. through negative effects on their natural enemies) Trait-mediated indirect effects (e.g. behavioural changes in the native hosts that promote transmission)	Direct predation on the native parasites or their free-living infective stages Concomitant predation Decreased abundance of the native host (e.g. through predation or competition) Trait-mediated indirect effects (e.g. behavioural changes in the native hosts that constrain transmission)

was significantly reduced because small insects on alien plants replaced large insects on native plants. Similarly, Carvalheiro et al. (2010) investigated how the invasive plant *Gaultheria shallon* integrated into the plant–herbivore–parasitoid network of a site dominated by *Calluna vulgaris* in the United Kingdom. They found a decrease in the abundance of all insects and a decrease in the richness of parasitoids, but these effects were confined to trophic specialists, not generalists. In terms of network structure, because specialist species dominated the undisturbed food web, *G. shallon* invasion caused a reduction of the importance of their trophic links, leading to an increase in the evenness of species abundances and interaction frequencies (Carvalheiro et al., 2010). Timms et al. (2012) found that invasion by gypsy moth (*Lymantria dispar*), one of the most harmful invasive forest insects in North America, had no significant effect on the structure of the native parasitoid–herbivore network. There was only a little sharing of parasitoids with native hosts and gypsy moth was the target of a generalist parasitoid that increased its specialization on gypsy moth at high gypsy moth abundances.

There is not enough empirical and theoretical works to infer a general pattern for the effects of invaders on the structure of recipient parasite–host networks, and it is likely that it depends on the structure of the pre-invasion food web. The works by Heleno et al. (2009), Carvalheiro et al. (2010) and Timms et al. (2012) emphasize the benefit of using the network approach with information on diet breadth to understand and predict the responses of native species to invasion. It permits the detection of bottom-up cascade effects that can lead to further top-down effects, via apparent competition (Carvalheiro et al., 2010).

By definition, invasive species proliferate in recipient ecosystems and can dominate invaded communities in terms of abundance or biomass. It follows that invasive species can become profitable prey for native predators. For instance, in North America, many freshwater predators have shifted their diet to include the round goby (*Neogobius melanostomus*) and/or the zebral mussel (*Dreissena polymorpha*), two of the most problematic freshwater invaders (reviewed in Bulté et al., 2012). In addition, a broad diet with opportunistic feeding is expected to be one of the attributes of successful invaders (Ricciardi and Rasmussen, 1998). Invasive species can thus become generalist consumers in the invaded food web, creating links with many native resources. Combining high generality with high vulnerability makes invasive species, and especially those at intermediate trophic levels, likely to occupy a central location in invaded food webs in that they are highly connected. Invasion might lead to food-web contraction with increased connectivity around invasive species and reduced connectivity among native species, without marked change in overall connectance. Several studies documented the preferential use by parasites of free-living species that occupy central locations in food webs and represent hubs (Amundsen et al., 2009; Anderson and Sukhdeo, 2011; Chen et al., 2008). These species constitute stable ecological and evolutionary units for parasites and offer transmission opportunities. Following the colonization time hypothesis (Guégan and Kennedy, 1993), one might thus predict that invasive species should acquire local parasites over time and should host a significant fraction of parasite diversify due to their central position. The work by Gendron et al. (2012) supports this prediction. As predicted by the enemy release hypothesis, they showed that the round goby was less infected than native fish in the early phase of its establishment in the St. Lawrence River. However, this advantage over native species is expected be of short duration as the parasite richness and abundance of the older population of round goby in Lake St. Clair has more than doubled within 15 years (Gendron et al. 2012).

7. QUALITATIVE VS QUANTITATIVE APPROACHES

The application of network analyses to predict how introduced free-living species and their parasites alter network properties has revealed their high utility in demonstrating the extent of the changes that can be incurred in native systems. However, trophic networks that are built on binary matrices provide only a qualitative perspective of food-web structure (Section 2.2). Whilst more quantitative approaches are possible that incorporate species abundance and interaction frequency using weighted metrics (e.g. Bersier et al., 2002), these are reliant on either using robust empirical data on predator–prey relationships or using heuristic approaches where these data are unavailable (e.g. Amundsen et al., 2013). These network approaches are therefore arguably unable to incorporate some of the more complex aspects of host–parasite trophic relationships that could result from processes such as parasite manipulation and infection-induced host phenotype alterations (Section 5). Thus, approaches that integrate the more qualitative network approaches discussed in this review with more quantitative approaches, such as stable isotope analyses, should provide greater insights into how parasites alter trophic relationships (Britton, 2013; Britton and Andreou, 2016; see also Kamenova et al., 2017). They should also help test how parasite loading affects trophic niche sizes and how infections modify energy flux (Britton and Andreou, 2016).

As the ratios of the stable isotopes of carbon ($^{13}C{:}^{12}C$) and nitrogen ($^{15}N{:}^{14}N$) vary predictably from resource to consumer (Fry, 2007), they enable reconstruction of the trophic structure and the analysis of the trophic niche sizes and the overall food-web structure (Grey, 2006). The carbon values ($\delta^{13}C$) of a consumer species indicates their energy source, with enrichment of approximately 1‰, indicating the move up a trophic position; the stable nitrogen isotope ($\delta^{15}N$) typically becomes enriched by 3–4‰ between prey and predator tissue and so is an indicator of consumer trophic position (Deniro and Epstein, 1981; Minagawa and Wada, 1984). The application of stable isotope techniques, using the predictable relationship between the isotopic composition of consumers and their diet, is then sufficiently powerful to detect long-term (e.g. 3–6 months) dietary differences between individuals of the same population (Fry, 2007), such as those that are parasitized with a specific parasite and those that are uninfected (Pegg et al., 2015b).

Examples of how parasites can affect the trophic niche of host populations are demonstrated in Figs. 6 and 7. In all cases, the host

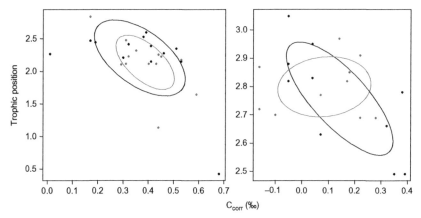

Fig. 6 Trophic niches of subgroups of infected (*grey ellipse*; individual data points *grey circles*) and uninfected (*black ellipse*; individual data points *black circles*) for populations of bullhead *Cottus gobio* (*left*) and minnow *Phoxinus phoxinus* (*right*) infected with *Pomphorhynchus laevis* from the River Loddon, Southern England, sampled in June 2015 (unpublished data). Both infected subgroups show strong patterns of niche constriction. Differences in values on the *X* and *Y* axis are for the purposes of presentation and clarity only.

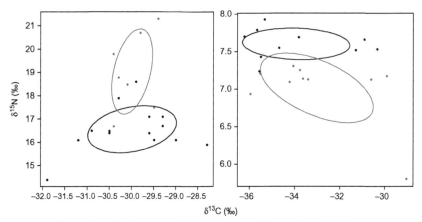

Fig. 7 Trophic niches of subgroups of infected (*grey ellipse*; individual data points *grey circles*) and uninfected (*black ellipse*; individual data points *black circles*) for lentic populations of roach *Rutilus rutilus* with the parasite *Ligula intestinalis* (*left*) and three-spined stickleback *Gasterosteus aculeatus* population infected with *Schistocephalus solidus* (*right*) in England (unpublished data). Both infected subgroups show strong patterns of niche divergence from their uninfected conspecifics. Differences in values of the *X* and *Y* axis between the plots are due to the fish being sampled from different waterbodies.

populations were sampled using electric fishing and seine netting, with dorsal muscle samples taken from a random selection of up to 30 fish per host population. Concomitantly, macroinvertebrate samples were collected by kick- and sweep-netting. All samples were dried at 50°C for 48 h before being sent to the Cornell Isotope Laboratory for analysis (Cornell University, New York, USA). The outputs were values of $\delta^{13}C$ and $\delta^{15}N$ for each individual fish and their putative food resources. Where $\delta^{15}N$ has been converted to trophic position, it used the formula $[(\delta^{15}N_i - \delta^{15}N_{baseline})/3.4] + 2$, where $N_{baseline}$ is the mean $\delta^{15}N$ of the putative food resources (macroinvertebrates). Where $\delta^{13}C$ was converted to C_{corr}, conversion was via $(\delta^{13}C_i - \delta^{13}C_{mean})/CR$, where $\delta^{13}C_{corr}$ is the corrected carbon isotope ratio of the fish, $\delta^{13}C_i$ is the uncorrected isotope ratio of the fish, $\delta^{13}C_{mean}$ is the mean macroinvertebrate isotope ratio and CR is the invertebrate carbon range $(\delta^{13}C_{max} - \delta^{13}C_{min})$. Irrespective of whether the stable isotope data were corrected or not, they were used to determine the trophic niche width of each population subgroup (infected/uninfected with the focal parasite) using the metric 'standard ellipse area' (SEA$_c$) (Jackson et al., 2011, 2012). These ellipses are based on the distribution of individuals in isotopic space as an estimate of each species core trophic niche using the 'siar' package (Jackson et al., 2011; Parnell et al., 2010) in the 'R' computing programme (R Core Development Team, 2012). The subscript 'c' in 'SEA$_c$' indicates that a small sample size correction factor was used, as sample size tended to be below 20 per species per subgroup (Jackson et al., 2011).

Fig. 6 shows the trophic niche sizes of subgroups of minnow *Phoxinus phoxinus* and bullhead *Cottus gobio* that are infected/uninfected with the focal parasite *P. laevis* in the River Loddon, Southern England. This is an intestinal parasite that has a complex life cycle involving *G. pulex* as the intermediate host. In both species, the infected fishes had a trophic niche that sat within the larger trophic niche of the uninfected fishes, suggesting that although both subgroups were consuming similar prey items, the infected individuals were specializing on only a proportion of these items, resulting in constriction of their niche. Fig. 7 revealed the trophic niche sizes of the infected/uninfected subgroups of a roach *Rutilus rutilus* population infected with *L. intestinalis*, and a *G. aculeatus* population infected with *S. solidus*. Both parasites have a complex life cycle where the fish represent the second intermediate host, where the final host is a fish-eating bird. For both host–parasite systems, the trophic niche of the infected fishes had almost completely diverged from that of the uninfected fishes. Thus, across these examples, stable isotope analysis revealed that trophic consequences of infection for the host population were marked, with these insights difficult to predict from qualitative network approaches. These outputs could then be used as the basis of weighting the interaction frequencies between infected population subgroups and their prey.

8. CONCLUSIONS AND FORWARD LOOK

Parasites can reach novel ecosystems with or without free-living species, and despite enemy release, the rate of parasite introductions is accelerating around the world. Identifying the structural features of recipient networks that constrain or promote the establishment success of introduced parasites has proven to be complex as antagonistic effects might be confounded. Particularly, the role of species diversity and connectance in invasibility by parasites has to be explored further. Less ambiguously, it seems that modular networks could more easily accommodate parasites experiencing strong ontogenetic shifts. Once established, the extent to which exotic parasites alter food-web structure mostly depends on their life strategies. For instance, pathogenic parasites causing emerging infectious diseases or acting as biological weapons for their cointroduced host are likely to make food webs more simple through the removal of nodes and links, whereas complex life cycle parasites with trophic transmission are likely to make food webs more complex through the creation of new links. At the individual and population levels, this involves various processes including parasite spill-over and spill-back, the modification of host phenotype and the mediation of biotic interactions. However, the ecological significance of such increased complexity in networks, and particularly whether it promotes or constrains stability, has to be further explored through theoretical and experimental investigations. For this, some outstanding methodological issues have to be resolved such as the problem of node resolution, or the question of how to deal with the distinct subwebs of a whole food web.

In addition to the traditional predator–prey links, including parasites in food webs adds new types of links that are often unequally documented. For instance, predator–parasite links, whereby free-living species prey directly on parasites at a free-living stage or indirectly through concomitant predation of their host, are not the most obvious and may be quite difficult to detect and quantify. However, concomitant predation could be what makes the topological role of parasites unique compared to free-living species, whilst predation on parasites may confer biotic resistance to native networks. Additional data are needed on predator–parasite links.

The network approach improves our understanding of invasive species impacts as it allows to track how they propagate throughout invaded communities through bottom-up and top-down effects involving density-dependent regulation and apparent competition. Nevertheless, some impacts may not be detected with a qualitative approach, which does not account for the strength of interactions. This is particularly true for parasites, which establish weak links with their hosts compared to predator–prey links. Accounting for this asymmetry of interaction strength through empirical data, heuristic approaches or stable isotopes, is also important when it comes to investigate the role of parasites in food-web stability. To go further in this way, it has been proposed to combine the network approach with an energetic approach (Sukhdeo, 2012). Under the energetic perspective, energy is the currency of biological interactions and can be used to characterize the role of parasites in food webs, for instance, by focusing on how host biomass and its stability constrains parasitism or how parasites drive energy fluxes in the web. This can be done by measuring and scaling up the costs associated with infection at the individual scale to the whole population based on parasite prevalence (see Lettini and Sukhdeo, 2010 for an example).

On the long run, free-living invaders are predicted to become important players in parasite diversity. As stated by the 'invasional meltdown hypothesis' (Simberloff and Von Holle, 1999) and illustrated in Amundsen et al. (2013), free-living invaders and the food-web reconfigurations they induce should facilitate subsequent invasions by the species with which they have coevolved, including parasites, whilst their central location in the web should facilitate the acquisition of local parasites.

APPENDIX. DATA FOR THE CASE STUDY OF ENEMY RELEASE IN INVASIVE CICHLID FISH

The table below shows a literature survey combining the reported Ancyrocephalid gill parasites species richness and the number of introduced host species per territory.

This table synthetizes available data on regional species richness of ancyrocephalid parasites in tilapia populations established out of Africa (their native range). For further details and list of references, see Firmat et al.'s (2016) supplementary material available at: https://static-content.springer.com/esm/art%3A10.1007%2Fs00436-016-5168-1/MediaObjects/436_2016_5168_MOESM1_ESM.doc.

Territory	Number Sampled Hosts Individuals Analysed	Number of African host species analyzed	Number of Established Host Species	Gill Monogenean Species Richness
Mexico	403	4	5	5
Cuba	—	1	4	2
Panama	80	2	3	3
Brazil	240	1	4	5
Bangladesh	—	2	2	4
Madagascar	—	8	8	12
Colombia	—	1	3	2
Viet Nam	—	3	4	3
Philippines	175	1	6	4
Thailand	—	1	5	5
China	—	2	6	5
Japan	212	3	9	3
Australia	668	2	4	1
New Caledonia	62	1	2	0

REFERENCES

Adamek, Z., Barus, V., Prokes, M., 1996. Summer diet of roach (*Rutilus rutilus*) infested by *Ligula infestinalis* (Cestoda) pleurocercoids in the Dalesice reservoir (Czech Republic). Folia Zool. 45, 347–354.

Albrecht, M., Duelli, P., Schmid, B., Müller, C.B., 2007. Interaction diversity within quantified insect food webs in restored and adjacent intensively managed meadow. J. Anim. Ecol. 76, 1015–1025.

Aliabadi, B.W., Juliano, S.A., 2002. Escape from gregarine parasites affects the competitive interactions of an invasive mosquito. Biol. Invasions 4, 283–297.

Amundsen, P.A., Lafferty, K.D., Knudsen, R., Primicerio, R., Klemetsen, A., Kuris, A.M., 2009. Food web topology and parasites in the pelagic zone of a subarctic lake. J. Anim. Ecol. 78, 563–572.

Amundsen, P.A., Lafferty, K.D., Knudsen, R., Primicerio, R., Kristoffersen, R., Klemetsen, A., Kuris, A.M., 2013. New parasites and predators follow the introduction of two fish species to a subarctic lake: implications for food-web structure and functioning. Oecologia 171, 993–1002.

Anderson, R.M., May, R.M., 1979. Population biology of infectious diseases: part I. Nature 280, 361–367.

Anderson, T.K., Sukhdeo, M.V.K., 2011. Host centrality in food web networks determines parasite diversity. PLoS One 6, e26798.

Baiser, B., Russell, G.J., Lockwood, J.L., 2010. Connectance determines invasion success via trophic interactions in model food webs. Oikos 119, 1970–1976.

Barber, I., Hoare, D.J., Krause, J., 2000. The effects of parasites on fish behaviour: an evolutionary perspective and review. Rev. Fish Biol. Fish. 10, 1–35.

Bascompte, J., Jordano, P., Melián, C.J., Olesen, J.M., 2003. The nested assembly of plant–animal mutualistic networks. Proc. Natl. Acad. Sci. U.S.A. 100, 9383–9387.

Bauer, A., Haine, E.R., Perrot-Minnot, M.J., Rigaud, T., 2005. The acanthocephalan parasite *Polymorphus minutus* alters the geotactic and clinging behaviours of two sympatric amphipod hosts: the native *Gammarus pulex* and the invasive *Gammarus roeseli*. J. Zool. 267, 39–43.

Beckerman, A.P., Petchey, O.L., 2009. Infectious food webs. J. Anim. Ecol. 78, 493–496.

Bellay, S., Lima, D.P., Takemoto, R.M., Luque, J.L., 2011. A host-endoparasite network of Neotropical marine fish: are there organizational patterns? Parasitology 138, 1945–1952.

Bellay, S., de Oliveira, E.F., Almeida-Neto, M., Lima Junior, D.P., Takemoto, R.M., Luque, J.L., 2013. Developmental stage of parasites influences the structure of fish-parasite networks. PLoS One 8, 6–11.

Bellay, S., Oliveira, E.F., Almeida-Neto, M., Mello, M.A.R., Takemoto, R.M., Luque, J.L., 2015. Ectoparasites and endoparasites of fish form networks with different structures. Parasitology 142, 901–909.

Benejam, L., Alcaraz, C., Sasal, P., Simon-Levert, G., García-Berthou, E., 2009. Life history and parasites of the invasive mosquitofish (*Gambusia holbrooki*) along a latitudinal gradient. Biol. Invasions 11, 2265–2277.

Benesh, D.P., Chubb, J.C., Parker, G.A., 2014. The trophic vacuum and the evolution of complex life cycles in trophically transmitted helminths. Proc. R. Soc. Lond. B Biol. Sci. 281. 20141462.

Bersier, L.F., Banašek-Richter, C., Cattin, M.F., 2002. Quantitative descriptors of food-web matrices. Ecology 83, 2394–2407.

Blackburn, T.M., Pyšek, P., Bacher, S., Carlton, J.T., Duncan, R.P., Jarošík, V., Wilson, J.R.U., Richardson, D.M., 2011. A proposed unified framework for biological invasions. Trends Ecol. Evol. 26, 333–339.

Blakeslee, A.M., Altman, I., Miller, A.W., Byers, J.E., Hamer, C.E., Ruiz, G.M., 2012. Parasites and invasions: a biogeographic examination of parasites and hosts in native and introduced ranges. J. Biogeogr. 39, 609–622.

Blanchet, S., Rey, O., Loot, G., 2010. Evidence for host variation in parasite tolerance in a wild fish population. Evol. Ecol. 24, 129–1139.

Boettner, G.H., Elkinton, J.S., Boettner, C.J., 2000. Effects of a biological control introduction on three nontarget native species of saturniid moths. Conserv. Biol. 14, 1798–1806.

Braga, M.P., Araújo, S.B.L., Boeger, W.A., 2014. Patterns of interaction between Neotropical freshwater fishes and their gill Monogenoidea (Platyhelminthes). Parasitol. Res. 113, 481–490.

Braga, M.P., Razzolini, E., Boeger, W.A., 2015. Drivers of parasite sharing among Neotropical freshwater fishes. J. Anim. Ecol. 84, 487–497.

Brito, S.V., Corso, G., Almeida, A.M., Ferreira, F.S., Almeida, W.O., Anjos, L.A., Mesquita, D.O., Vasconcellos, A., 2014. Phylogeny and micro-habitats utilized by lizards determine the composition of their endoparasites in the semiarid Caatinga of Northeast Brazil. Parasitol. Res. 113, 3963–3972.

Britton, J.R., 2013. Introduced parasites in food webs: new species, shifting structures? Trends Ecol. Evol. 28, 93–99.

Britton, J.R., Andreou, D., 2016. Parasitism as a driver of trophic niche specialisation. Trends Parasitol. 32, 437–445. http://dx.doi.org/10.1016/j.pt.2016.02.007.

Britton, J.R., Jackson, M.C., Harper, D.M., 2009. *Ligula intestinalis* (Cestoda: Diphyllobothriidae) in Kenya: a field investigation into host specificity and behavioural alterations. Parasitology 136, 1367–1373.

Britton, J.R., Pegg, J., Williams, C.F., 2011. Pathological and ecological host consequences of infection by an introduced fish parasite. PLoS One 6 (10), e26365.

Britton, J.R., Pegg, J., Baker, D., Williams, C.F., 2012. Do lower feeding rates result in reduced growth of a cyprinid fish infected with the Asian tapeworm? Ecol. Freshw. Fish 21, 172–175.

Bulté, G., Robinson, S.a., Forbes, M.R., Marcogliese, D.J., 2012. Is there such thing as a parasite free lunch? The direct and indirect consequences of eating invasive prey. Ecohealth 9, 6–16.

Byers, J.E., 2008. Including parasites in food webs. Trends Parasitol. 25, 55–57.

Campião, K.M., Ribas, A., Tavares, L.E.R., 2015. Diversity and patterns of interaction of an anuran–parasite network in a neotropical wetland. Parasitology 142, 1751–1757.

Canonico, G.C., Arthington, A., McCrary, J.K., Thieme, M.L., 2005. The effects of introduced tilapias on native biodiversity. Aquat. Conserv. Mar. Freshw. Ecosyst. 15, 463–483.

Carvalheiro, L.G., Buckley, Y.M., Memmott, J., 2010. Diet breadth influences how the impact of invasive plants is propagated through food webs. Ecology 91, 1063–1074.

Chen, H.W., Liu, W.C., Davis, A.J., Jordán, F., Hwang, M.J., Shao, K.T., 2008. Network position of hosts in food webs and their parasite diversity. Oikos 117, 1847–1855.

Chen, H.W., Shao, K.T., Liu, C.W.J., Lin, W.H., Liu, W.C., 2011. The reduction of food web robustness by parasitism: fact and artefact. Int. J. Parasitol. 41, 627–634.

Cirtwill, A.R., Stouffer, D.B., 2015. Concomitant predation on parasites is highly variable but constrains the ways in which parasites contribute to food web structure. J. Anim. Ecol. 84, 734–744.

Cohen, J.E., Beaver, R.A., Cousins, S.H., DeAngelis, D.L., Goldwasser, L., Heong, K.L., Holt, R.D., Kohn, A.J., Lawton, J.H., Martinez, N., O'Malley, R., Page, L.M., Patten, B.C., Pimm, S.L., Polis, G.A., Rejmánek, M., Schoener, T.W., Schoenly, K., Sprules, W.G., Teal, J.M., Ulanowicz, R.E., Warren, P.H., Wilbur, H.M., Yodzis, P., 1993. Improving food webs. Ecology 74, 252–258.

Cunningham, E.J., Tierney, J.F., Huntingford, F.A., 1994. Effects of the cestode *Schistocephalus solidus* on food intake and foraging decisions in the three-spined stickleback *Gasterosteus aculeatus*. Ethology 97, 65–75.

Darwin, C., 1859. On the Origin of Species by Means of Natural Selection. Murray, London, UK.

Daszak, P., Cunningham, A.A., Hyatt, A.D., 2000. Emerging infectious diseases of wildlife—threats to biodiversity and human health. Science 287, 443–449.

David, P., Thébault, E., Anneville, O., Duyck, P.-F., Chapuis, E., Loeuille, N., 2017. Impacts of invasive species on food webs: a review of empirical data. Adv. Ecol. Res. 56, 1–60.

DeNiro, M.J., Epstein, S., 1981. Influence of diet on the distribution of nitrogen isotopes in animals. Geochim. Cosmochim. Acta 45, 341–351.

Dunn, A.M., Dick, J.T., 1998. Parasitism and epibiosis in native and non-native gammarids in freshwater in Ireland. Ecography 21, 593–598.

Dunn, A.M., Hatcher, M.J., 2015. Parasites and biological invasions: parallels, interactions, and control. Trends Parasitol. 31, 189–199.

Dunn, J.C., McClymont, H.E., Christmas, M., Dunn, A.M., 2009. Competition and parasitism in the native White Clawed Crayfish *Austropotamobius pallipes* and the invasive Signal Crayfish *Pacifastacus leniusculus* in the UK. Biol. Invasions 11, 315–324.

Dunn, A.M., Torchin, M.E., Hatcher, M.J., Kotanen, P.M., Blumenthal, D.M., Byers, J.E., Coon, C.A.C., Frankel, V.M., Holt, R.D., Hufbauer, R.A., Kanarek, A.R., Schierenbeck, K.A., Wolfe, L.M., Perkins, S.E., 2012. Indirect effects of parasites in invasions. Funct. Ecol. 26, 1262–1274.

Dunne, J.A., Williams, R.J., Martinez, N.D., 2002. Network structure and biodiversity loss in food webs: robustness increases with connectance. Ecol. Lett. 5, 558–567.

Dunne, J.A., Lafferty, K.D., Dobson, A.P., Hechinger, R.F., Kuris, A.M., Martinez, N.D., McLaughlin, J.P., Mouritsen, K.N., Poulin, R., Reise, K., Stouffer, D.B., Thieltges, D.W., Williams, R.J., Zander, C.D., 2013. Parasites affect food web structure primarily through increased diversity and complexity. PLoS Biol. 11, e1001579.

Elton, C.S., 1927. Animal Ecology. Sidgwick & Jackson, London.

Emmerson, M.C., Raffaelli, D., 2004. Predator–prey body size, interaction strength and the stability of a real food web. J. Anim. Ecol. 73, 399–409.

Firmat, C., Alibert, P., Mutin, G., Losseau, M., Pariselle, A., Sasal, P., 2016. A case of complete loss of gill parasites in the invasive cichlid *Oreochromis mossambicus*. Parasitol. Res. 115 (9), 3657–3661.

Fonseca, C.R., Prado, P.I., Almeida-Neto, M., Kubota, U., Lewinsohn, T.M., 2005. Flower-heads, herbivores, and their parasitoids: food web structure along a fertility gradient. Ecol. Entomol. 30, 36–46.

Fontaine, C., Guimarães, P.R., Kéfi, S., Loeuille, N., Memmott, J., van der Putten, W.H., van Veen, F.J.F., Thébault, E., 2011. The ecological and evolutionary implications of merging different types of networks. Ecol. Lett. 14, 1170–1181.

Food and Agriculture Organization, 2004. Tilapias as alien aquatics in Asia and the Pacific: a review FAO Fisheries Technical Paper. http://www.fao.org/docrep/007/y5728e/y5728e04.htm#bm4.

Fortuna, M.A., Stouffer, D.B., Olesen, J.M., Jordano, P., Mouillot, D., Krasnov, B.R., Poulin, R., Bascompte, J., 2010. Nestedness versus modularity in ecological networks: two sides of the same coin? J. Anim. Ecol. 79, 811–817.

Frick, W.F., Pollock, J.F., Hicks, A.C., Langwig, K.E., Reynolds, D.S., Turner, G.G., Butchkoski, C.M., Kunz, T.H., 2010. An emerging disease causes regional population collapse of a common North American bat species. Science 329, 679–682.

Froese, R., Pauly, D., 2011. FishBase. World Wide Web Electronic Publication. www.fishbase.org.

Fry, B., 2007. Stable Isotope Ecology. Springer Science & Business Media, New York, NY.

Gagic, V., Tscharntke, T., Dormann, C.F., Gruber, B., Wilstermann, A., Thies, C., 2011. Food web structure and biocontrol in a four-trophic level system across a landscape complexity gradient. Proc. R. Soc. Lond. B Biol. Sci. 278, 2946–2953.

Gendron, A.D., Marcogliese, D.J., Thomas, M., 2012. Invasive species are less parasitized than native competitors, but for how long? The case of the round goby in the Great Lakes-St. Lawrence Basin. Biol. Invasions 14, 367–384.

Goedknegt, M.A., Welsh, J.E., Drent, J., Thieltges, D.W., 2015. Climate change and parasite transmission: how temperature affects parasite infectivity via predation on infective stages. Ecosphere 6, 1–9.

Graham, S.P., Hassan, H.K., Burkett-Cadena, N.D., Guyer, C., Unnasch, T.R., 2009. Nestedness of ectoparasite-vertebrate host networks. PLoS One 4, e7873.

Grey, J., 2006. The use of stable isotope analyses in freshwater ecology: current awareness. Pol. J. Ecol. 54, 563–584.

Guégan, J.F., Kennedy, C.R., 1993. Maximum local helminth parasite community richness in British freshwater fish: a test of the colonization time hypothesis. Parasitology 106, 91–100.

Hartigan, A., Fiala, I., Dyková, I., Jirků, M., Okimoto, B., Rose, K., Phalen, D.N., Šlapeta, J., 2011. A suspected parasite spill-back of two novel Myxidium spp.

(Myxosporea) causing disease in Australian endemic frogs found in the invasive cane toad. PLoS One 6, e18871.

Hatcher, M.J., Dunn, A.M., 2011. Parasites in Ecological Communities: From Interactions to Ecosystems. Cambridge University Press, Cambridge, UK.

Hatcher, M.J., Dick, J.T.A., Dunn, A.M., 2006. How parasites affect interactions between competitors and predators. Ecol. Lett. 9, 1253–1271.

Hatcher, M.J., Dick, J.T.A., Dunn, A.M., 2012a. Diverse effects of parasites in ecosystems: linking interdependent processes. Front. Ecol. Environ. 10, 186–194.

Hatcher, M.J., Dick, J.T.A., Dunn, A.M., 2012b. Disease emergence and invasions. Funct. Ecol. 26, 1275–1287.

Hatcher, M.J., Dick, J.T.A., Dunn, A.M., 2014. Parasites that change predator or prey behaviour can have keystone effects on community composition. Biol. Lett. 10. 20130879.

Hawkins, B.A., Marino, P.C., 1997. The colonization of native phytophagous insects in North America by exotic parasitoids. Oecologia 112, 566–571.

Hawley, D.M., Dhondt, K.V., Dobson, A.P., Grodio, J.L., Hochachka, W.M., Ley, D.H., Osnas, E.E., Schat, K.A., Dhondt, A.A., 2010. Common garden experiment reveals pathogen isolate but no host genetic diversity effect on the dynamics of an emerging wildlife disease. J. Evol. Biol. 23, 1680–1688.

Hechinger, R.F., Lafferty, K.D., McLaughlin, J.P., Fredensborg, B.L., Huspeni, T.C., Lorda, J., Sandhu, P.K., Shaw, J.C., Torchin, M.E., Whitney, K.L., Kuris, A.M., 2011. Food webs including parasites, biomass, body sizes, and life stages for three California/BajaCalifornia Estuaries. Ecology 92, 791.

Heleno, R.H., Ceia, R.S., Ramos, J.A., Memmott, J., 2009. Effects of alien plants on insect abundance and biomass: a food-web approach. Conserv. Biol. 23, 410–419.

Henneman, M.L., Memmott, J., 2001. Infiltration of a Hawaiian community by introduced biological control agents. Science 293, 1314–1316.

Hernandez, A.D., Sukhdeo, M.V.K., 2008. Parasites alter the topology of a stream food web across seasons. Oecologia 156, 613–624.

Hine, P.M., Jones, J.B., Diggles, B.K., 2000. A checklist of the parasites of New Zealand fishes, including previously unpublished records: NIWA.

Holt, R.D., 1977. Predation, apparent competition, and the structure of prey communities. Theor. Popul. Biol. 12, 197–229.

Holt, R.D., Dobson, A.P., 2006. Extending the principles of community ecology to address the epidemiology of host-pathogen systems. In: Collinge, S.K., Ray, C. (Eds.), Disease Ecology: Community Structure and Pathogen Dynamics, Oxford University Press, Oxford, pp. 6–27.

Holt, R.D., Pickering, J., 1985. Infectious disease and species coexistence: a model of Lotka-Volterra form. Am. Nat. 126, 196–211.

Hui, C., Richardson, D.M., Landi, P., Minoarivelo, H.O., Garnas, J., Roy, H.E., 2016. Defining invasiveness and invasibility in ecological networks. Biol. Invasions 18, 971–983.

Hulme, P.E., Pyšek, P., Nentwig, W., Vilà, M., 2009. Will threat of biological invasions unite the European Union. Science 324, 40–41.

Huxham, M., Raffaelli, D., Pike, A., 1995. Parasites and food web patterns. J. Anim. Ecol. 64, 168–176.

Jackson, A.L., Inger, R., Parnell, A.C., Bearhop, S., 2011. Comparing isotopic niche widths among and within communities: SIBER–Stable Isotope Bayesian Ellipses in R. J. Anim. Ecol. 80, 595–602.

Jackson, M.C., Donohue, I., Jackson, A.L., Britton, J.R., Harper, D.M., Grey, J., 2012. Population-level metrics of trophic structure based on stable isotopes and their application to invasion ecology. PLoS One 7, e31757.

Jacobs, A.Z., Dunne, J.A., Moore, C., Clauset, A., 2015. Untangling the roles of parasites in food webs with generative network models. arXiv: 1505.04741.

Jacquin, L., Mori, Q., Pause, M., Steffen, M., Médoc, V., 2014. Non-specific manipulation of gammarid behaviour by *P. minutus* parasite enhances their predation by definitive bird hosts. PLoS One 9, e101684.

Jansen, P.A., Bakke, T.A., 1993. Regulatory processes in the monogenean *Gyrodactylus salaris* Malmberg—Atlantic salmon (*Salmo salar* L.) association. I. Field studies in southeast Norway. Fish. Res. 17, 87–101.

Jephcott, T.G., Sime-Ngando, T., Gleason, F.H., Macarthur, D.J., 2016. Host-parasite interactions in food webs: diversity, stability, and coevolution. Food Webs 6, 1–8.

Jimenez-Garcia, M.I., Vidal-Martinez, V.M., Lopez-Jimenez, S., 2001. Monogeneans in introduced and native cichlids in Mexico: evidence for transfer. J. Parasitol. 87, 907–909.

Johnson, P.T., Hoverman, J.T., 2012. Parasite diversity and coinfection determine pathogen infection success and host fitness. Proc. Natl. Acad. Sci. U.S.A. 109, 9006–9011.

Johnson, P.T., Hartson, R.B., Larson, D.J., Sutherland, D.R., 2008. Diversity and disease: community structure drives parasite transmission and host fitness. Ecol. Lett. 11, 1017–1026.

Johnson, P.T.J., Dobson, A., Lafferty, K.D., Marcogliese, D.J., Memmott, J., Orlofske, S.A., Poulin, R., Thieltges, D.W., 2010. When parasites become prey: ecological and epidemiological significance of eating parasites. Trends Ecol. Evol. 25, 362–371.

Kaartinen, R., Roslin, T., 2011. Shrinking by numbers: landscape context affects the species composition but not the quantitative structure of local food webs. J. Anim. Ecol. 80, 622–631.

Kamenova, S., Bartley, T.J., Bohan, D.A., Boutain, J.R., Colautti, R.I., Domaizon, I., Fontaine, C., Lemainque, A., Le Viol, I., Mollot, G., Perga, M.-E., Ravigné, V., Massol, F., 2017. Invasions toolkit: current methods for tracking the spread and impact of invasive species. Adv. Ecol. Res. 56, 85–182.

Kamiya, T., O'Dwyer, K., Nakagawa, S., Poulin, R., 2014. Host diversity drives parasite diversity: meta-analytical insights into patterns and causal mechanisms. Ecography 37, 689–697.

Keane, R.M., Crawley, M.J., 2002. Exotic plant invasions and the enemy release hypothesis. Trends Ecol. Evol. 17, 164–170.

Kelly, D.W., Paterson, R.A., Townsend, C.R., Poulin, R., Tompkins, D.M., 2009. Parasite spillback: a neglected concept in invasion ecology? Ecology 90, 2047–2056.

Kennedy, T.A., Naeem, S., Howe, K.M., Knops, J.M., Tilman, D., Reich, P., 2002. Biodiversity as a barrier to ecological invasion. Nature 417, 636–638.

Kirk, R.S., 2003. The impact of Anguillicola crassus on European eels. Fish. Manag. Ecol. 10, 385–394.

Knudsen, R., Curtis, M.A., Kristoffersen, R., 2004. Aggregation of helminths: the role of feeding behavior of fish hosts. J. Parasitol. 90, 1–7.

Krasnov, B.R., Fortuna, M.a., Mouillot, D., Khokhlova, I.S., Shenbrot, G.I., Poulin, R., 2012. Phylogenetic signal in module composition and species connectivity in compartmentalized host-parasite networks. Am. Nat. 179, 501–511.

Kuang, W., Zhang, W., 2011. Some effects of parasitism on food web structure: a topological analysis. Netw. Biol. 1, 171–185.

Kuris, A.M., Lafferty, K.D., 2000. Parasite–host modelling meets reality: adaptive peaks and their ecological attributes. In: Poulin, R., Morand, S., Skorping, A. (Eds.), Evolutionary Biology of Host-Parasite Relationships: Theory Meets Reality. Elsevier Science, Amsterdam, pp. 9–26.

Kuris, A.M., Hechinger, R.F., Shaw, J.C., Whitney, K.L., Aguirre-Macedo, L., Boch, C.A., Dobson, A.P., Dunham, E.J., Fredensborg, B.L., Huspeni, T.C., Lorda, J., Mababa, L., Mancini, F.T., Mora, A.B., Pickering, M., Talhouk, N.L., Torchin, M.E.,

Lafferty, K.D., 2008. Ecosystem energetic implications of parasite and free-living biomass in three estuaries. Nature 454, 515–518.

Lafferty, K.D., 2009. The ecology of climate change and infectious diseases. Ecology 90, 888–900.

Lafferty, K.D., 2012. Biodiversity loss decreases parasite diversity: theory and patterns. Philos. Trans. R. Soc. Lond. B Biol. Sci. 367, 2814–2827.

Lafferty, K.D., Kuris, A.M., 2002. Trophic strategies, animal diversity and body size. Trends Ecol. Evol. 17, 507–513.

Lafferty, K.D., Kuris, A.M., 2009. Parasites reduce food web robustness because they are sensitive to secondary extinction as illustrated by an invasive estuarine snail. Philos. Trans. R. Soc. Lond. B Biol. Sci. 364, 1659–1663.

Lafferty, K.D., Dobson, A.P., Kuris, A.M., 2006. Parasites dominate food web links. Proc. Natl. Acad. Sci. U.S.A. 103, 11211–11216.

Lafferty, K.D., Allesina, S., Arim, M., Briggs, C.J., De Leo, G., Dobson, A.P., Dunne, J.A., Johnson, P.T.J., Kuris, A.M., Marcogliese, D.J., Martinez, N.D., Memmott, J., Marquet, P.A., McLaughlin, J.P., Mordecai, E.A., Pascual, M., Poulin, R., Thieltges, D.W., 2008. Parasites in food webs: the ultimate missing links. Ecol. Lett. 11, 533–546.

Lafferty, K.D., Torchin, M.E., Kuris, A.M., 2010. The geography of host and parasite invasions. In: Morand, S., Krasnov, B.R. (Eds.), The Biogeography of Host–Parasite Interactions. Oxford University Press, New York, pp. 191–203.

Lettini, S.E., Sukhdeo, M.K., 2010. The energetic cost of parasitism in a population of isopods. Ecoscience 17, 1–8.

Lewinsohn, T., Prado, P.I., Jordano, P., Bascompte, J., Olesen, J., 2006. Structure in plant–animal interaction assemblages. Oikos 113, 174–184.

Lima, D.P., Giacomini, H.C., Takemoto, R.M., Agostinho, A.A., Bini, L.M., 2012. Patterns of interactions of a large fish-parasite network in a tropical floodplain. J. Anim. Ecol. 81, 905–913.

Locke, S.A., Marcogliese, D.J., Tellervo Valtonen, E., 2014. Vulnerability and diet breadth predict larval and adult parasite diversity in fish of the Bothnian Bay. Oecologia 174, 253–262.

Lohaus, K., Vidal, S., Thies, C., 2013. Farming practices change food web structures in cereal aphid–parasitoid–hyperparasitoid communities. Oecologia 171, 249–259.

Lonsdale, W.M., 1999. Global patterns of plant invasions and the concept of invasibility. Ecology 80, 1522–1536.

Loot, G., Brosse, S., Lek, S., Guégan, J.F., 2001. Behaviour of roach (*Rutilus rutilus* L.) altered by *Ligula intestinalis* (Cestoda: Pseudophyllidea): a field demonstration. Freshw. Biol. 46, 1219–1227.

Luque, J.L., Poulin, R., 2007. Metazoan parasite species richness in Neotropical fishes: hotspots and the geography of biodiversity. Parasitology 134, 865–878.

Lurgi, M., Galiana, N., López, B.C., Joppa, L.N., Montoya, J.M., 2014. Network complexity and species traits mediate the effects of biological invasions on dynamic food webs. Front. Ecol. Evol. 2, 1–11.

Lymbery, A.J., Morine, M., Kanani, H.G., Beatty, S.J., Morgan, D.L., 2014. Co-invaders: the effects of alien parasites on native hosts. Int. J. Parasitol. Parasites Wildl. 3, 171–177.

MacLeod, C.J., Paterson, A.M., Tompkins, D.M., Duncan, R.P., 2010. Parasites lost—do invaders miss the boat or drown on arrival? Ecol. Lett. 13, 516–527.

Macnab, V., Barber, I., 2012. Some (worms) like it hot: fish parasites grow faster in warmer water, and alter host thermal preferences. Glob. Chang. Biol. 18, 1540–1548.

MacNeil, C., Fielding, N.J., Hume, K.D., Dick, J.T., Elwood, R.W., Hatcher, M.J., Dunn, A.M., 2003. Parasite altered micro-distribution of *Gammarus pulex* (Crustacea: Amphipoda). Int. J. Parasitol. 33, 57–64.

Madanire-Moyo, G.N., Matla, M.M., Olivier, P.A.S., Luus-Powell, W.J., 2011. Population dynamics and spatial distribution of monogeneans on the gills of *Oreochromis mossambicus* (Peters, 1852) from two lakes of the Limpopo River System, South Africa. J. Helminthol. 85, 146–152.

Marcogliese, D.J., 2003. Food webs and biodiversity: are parasites the missing link. J. Parasitol. 89, 106–113.

Marcogliese, D.J., Cone, D.K., 1997. Food webs: a plea for parasites. Trends Ecol. Evol. 12, 320–325.

Marr, S.R., Mautz, W.J., Hara, A.H., 2008. Parasite loss and introduced species: a comparison of the parasites of the Puerto Rican tree frog (*Eleutherodactylus coqui*), in its native and introduced ranges. Biol. Invasions 10, 1289–1298.

Marrero, H.J., Torretta, J.P., Pompozzi, G., 2013. Triple interaction network among flowers, flower visitors and crab spiders in a grassland ecosystem. Stud. Neotropical Fauna Environ. 48, 153–164.

Maunsell, S.C., Kitching, R.L., Burwell, C.J., Morris, R.J., 2015. Changes in host-parasitoid food web structure with elevation. J. Anim. Ecol. 84, 353–363.

May, R.M., Anderson, R.M., 1979. Population biology of infectious diseases: part II. Nature 280, 455–461.

McCann, K., Hastings, A., Huxel, G.R., 1998. Weak trophic interactions and the balance of nature. Nature 395, 794–798.

Médoc, V., Beisel, J.N., 2009. Field evidence for non-host predator avoidance in a manipulated amphipod. Naturwissenschaften 96, 513–523.

Médoc, V., Beisel, J.N., 2011. When trophically-transmitted parasites combine predation enhancement with predation suppression to optimize their transmission. Oikos 120, 1452–1458.

Médoc, V., Bollache, L., Beisel, J.N., 2006. Host manipulation of a freshwater crustacean (*Gammarus roeseli*) by an acanthocephalan parasite (*Polymorphus minutus*) in a biological invasion context. Int. J. Parasitol. 36, 1351–1358.

Médoc, V., Rigaud, T., Bollache, L., Beisel, J.N., 2009. A manipulative parasite increasing an antipredator response decreases its vulnerability to a nonhost predator. Anim. Behav. 77, 1235–1241.

Memmott, J., Martinez, N.D., Cohen, J.E., 2000. Predators, parasitoids and pathogens: species richness, trophic generality and body sizes in a natural food web. J. Anim. Ecol. 69, 1–15.

Mendlova, M., Simkova, A., 2014. Evolution of host specificity in monogeneans parasitizing African cichlid fish. Parasit. Vectors 7, 69.

Milinski, M., 1984. Parasites determine a predators optimal feeding strategy. Behav. Ecol. Sociobiol. 15, 35–37.

Milinski, M., 1985. Risk of predation of parasitized sticklebacks (*Gasterosteus aculeatus* L.) under competition for food. Behaviour 93, 203–216.

Minagawa, M., Wada, E., 1984. Stepwise enrichment of 15 N along food chains: further evidence and the relation between δ 15 N and animal age. Geochim. Cosmochim. Acta 48, 1135–1140.

Mitchell, C.E., Power, A.G., 2003. Release of invasive plants from fungal and viral pathogens. Nature 421, 625–627.

Morris, R.J., Sinclair, F.H., Burwell, C.J., 2015. Food web structure changes with elevation but not rainforest stratum. Ecography 38, 792–802.

Mouillot, D., Krasnov, B.R., Poulin, R., 2008. High intervality explained by phylogenetic constraints in host–parasite webs. Ecology 89, 2043–2051.

Mouritsen, K.N., Poulin, R., McLaughlin, J.P., Thieltges, D.W., 2011. Food web including metazoan parasites for an intertidal ecosystem in New Zealand. Ecology 92, 2006.

Murakami, M., Hirao, T., Kasei, A., 2008. Effects of habitat configuration on host-parasitoid food web structure. Ecol. Res. 23, 1039–1049.

O'Hara, R.B., Kotze, D.J., 2010. Do not log-transform count data. Methods Ecol. Evol. 1, 118–122.

Olesen, J.M., Bascompte, J., Dupont, Y.L., Jordano, P., 2007. The modularity of pollination networks. Proc. Natl. Acad. Sci. U.S.A. 104, 19891–19896.

Ortubay, S., Semenas, L., Ubeda, C., Quaggiotto, A., Viozzi, G., 1994. Catalogo de peces dulceacuícolas de la Patagonia argentina y sus parásitos metazoos. Subsecretaría de Recursos Naturales, Río Negro.

Pariselle, A., Euzet, L., 2009. Systematic revision of dactylogyridean parasites (Monogenea) from cichlid fishes in Africa, the Levant and Madagascar. Zoosystema 31, 849–898.

Parker, G.A., Ball, M.A., Chubb, J.C., 2015. Evolution of complex life cycles in trophically transmitted helminths. I. Host incorporation and trophic ascent. J. Evol. Biol. 28 (2), 267–291.

Parnell, A.C., Inger, R., Bearhop, S., Jackson, A.L., 2010. Source partitioning using stable isotopes: coping with too much variation. PLoS One 5, e9672.

Parry, D., 2009. Beyond Pandora's box: quantitatively evaluating non-target effects of parasitoids in classical biological control. Biol. Invasions 11, 47–58.

Paterson, R.A., Townsend, C.R., Tompkins, D.M., Poulin, R., 2012. Ecological determinants of parasite acquisition by exotic fish species. Oikos 121, 1889–1895.

Pegg, J., Andreou, D., Williams, C.F., Britton, J.R., 2015a. Head morphology and piscivory of European eels, *Anguilla anguilla*, predict their probability of infection by the invasive parasitic nematode *Anguillicoloides crassus*. Freshw. Biol. 60, 1977–1987.

Pegg, J., Andreou, D., Williams, C.F., Britton, J.R., 2015b. Temporal changes in growth, condition and trophic niche in juvenile *Cyprinus carpio* infected with a non-native parasite. Parasitology 142, 1579–1587.

Penczykowski, R.M., Forde, S.E., Duffy, M.A., 2011. Rapid evolution as a possible constraint on emerging infectious diseases. Freshw. Biol. 56, 689–704.

Pilosof, S., Morand, S., Krasnov, B.R., Nunn, C.L., 2015. Potential parasite transmission in multi-host networks based on parasite sharing. PLoS One 10, 1–19.

Pocock, M.J., Evans, D.M., Fontaine, C., Harvey, M., Julliard, R., McLaughlin, Ó., Silvertown, J., Tamaddoni-Nezhad, A., White, P.C.L., Bohan, D.A., 2016. The visualisation of ecological networks, and their use as a tool for engagement, advocacy and management. Adv. Ecol. Res. 54, 41–85.

Ponton, F., Biron, D.G., Joly, C., Helluy, S., Duneau, D., Thomas, F., 2005. Ecology of parasitically modified populations: a case study from a gammarid-trematode system. Mar. Ecol. Prog. Ser. 299, 205–215.

Poulin, R., 2010. Network analysis shining light on parasite ecology and diversity. Trends Parasitol. 26, 492–498.

Poulin, R., 2011. The many roads to parasitism: a tale of convergence. Adv. Parasitol. 74, 40.

Poulin, R., Maure, F., 2015. Host manipulation by parasites: a look back before moving forward. Trends Parasitol. 31, 563–570.

Poulin, R., Paterson, R.A., Townsend, C.R., Tompkins, D.M., Kelly, D.W., 2011. Biological invasions and the dynamics of endemic diseases in freshwater ecosystems. Freshw. Biol. 56, 676–688.

Prenter, J., MacNeil, C., Dick, J.T.A., Dunn, A.M., 2004. Roles of parasites in animal invasions. Trends Ecol. Evol. 19, 385–390.

Preston, D.L., Orlofske, S.A., McLaughlin, J.P., Johnson, P.T.J., 2012. Food web including infectious agents for a California freshwater pond. Ecology 93, 1760.

Preston, D.L., Jacobs, A.Z., Orlofske, S.A., Johnson, P.T.J., 2014. Complex life cycles in a pond food web: effects of life stage structure and parasites on network properties, trophic positions and the fit of a probabilistic niche model. Oecologia 174, 953–965.

Proulx, S.R., Promislow, D.E.L., Phillips, P.C., 2005. Network thinking in ecology and evolution. Trends Ecol. Evol. 20, 345–353.

Pyšek, P., Jarošík, V., Hulme, P., Kühn, I., Wild, J., Arianoutsou, M., Bacher, S., Chiron, F., Didžiulis, V., Essl, F., Genovesi, P., Gherardi, F., Hejda, M., Kark, S., Lambdon, P.W., Desprez-Loustau, M.L., Nentwig, W., Pergl, J., Poboljšaj, K., Rabitsch, W., Roques, A., Roy, D.B., Shirley, S., Solarz, W., Montserrat, V., Winter, M., 2010. Disentangling the role of environmental and human pressures on biological invasions across Europe. Proc. Natl. Acad. Sci. U.S.A. 107, 12157–12162.

Ranta, E., 1995. *Schistocephalus* infestation improves prey-size selection by 3-spined sticklebacks, *Gasterosteus aculeatus*. J. Fish Biol. 46, 156–158.

R Core Development Team, 2012. R: A Language and Environment for Statistical Computing. R Foundation for Statistical Computing, Vienna, Austria. ISBN: 3-900051-07-0, URL Available at: http://www.R-project.org/.

Ricciardi, A., Rasmussen, J.B., 1998. Predicting the identity and impact of future biological invaders: a priority for aquatic resource management. Can. J. Fish. Aquat. Sci. 55, 1759–1765.

Richardson, D.M., Pyšek, P., 2006. Plant invasions: merging the concepts of species invasiveness and community invasibility. Prog. Phys. Geogr. 30, 409–431.

Riede, J.O., Rall, B.C., Banasek-Richter, C., Navarrete, S.A., Wieters, E.A., Emmerson, M.C., Jacob, U., Brose, U., 2010. Scaling of food-web properties with diversity and complexity across ecosystems. Adv. Ecol. Res. 42, 139–170.

Rohde, K., 1999. Latitudinal gradients in species diversity and Rapoport's rule revisited: a review of recent work and what can parasites teach us about the causes of the gradients? Ecography 22, 593–613.

Romanuk, T.N., Zhou, Y., Brose, U., Berlow, E.L., Williams, R.J., Martinez, N.D., 2009. Predicting invasion success in complex ecological networks. Philos. Trans. R. Soc. Lond. B Biol. Sci. 364, 1743–1754.

Rooney, N., McCann, K.S., 2012. Integrating food web diversity, structure and stability. Trends Ecol. Evol. 27, 40–45.

Ross, J.L., Ivanova, E.S., Severns, P.M., Wilson, M.J., 2010. The role of parasite release in invasion of the USA by European slugs. Biol. Invasions 12, 603–610.

Sato, T., Egusa, T., Fukushima, K., Oda, T., Ohte, N., Tokuchi, N., Watanabe, K., Kanaiwa, M., Murakami, I., Lafferty, K.D., 2012. Nematomorph parasites indirectly alter the food web and ecosystem function of streams through behavioural manipulation of their cricket hosts. Ecol. Lett. 15, 786–793.

Schall, J.J., 1992. Parasite-mediated competition in Anolis lizards. Oecologia 92, 58–64.

Scott, J., 2000. Social Network Analysis: A Handbook, second ed. Sage Publications, Newberry Park, CA.

Sheath, D.J., Williams, C.F., Reading, A.J., Britton, J.R., 2015. Parasites of non-native freshwater fishes introduced into England and Wales suggest enemy release and parasite acquisition. Biol. Invasions 17, 2235–2246.

Sheath, D.J., Andreou, D., Britton, J.R., 2016. Interactions of warming and exposure affect susceptibility to parasite infection in a temperate fish species. Parasitology 143, 1340–1346.

Simberloff, D., Von Holle, B., 1999. Positive interactions of nonindigenous species: invasional meltdown? Biol. Invasions 1, 21–32.

Sousa, W.P., 1992. Interspecific interactions among larval trematode parasites of freshwater and marine snails. Am. Zool. 32, 583–592.

Strauss, A., White, A., Boots, M., 2012. Invading with biological weapons: the importance of disease-mediated invasions. Funct. Ecol. 26, 1249–1261.

Sukhdeo, M.V.K., 2010. Food webs for parasitologists: a review. J. Parasitol. 96, 273–284.

Sukhdeo, M.V.K., 2012. Where are the parasites in food webs? Parasit. Vectors 5, 239.

Svanbäck, R., Quevedo, M., Olsson, J., Eklöv, P., 2015. Individuals in food webs: the relationships between trophic position, omnivory and among-individual diet variation. Oecologia 178, 103–114.

Taraschewski, H., 2006. Hosts and parasites as aliens. J. Helminthol. 80, 99–128.

Telfer, S., Bown, K.J., Sekules, R., Begon, M., Hayden, T., Birtles, R., 2005. Disruption of a host-parasite system following the introduction of an exotic host species. Parasitology 130, 661–668.

Thébault, E., Fontaine, C., 2010. Stability of ecological communities and the architecture of mutualistic and trophic networks. Science 80, 853–856.

Thieltges, D.W., Jensen, K.T., Poulin, R., 2008. The role of biotic factors in the transmission of free-living endohelminth stages. Parasitology 135, 407–426.

Thieltges, D.W., Reise, K., Prinz, K., Jensen, K.T., 2009. Invaders interfere with native parasite–host interactions. Biol. Invasions 11, 1421–1429.

Thieltges, D.W., Reise, K., Mouritsen, K.N., McLaughlin, J.P., Poulin, R., 2011. Food web including metazoan parasites for a tidal basin in Germany and Denmark: ecological archives E092-172. Ecology 92, 2005.

Thieltges, D.W., Amundsen, P.A., Hechinger, R.F., Johnson, P.T.J., Lafferty, K.D., Mouritsen, K.N., Preston, D.L., Reise, K., Zander, C.D., Poulin, R., 2013. Parasites as prey in aquatic food webs: implications for predator infection and parasite transmission. Oikos 122, 1473–1482.

Thomas, F., Schmidt-Rhaesa, A., Martin, G., Manu, C., Durand, P., Renaud, F., 2002. Do hairworms (Nematomorpha) manipulate the water seeking behaviour of their terrestrial hosts? J. Evol. Biol. 15, 356–361.

Thompson, R.A., 2013. Parasite zoonoses and wildlife: one health, spillover and human activity. Int. J. Parasitol. 43, 1079–1088.

Thompson, R.M., Mouritsen, K.N., Poulin, R., 2005. Importance of parasites and their life cycle characteristics in determining the structure of a large marine food web. J. Anim. Ecol. 74, 77–85.

Timms, L.L., Walker, S.C., Smith, S.M., 2012. Establishment and dominance of an introduced herbivore has limited impact on native host-parasitoid food webs. Biol. Invasions 14, 229–244.

Tompkins, D.M., Poulin, R., 2006. Parasites and biological invasions. In: Allen, R.B., Lee, W.G. (Eds.), Biological invasions in New Zealand. Ecological Studies, 168. Springer, Berlin and Heidelberg, pp. 67–84.

Tompkins, D.M., White, A.R., Boots, M., 2003. Ecological replacement of native red squirrels by invasive greys driven by disease. Ecol. Lett. 6, 189–196.

Tompkins, D.M., Dunn, A.M., Smith, M.J., Telfer, S., 2011. Wildlife diseases: from individuals to ecosystems. J. Anim. Ecol. 80, 19–38.

Torchin, M.E., Mitchell, C.E., 2004. Parasites, pathogens, and invasions by plants and animals. Front. Ecol. Environ. 2, 183–190.

Torchin, M.E., Lafferty, K.D., Kuris, A.M., 2001. Release from parasites as natural enemies: increased performance of a globally introduced marine crab. Biol. Invasions 3, 333–345.

Torchin, M.E., Lafferty, K.D., Dobson, A.P., McKenzie, V.J., Kuris, A.M., 2003. Introduced species and their missing parasites. Nature 421, 628–630.

Torchin, M.E., Byers, J.E., Huspeni, T.C., 2005. Differential parasitism of native and introduced snails: replacement of a parasite fauna. Biol. Invasions 7, 885–894.

Tylianakis, J.M., Tscharntke, T., Lewis, O.T., 2007. Habitat modification alters the structure of tropical host-parasitoid food webs. Nature 445, 202–205.

Tylianakis, J.M., Laliberté, E., Nielsen, A., Bascompte, J., 2010. Conservation of species interaction networks. Biol. Conserv. 143, 2270–2279.

Valladares, G., Cagnolo, L., Salvo, A., 2012. Forest fragmentation leads to food web contraction. Oikos 121, 299–305.

Vanhove, M.P.M., Hablützel, P.I., Pariselle, A., Šimková, A., Huyse, T., Raeymaekers, J.A.M., 2016. Cichlids: a host of opportunities for evolutionary parasitology. Trends Parasitol. 32, 820–832.

Van Kleunen, M., Dawson, W., Schlaepfer, D., Jeschke, J.M., Fischer, M., 2010. Are invaders different? A conceptual framework of comparative approaches for assessing determinants of invasiveness. Ecol. Lett. 13, 947–958.

Van Steenberge, M., Pariselle, A., Huyse, T., Volckaert, F.A.M., Snoeks, J., Vanhove, M.P.M., 2015. Morphology, molecules, and monogenean parasites: an example of an integrative approach to cichlid biodiversity. PLoS One 10, e0124474.

Vázquez, D.P., Poulin, R., Krasnov, B.R., Shenbrot, G.I., 2005. Species abundance and the distribution of specialization in host-parasite interaction networks. J. Anim. Ecol. 74, 946–955.

Vermaat, J.E., Dunne, J.A., Gilbert, A.J., 2009. Major dimensions in food-web structure properties. Ecology 90, 278–282.

Warren, P.H., Lawton, J.H., 1987. Invertebrate predator–prey body size relationships: an explanation for upper triangular food webs and patterns in food web structure? Oecologia 74 (2), 231–235.

Wei, Z., Yang, T., Friman, V.-P., Xu, Y., Shen, Q., Jousset, A., 2015. Trophic network architecture of root-associated bacterial communities determines pathogen invasion and plant health. Nat. Commun. 6, 8413.

Wood, C.L., Byers, J.E., Cottingham, K.L., Altman, I., Donahue, M.J., Blakeslee, A.M., 2007. Parasites alter community structure. Proc. Natl. Acad. Sci. U.S.A. 104, 9335–9339.

Zander, C.D., Josten, N., Detloff, K.C., Poulin, R., McLaughlin, J.P., Thieltges, D.W., 2011. Food web including metazoan parasites for a brackish shallow water ecosystem in Germany and Denmark: ecological archives E092-174. Ecology 92, 2007.

CHAPTER TWO

Novel and Disrupted Trophic Links Following Invasion in Freshwater Ecosystems

M.C. Jackson*,†,1, R.J. Wasserman‡,§, J. Grey¶, A. Ricciardi‖, J.T.A. Dick#, M.E. Alexander**

*Centre for Invasion Biology, University of Pretoria, Hatfield, South Africa
†Life Sciences, Imperial College London, Ascot, Berkshire, United Kingdom
‡South African Institute for Aquatic Biodiversity, Grahamstown, South Africa
§School of Science, Monash University, Jalan Lagoon Selatan, Selangor, Malaysia
¶Lancaster Environment Centre, Lancaster University, Lancaster, United Kingdom
‖Redpath Museum, McGill University, Montreal, QC, Canada
#Institute for Global Food Security, School of Biological Sciences, Queen's University Belfast, MBC, Belfast, United Kingdom
**Institute for Biomedical and Environmental Health Research, School of Science and Sport, University of the West of Scotland, Paisley, United Kingdom
¹Corresponding author: e-mail address: m.jackson@imperial.ac.uk

Contents

Advances in Ecological Research, Volume 57
ISSN 0065-2504
http://dx.doi.org/10.1016/bs.aecr.2016.10.006

Abstract

When invasive species become integrated within a food web, they may have numerous direct and indirect impacts on the native community by creating novel trophic links, and modifying or disrupting existing ones. Here we discuss these impacts by drawing on examples from freshwater ecosystems, and argue that future research should quantify changes in such trophic interactions (i.e. the links in a food web), rather than simply focusing on traditional measures of diversity or abundance (i.e. the nodes in a food web). We conceptualise the impacts of invaders on trophic links as either direct consumption, indirect trophic effects (e.g. cascading interactions, competition) or indirect nontrophic effects (e.g. behaviour mediated). We then discuss how invader impacts on trophic links are context-dependent, varying with invader traits (e.g. feeding rates), abiotic variables (e.g. temperature, pH) and the traits of the receiving community (e.g. predators or competitors). Co-occurring invasive species and other environmental stressors, such as climate change, will also influence invader impacts on trophic links. Finally, we discuss the available methods to identify new food web interactions following invasion and to quantify how invasive species disrupt existing feeding links. Methods include direct observations in the field, laboratory trials (e.g. to quantify functional responses) and controlled mesocosm experiments to elucidate impacts on food webs. Field studies which use tracer techniques, such as stable isotope analyses, allow diet characterisation of both invaders and interacting native species in the wild. We conclude that invasive species often drastically alter food webs by creating and disrupting trophic links, and future research should be directed particularly towards disentangling the effects of invaders from other environmental stressors.

1. INTRODUCTION

Food webs are networks of organisms that interact through both direct and indirect links (Heath et al., 2014). Under natural conditions, the communities that comprise these food webs are a result of the evolution of species interactions within the physicochemical constraints of their surroundings (Smetacek et al., 2004). Biological invasions (i.e. the establishment of nonnative species), however, inevitably result in food web alterations through node and link augmentation (David et al., 2017). This can affect ecosystems in numerous (sometimes disruptive) ways through various mechanisms. As such, biological invasions are now considered one of the most important drivers of biodiversity loss, with both ecological and economic impacts (Clavero and Garcia-Berthou, 2005; Mollot et al., 2017;

Pejchar and Mooney, 2009; Simberloff et al., 2013). Increasing globalisation is continuing to create a greater range of pathways for species to invade (Hulme, 2009) and as a result, rates of invasion are accelerating, especially in aquatic systems (Cohen and Carlton, 1998; Jackson and Grey, 2013; Ricciardi, 2006).

Invasive species may have impacts across multiple levels of organisation and trophic levels (e.g. Brennan et al., 2014; Cameron et al., 2016; Jackson et al., 2014). Compared with other environmental stressors such as climate change or habitat destruction, the effects of invasive species can be considered more direct, as these novel species become integrated within the community food web. Some of the most conspicuous impacts of invaders are through direct predator–prey links, which is exemplified by the damage to native cichlid populations following the introduction of Nile Perch *Lates niloticus* in Lake Victoria, East Africa (Witte et al., 1992). It is clear that some of the most severe impacts of invaders are mediated by these direct trophic interactions or the resulting indirect trophic cascades (e.g. a predator reduces herbivore abundance, in turn promoting plant abundance), yet studies typically quantify invader impacts on species or communities without clarifying changes in species interactions. Quantifying changes in trophic links as a result of invasions will account for impacts across different levels of organisation (individual to network) *and* different trophic levels (producer to predator), revealing potentially subtle or otherwise unexpected effects which could have cascading consequences for ecosystem processes.

The impacts of invasive species are particularly pronounced in insular systems, such as islands and lakes, owing largely to their typically depauperate food webs and ecological naïveté of their biota (D'Antonio and Dudley, 1995; Ricciardi and MacIsaac, 2011; Spencer et al., 2016). Furthermore, the growth of ecological theory on food webs and trophic cascades in lakes and rivers demonstrates their value as model systems to explore trophic ecology. For example, the now well-known trophic cascade 'producer–herbivore–predator' was first identified in lake systems when Carpenter et al. (1985) described how fish biomass indirectly controls phytoplankton via their direct trophic link with zooplankton. This idea has since been developed into more complex food web theory, including mass–abundance relationships (Cohen et al., 2003; O'Gorman and Emmerson, 2011). Therefore, here we: (1) conceptualise how free-living invasive species alter trophic links by drawing on examples from freshwater ecosystems, (2) examine the context dependency of such impacts and (3) outline methods to quantify changes in trophic links or food webs following invasion.

2. WHAT ARE THE IMPACTS OF INVASIVE SPECIES ON TROPHIC LINKS IN FRESHWATER ECOSYSTEMS?

Freshwater invasions may alter food web structure by creating new trophic links or by disrupting, altering or reducing the number of existing links (e.g. Simon and Townsend, 2003). This can be through direct trophic interactions, trophic cascades or other indirect effects such as alterations in behaviour (Box 1). Here we have outlined these trophic impacts drawing on examples from freshwater ecosystems.

BOX 1 Impacts of Invaders on Trophic Links

The possible effects of an invasive population (I) on trophic links in a simplified food chain involving two native species, one of which is prey or a resource (R) to a consumer species (C). Red arrows represent new trophic links created by the invader and black arrows represent indirectly disrupted trophic links as a result of the invasion.

a. Direct links	b. Indirect links: trophic cascades	c. Indirect links: competition	d. Indirect nontrophic effects
(i) I (ii) C C I R R	(i) C (ii) I R C I R	(i) C I R	(i) C I R
To be successful, invaders must consume resources in the recipient ecosystem. Many are omnivores, feeding across multiple trophic levels (i). Additionally, invaders can become a resource to native consumers (ii), often creating novel trophic pathways.	When an invasive species acts as a resource for a native species, it can cause a cascade where the newly more abundant native species is consumed by upper trophic levels (i). Alternatively, a new invader at the top of the food chain will not only affect its direct prey but also lower trophic levels because of a release from predation or herbivory pressure (ii).	An invader may compete with a native consumer for the same resource due to limited resource availability (apparent competition) or antagonistic interactions (interference competition).	Some invasive species alter the avoidance behaviour of native prey, and therefore its interactions with a native predator. In addition, invaders can alter community structure or habitat complexity with implications for native species foraging behaviour.

2.1 Direct and Indirect Effects From the Bottom-Up

The presence of invasive prey can represent an abundant novel food source for native predators (Box 1a and b). For example, invasive zebra mussels (*Dreissena polymorpha*) are important in the diet of blue crabs (*Callinectes sapidus*) in the Hudson River (Molloy et al., 1994) and invasive round gobies (*Neogobius melanostomus*) are consumed by the Lake Erie water snake (*Nerodia sipedon insularum*) in regions of Canada (King et al., 2006). Invasive plants may also sustain native populations of herbivores or detritivores which, in turn, can alter resource availability to higher trophic levels; therefore, energy derived from invasive species is transferred up the food chain (Box 1b). Some native freshwater herbivores show a preference for nonnative plants in their diet (Parker and Hay, 2005), adding a new link to the food web following invasion. Alternatively, nonnative plants may be unpalatable to native herbivores (Burlakova et al., 2009; Cano et al., 2009; Keane and Crawley, 2002) and remain disconnected from the food chain as a result (i.e. the enemy release hypothesis). Trees which invade the riparian zone of rivers and lakes can also act as a new resource to aquatic primary consumers when their leaves fall into the water (Reinhart and VandeVoort, 2006; Swan et al., 2008; Box 1a). In addition, these new resources can ultimately be made available to organisms further up the food web. For example, in Hawaii leaf litter from the invasive albizia tree *Falcataria moluccana* is an indirect nutrient source for consumers higher in the food chain, such as the green swordtail *Xiphophorus helleri* and guppy *Poecilia reticulata*, because albizia is the major dietary component of their invertebrate prey (Atwood et al., 2010). There are instances, however, where such invasive resources are unpalatable and are therefore left largely unexploited (Hladyz et al., 2011). Stoichiometry of the leaf litter (relative concentrations of nutrients) will affect palatability, with those high in nitrogen content usually consumed by invertebrates at faster rates (Hladyz et al., 2009; Martínez et al., 2013). Many introduced commercial plants (e.g. *Pinus radiata*, *Rhododendron ponticum*, *Eucalyptus* spp.) are low in nitrogen content and, therefore, rarely eaten by aquatic invertebrates (Hladyz et al., 2011; Larrañaga et al., 2014; Martínez et al., 2013). This can have implications for decomposition rates and other trophic links in the food web through cascading interactions (Hladyz et al., 2011; Swan et al., 2008).

Leaf litter from invasive trees such as albizia (Atwood et al., 2010) often become an alternate resource for consumers, which might also result in previously preferred resources being released from herbivory or predation pressure (Box 1a). For instance, on islands in New Zealand, invasive rats *Rattus spp.* are consumed by top predators which releases an endangered parrot *Strigops habroptilus* from predation (Zavaleta et al., 2001). Native snakes (*Dinodon rufozonatum*) preferentially consume invasive North American bullfrogs *Lithobates catesbeianus* in parts of China, reducing predation by the snakes on native anuran species (Li et al., 2011). As alternate food sources, invasive species can also indirectly affect trophic interactions by providing native predators a fitness advantage over competitors unable to exploit the novel resource. Although this aspect of invasion ecology has been largely understudied, an extreme example is seen in the consumption of invasive cane toads *Rhinella marina* by a variety of snake species in Australia. Cane toads represent a novel potential food source for snakes and while many native snake species are able to consume the toads (Phillips et al., 2003), some are susceptible to the toad lethal toxin where consumption may result in death (Shine, 2010). In this way, the invasive toads represent not only an additional food source but also a means through which potential snake predator competition is mediated (Brown et al., 2013).

2.2 Invading the Middle of the Food Chain

Introduced animals can subsidise secondary consumers as an intermediate link in the food chain (Box 1a; Rodriguez, 2006). Invasive red swamp crayfish are a resource for many native mammals and birds in Portugal (Correia, 2001) and as a result can have far-reaching trophic consequences for terrestrial ecosystems. At the same time, red swamp crayfish consume native macroinvertebrates with negative implications for community assemblages (Gherardi and Acquistapace, 2007; Jackson et al., 2014). Similarly, the invasive amphipods *Corophium curvispinum, Dikerogammarus villosus* and *Gammarus tigrinus* in the River Rhine, Germany, consume native invertebrates voraciously with negative consequences for diversity (van Riel et al., 2006), while acting as significant resources themselves for European perch *Perca fluviatilis* and European eel *Anguilla anguilla* (Kelleher et al., 1998). Introduced bream *Abramis brama* in Northern Ireland are consumed by native river lamprey *Lamperta fluviatilis* and have been linked to elevated population sizes of the latter (Inger et al., 2010). In South America, the freshwater bivalves *Corbicula fluminea* and *Limnoperna fortunei* are further examples of intermediate food web invaders. Both species accumulate pollutants through their filter feeding activity and they are also consumed by numerous species of native fish (e.g. *Pterodoras granulosus* and *Brochiloricaria chauliodon*, Garcia and Protogino, 2005). By creating new trophic links with both upper and lower trophic levels, the invasive bivalves transfer pollutants through the food web.

A key trait of many invasive species is a generalist diet across numerous trophic levels, often comprising both plant and animal material (Crowder and Snyder, 2010). This omnivorous behaviour means that many invaders sit in the middle of the food web with direct dietary links with a number of trophic levels (Box 1a), sometimes a beneficial trait given the diversity in trophic states among receiving environments. Invasive crayfish, of which several species now have a global distribution (Capinha et al., 2011), tend to be opportunistic omnivores and are often associated with the decoupling of trophic cascades as a result (Jackson et al., 2014). For example, a trophic cascade where the consumption of native invertebrate shredders by invertebrate predators caused a reduction in net leaf litter decomposition was decoupled by invasive red swamp (*Procambarus clarkii*) and signal crayfish (*Pacifastacus leniusculus*), since they consumed both shredders and leaf litter directly (Jackson et al., 2014). While there is little information available for freshwater omnivores other than crayfish, invasive bighead (*Hypophthalmichthys nobilis*) and silver carp (*Hypophthalmichthys molitrix*) in North America have the potential to directly influence both phytoplankton and zooplankton biomass (Radke and Kahl, 2002; Sampson et al., 2009).

2.3 Direct and Indirect Effects From the Top-Down

The most evident impacts of invaders are through direct predator–prey links, where the invasive species assumes a predatory role within a food web (Box 1a and b; Roemer et al., 2002). Invasive predators can have elevated consumption rates when compared to native equivalents, for instance in invasive fish (Alexander et al., 2014), mysids (Bohmann et al., 2014) and amphipod species (Laverty et al., 2015). Similarly, native and invasive hybrids can affect native species directly through cannibalism as shown in salamanders, whereby hybrids (*Ambystoma californiense* × *A. tigrinum mavortium*) reduce native salamander (*A. californiense*) numbers through predation (Ryan et al., 2009). Species invasions greatly enhance the chance of new, and potentially damaging, hybridisations in the future (Brennan et al., 2014). In addition to the direct effects of predation, the effects of novel predators can also have further reaching indirect implications through trophic cascades (Box 1b). Classic examples of this from freshwater ecosystems involve introduced apex predatory fish, including the Nile Perch in Kenya (Goudswaard et al., 2008; Witte et al., 1992), and both brown trout *Salmo trutta* and rainbow trout *Oncorhynchus mykiss* in New Zealand (McIntosh et al., 2010; Simon and Townsend, 2003). However, in some of these cases, the impact of the invader is through behavioural shifts in the prey rather than direct consumption (Section 2.5). Largemouth bass *Micropterus salmoides* has also been introduced to many countries where it invades the top of the food web, with cascading consequences for lower trophic levels through dietary interactions (Box 1d; Maezono and Miyashita, 2003; Vander Zanden et al., 1999). For instance, in Japanese lakes, largemouth bass consumes native fish species, thereby elevating the abundance of native invertebrate species released from predation pressure (Maezono and Miyashita, 2003).

2.4 Competition

One of the most well-described impacts of invaders is through their competition with native species for resources (i.e. indirect trophic effects; Box 1c) and there is a wealth of evidence demonstrating this following invasion by fish, amphibians and invertebrates (Bergstrom and Mensinger, 2009; Jackson et al., 2016a; Ryan et al., 2009). For example, invasive smallmouth bass *Micropterus dolomieu* and rock bass *Ambloplites rupestris* compete with native lake trout *Salvelinus namaycush* for fish prey in lakes in North America, causing the native fish to start foraging on macroinvertebrates lower down the food web (Lepak et al., 2006; Vander Zanden et al., 1999). This results from a decline in resource availability (i.e. exploitative competition), but antagonistic interactions between natives and invaders can also play a role (i.e. interference competition). The native Spanish terrapin *Mauremys leprosa*, for instance, consumes less food (i.e. reduced link strength) in the presence of invasive red-eared sliders *Trachemys scripta elegans*, because of the latter species' aggressive competitive behaviour (Polo-Cavia et al., 2011). There is recent evidence suggesting that invasive species in freshwaters can even compete with terrestrial consumers—including spiders, birds and bats—for emerging insects (Baxter et al., 2004; Benjamin et al., 2013; Jackson et al., 2016a). Invasive brown trout have overlapping diets with native riparian spiders in South Africa (Jackson et al., 2016a) and spiders which are found adjacent to streams invaded by the amphipod *D. villosus* in Germany have significantly less aquatic resources in their diet than those at uninvaded sites (Gergs et al., 2014). These processes could also lead to 'apparent trophic cascades', where bottom-up-controlled detritus-based food chains are linked to parallel top-down-controlled algae food channels by generalist consumers (i.e. the fish and spiders; Wollrab et al., 2012).

Community assembly theory suggest that two species occupying the same niche will not occur in the same place at the same time (Moyle and Light, 1996) and there are documented cases in the invasion literature where an invasive species has resulted in population crashes of functionally similar natives (Cucherousset and Olden, 2011; Simberloff et al., 2013). For example, invasive species of crayfish regularly out-compete native crayfish or previously established invaders (e.g. Hill and Lodge, 1999; James et al., 2016). Successful introduced crayfish commonly exhibit faster growth rates and achieve larger sizes than their native counterparts, which gives them an advantage by increasing their fecundity and success in competition scenarios (Alonso and Martínez, 2006). However, evidence also suggests that competitive exclusion by invaders is rare in freshwater ecosystems due to high levels of generalism (Covich, 2010; Jackson and Britton, 2014). For example, Westhoff and Rabeni (2013) found that a native and invasive crayfish shared the same habitat with no competitive exclusion, and Jackson et al. (2014) showed that multiple crayfish invaders were able to coexist by niche partitioning. Furthermore, the assumption that invasive and native species engage in competition, and that this drives native coexistence/exclusion, can be erroneous, especially where intraguild predation operates—for example, patterns of invader and native amphipod exclusion are driven by direct intraguild predation, with competition as a very weak interaction (e.g. Dick, 2008).

2.5 Indirect Nontrophic Effects

Invasive species can alter food web dynamics without directly interacting with biota through various behaviourally mediated mechanisms (Box 1d). For example, the mere presence of a new nonnative species may cause some resident species to shift their feeding behaviour. This is particularly true when the new species is a predator which creates a 'landscape of fear' due to the perceived predation risk. This has been well documented in Yellowstone National Park, North America, where the re-introduction of wolves *Canis lupus* caused native elk *Cervus elaphus* to shift their diet to less profitable resources (Hernández and Laundré, 2005), but is less well understood in true invasive species. However, evidence from freshwater systems suggests that native tadpoles (*Pseudacris regilla* and *Rana aurora aurora*) will respond to predator cues (i.e. waterborne cues) alone when they are associated with novel invasive red swamp crayfish *P. clarkii* or bluegill sunfish *Lepomis macrochirus* by increasing refuge use (Pearl et al., 2003), which may then alter resource availability and diet. Similarly, the presence of an invasive species might also have implications for competitors. The chemical cues from the invasive gastropod *Tarebia granifera* resulted in the spatial displacement of three native snail species in St Lucia Estuary, South Africa (Raw et al., 2013). These trends are not, however, universal as other studies have found a lack of response of native prey to chemical cues from invasive species as a possible result of naiveté to the predation risk (Pearl et al., 2003; Smith et al., 2007).

A second impact of invasive species which involves no direct trophic interaction involves that of habitat alteration by the invader (Box 1d). Habitat alteration can have implications for predator–prey dynamics (Alexander et al., 2015; Turesson and Brönmark, 2007; Wasserman et al., 2016a) and the presence of invasive species can alter the physical environment in such a way as to have indirect nontrophic effects on food web dynamics. For example, dense colonies of invasive zebra mussels (*D. polymorpha*) can promote higher local densities of benthic invertebrates, in part because of enhanced benthic structural complexity that provides refugia against fish predation (Mayer et al., 2001; Valley and Bremigan, 2002). Similarly, invasive water plants can indirectly alter the diet of native consumers by forming dense mats which alter light availability and habitat complexity (Schultz and Dibble, 2011). Invasive macrophytes may also increase habitat availability for certain native species and subsequently elevate their abundance (Kelly and Hawes, 2005; Strayer et al., 2003), causing other native consumers to switch their diet to the newly abundant prey (Box 1a). For example, in Lake Victoria the diet of Nile tilapia *Oreochromis niloticus* changed from being algal-dominated to constituting primarily of invertebrates following the invasion of water hyacinth *Eichhornia crassipes* (Njiru et al., 2004).

3. WHAT INFLUENCES THE IMPACTS OF INVADERS ON FRESHWATER FOOD WEBS?

The impact of an invasive species is mediated by factors related to the invader itself (i.e. autoecological traits such as body size and seasonal life history), the abiotic environment, and direct and indirect effects involving the resident community (i.e. synecological traits including competitors, parasites, predators and prey) (Hatcher et al., 2006; Ricciardi et al., 2013; Strayer et al., 2006). Anthropogenic stressors such as climate change, nutrient enrichment and habitat alteration, both individually and interactively, also alter invader impacts (e.g. Jackson et al., 2016b; Vye et al., 2015). As we show here, these numerous context dependencies challenge our understanding and prediction of the effects of invasions and must be considered when developing models of impact on food webs. However, we also show that recent advances in experimental approaches to invasions can capture such complexities and improve our predictive capability.

3.1 Autoecological Traits

The search for specific invader traits that explain invasion success or predict invader impacts on food webs has been generally unsuccessful, although some taxon-specific traits are valuable (Dick et al., 2014). Impacts of invaders in freshwater ecosystems will vary with: (1) species traits, such as feeding rates, dispersal mechanism, reproductive system and range; (2) population traits or propagule pressure and (3) traits which can vary between individuals, including body size and growth rate (Kolar and Lodge, 2001).

Traits associated with high-impact invaders include greater reproductive potential and rapid exploitation of resources (Dick et al., 2014; Ricciardi et al., 2013). High feeding rates, in particular, are thought to be associated with severe food web impacts (Dick et al., 2013a, 2014; Johnson et al., 2008; Morrison and Hay, 2011). Several recent studies indicate that per capita feeding (i.e. interaction strength) is higher in invasive species (both predators and herbivores) which have severe ecological impacts (Alexander et al., 2014; Xu et al., 2016). Interaction strength represents the per capita effect of one species on another species' density and, alongside complexity, is important in maintaining food web stability (May, 1973). Invasive species can, therefore, directly affect stability by both creating and disrupting trophic links.

Feeding rates and other per capita effects may vary spatiotemporally within an expanding population as a result of trait-based sorting and segregation of individuals during the dispersal process that leads to differences in character traits such as heightened boldness and voracity at an invasion front (Brandner et al., 2013a; Iacarella et al., 2015a; Phillips, 2009). Using a combination of reciprocal transplant mesocosms and functional response experiments, Iacarella et al. (2015a) showed that invasion front mysids had higher predation rates, particularly a more efficient ability to locate and capture

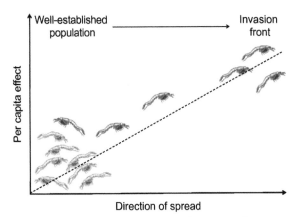

Fig. 1 The invasion front hypothesis (Iacarella et al., 2015a): The per capita effects of invading species may vary spatially as a result of trait-based sorting of individuals through dispersal. Individuals aggregating at invasion fronts tend to be more aggressive and voracious compared to those in well-established, higher density populations, where traits leading to lower intraspecific competition are selected. This gradient is reflected in higher per capita effects (e.g. predation rates) at the invasion front.

prey, compared with those from a well-established population; this resulted in a community-level impact on native zooplankton at the invasion front (Fig. 1).

The greatest impacts of invaders are hypothesised to be caused by generalist consumers, both apex and mesopredators, introduced to naturally simple food webs (Ricciardi et al., 2013). Native omnivores can potentially stabilise food webs, if their trophic interactions are weak (McCann, 2000); however, invasive omnivores have other ecosystem-level impacts (e.g. Bobeldyk and Lamberti, 2010). Ecosystem engineers, species capable of modifying their invaded habitat, can create more favourable abiotic conditions for themselves, and for certain co-occurring invaders and natives (especially in highly stressful environments; Crain and Bertness, 2006), while also reducing populations of native species (Gonzalez et al., 2008).

3.2 Abiotic Drivers

The outcome of trophic interactions involving invaders can vary with abiotic factors such as water chemistry, owing to differences in the physiological tolerances of the species involved. For example, field and lab experiments on intraguild predation between native and invasive amphipods in the St. Lawrence River, North America, revealed that magnitude and direction of this predation was reversed along a natural conductivity gradient (Kestrup and Ricciardi, 2009). Intraguild predation is a ubiquitous and powerful species displacement mechanism in crustaceans (Dick, 2008) and other groups (Polis et al., 1989) that is likely to override the effects of interspecific competition and, unlike cannibalism, lead to decreased stability in the populations of the interactors (Dick and Platvoet, 1996; Polis et al., 1989). The outcome of this interaction following invasion of a novel species may be sensitive to the influence of conductivity on crustacean moulting, which is a vulnerable predation window, even leading to exclusion of larger species by smaller species (e.g. Dick and Platvoet, 1996; Kestrup et al., 2011).

Spatial and temporal heterogeneity of the physical habitat can modify impacts over landscapes, thus simultaneously allowing native-invader coexistence at some sites and suppression (or exclusion) of either the native or the invader at others (MacNeil et al., 2004). This can create prey refugia within systems where invasive predator impacts are otherwise devastating (e.g. Reid et al., 2013). As such, habitat simplification (e.g. through river channelization or wetland reduction) can increase an invader's capacity for top-down control on native populations (Alexander et al., 2015). Similarly, variations in the heterogeneity in terms of water chemistry, temperature and oxygen concentration can also affect prey handling and consumption rates, and therefore the impacts of invasive species (Dick and Platvoet, 1996; Iacarella and Ricciardi, 2015; Laverty et al., 2015; Sardiña et al., 2015). A meta-analysis (Iacarella et al., 2015b) revealed that the per capita effects of invasive aquatic predators tend to vary unimodally with water temperature, such that their interaction strengths are reduced with distance from the predator's thermal optimum. This evidence supports the Environmental Matching Hypothesis that the consumptive impacts of invasive species are maximal in habitats whose abiotic conditions most closely match their physiological optima (Iacarella et al., 2015b; Ricciardi et al., 2013).

Water pH is a powerful predictor of community structure and the recovery of freshwater ecosystems from acidification since the 1970s has resulted in 'invasions' of progressively larger species which would have been part of the food web prior to the industrial revolution (Layer et al., 2011). Although these are not invasions sensu stricto, they provide some interesting insight. For instance, a recovering stream food web in the United Kingdom experienced increased food chain length and network complexity following establishment of the large dragonfly *Cordulegaster boltonii* and brown trout *S. trutta* (Layer et al., 2011; Woodward and Hildrew, 2001). Both of these changes following invasion could result in reduced network stability. Temperature can also alter food web connectance (Petchey et al., 2010), perhaps increasing the number of direct links between an invasive species and native community members, and therefore providing more opportunities for cascading indirect effects. Body size, attack rates and handling times will all be key in determining the strength of temperature effects on food webs (Petchey et al., 2010).

3.3 Synecological Traits

Depending on its composition and structure, the receiving community may either suppress or enhance the effects of an invader. Logically, a more diverse receiving community is more likely to contain strong competitors and predators that could limit the behaviour, population growth and impact of invaders (e.g. Levine et al., 2004; Robinson and Wellborn, 1988). Parasites in recipient communities can also have a demonstrably major influence on the outcome of invasions (Prenter et al., 2004). Invasive *Gammarus pulex*, for example, is a poorer intraguild predator when carrying parasites, but simultaneously a better predator of other trophic groups (MacNeil et al., 2003). However, both simple and diverse communities have been heavily impacted by invaders and no clear pattern has emerged in the diversity–invasibility debate (Belote et al., 2008). Food web topology will also influence invasion success and subsequent impact. There is no universal rule for food web structure but only a few display small-world (e.g. clustered nodes) and scale-free (uneven distribution of links per node) topologies (Dunne et al., 2002). Food webs tend to have low clustering and short pathways between any two nodes (Williams et al., 2002) and the degree of overall connectance between these nodes can influence invasion success (Baiser et al., 2010). For instance, models suggest that invasive predator success will increase with food web connectance because of an increase in intermediate prey (Baiser et al., 2010).

Resistance to the potential impact of an invader following establishment could be more frequent than resistance to establishment itself (Levine et al., 2004). Nevertheless, even species-rich communities can be dramatically altered by an invader (e.g. Boudreau and Yan, 2003; Kaufman, 1992), and a species-poor community can be quite resistant to being disrupted by invasion (e.g. Baltz and Moyle, 1993). Generalist predators in the receiving community may not necessarily prevent invasion, but can still reduce the abundance and distribution of an established invader (Dick et al., 2013a; Parker et al., 2007; Robinson and Wellborn, 1988) and thus limit its influence on food webs. However, such interactions may nonetheless lead to higher impacts on native prey populations through native predator enhancement (Noonburg and Byers, 2005). In contrast to resistance, invader facilitation by native species is rarely studied (but see Michelan et al., 2014; Rodriguez, 2006; Taniguchi et al., 2002), especially in terms of its moderating effects on impact.

A major driver of context dependency of an invader's impact on food webs is the receiving community's naiveté, or lack of evolutionary exposure, to functionally similar species (Cox and Lima, 2006). In terms of their insularity, freshwater systems are analogous to islands (Ricciardi and MacIsaac, 2011) and are thus naïve to the effects of a broad range of consumers, which perhaps explains why a larger proportion of introduced species appear to cause native population declines in freshwater versus marine systems (Ricciardi and Kipp, 2008). The greatest trophic disruptions will likely be caused by introductions of species to lakes or river basins in which they are functionally (or phylogenetically) novel—i.e. when they exploit key resources in novel ways (Ricciardi and Atkinson, 2004). Naiveté and 'ethological traps' can also lead to major impacts on native species where per capita effects of invasive predators are extreme; for example, fox predation of turtle nests (Spencer et al., 2016).

3.4 Multiple Invaders

Freshwater ecosystems are accumulating burgeoning numbers of invasive species (Cohen and Carlton, 1998; Jackson and Grey, 2013; Ricciardi, 2006), which may interact with each other and thus create new food web links and alter existing ones (Jackson, 2015). For instance, in Lake Naivasha (Kenya), invasive bass *M. salmoides* consume invasive crayfish *P. clarkii*, thereby reducing competitive pressure on invasive carp *Cyprinus carpio* (Britton et al., 2010; Jackson et al., 2012). In the North American Great Lakes, the invasive sea lamprey *Petromyzon marinus* nearly eliminated a dominant native piscivore, lake trout *S. namaycush*, thereby facilitating the expansion of an invasive planktivore, alewife *Alosa pseudoharengus*, which ultimately resulted in the competitive reduction of native planktivores and an overall loss of fishery productivity (Ricciardi, 2001). Alternatively, invaders may only be connected through links with native species, resulting in impacts on such species through multiple processes. For example, in lakes in North America, native snail communities are simultaneously and negatively impacted by invasive Chinese mystery snails *Bellamya chinensis* and invasive rusty crayfish *Orconectes rusticus* through competition and predation, respectively (Johnson et al., 2009).

Owing to the complexity of interactions between invaders, which depend on, inter alia, species identity, the receiving environment and the native community, the combined effect of multiple invaders is often unpredictable. Opposing theories suggest that invaders will generally benefit other invaders (e.g. 'Invasional Meltdown'), or, alternatively, inhibit and exclude other invaders (e.g. 'Biotic Resistance', 'Invasion Treadmill'; Jackson, 2015). Invasion meltdown predicts enhanced invasion success and impacts of invaders through interspecific facilitation (Simberloff and von Holle, 1999), and examples of this phenomenon are documented for freshwater systems (Adams et al., 2003; Ricciardi, 2001, 2005). In contrast, biotic resistance suggests that invaders will compete with one another for resources, often resulting in an invasion treadmill where a newly successful invader will replace previously established invasive competitors (e.g. Jackson et al., 2012). A number of factors, including evolutionary background, functional traits and niche opportunities, are likely to play a role in the outcome of such interactions, but a recent meta-analysis indicated that neutral interactions (i.e. where the presence of another invader had no influence on invasion success) were the most common (Jackson, 2015).

Once multiple invaders become established, their combined impacts can be additive or nonadditive (Box 2). Nonadditive effects are those which are less than, or more than, the sum of the impacts of a single invasive species (Jackson, 2015). The outcome will likely depend on where the invaders are positioned in a food web in relation to one another but overall additive effects tend to be more common (Box 2; Jackson, 2015). Resource partitioning and facilitation among invasive predators may lead to prey reductions, whereas interference or other antagonisms can increase prey populations (Trumpickas et al., 2011). In some cases, multiple predatory species have no additional effect on native prey assemblages, compared with single predators (Trumpickas et al., 2011). Reponses of fish assemblages to multiple invasive predators show no consistent pattern (Findlay et al., 2000; Trumpickas et al., 2011; Young et al., 2009), suggesting highly context-dependent effects or chaotic food web dynamics.

BOX 2 Multiple Invaders

Multiple invasive species can interact with one another to cause impacts which are additive (the sum of their single effects), antagonistic (less than the sum of their single effects) or synergistic (more than the sum of their single effects). The type of interaction will depend on the trophic ecology of each invasive species. For instance, two functionally similar invaders might have overlapping niches and therefore exert pressure on the same part of the food web. If the two invaders compete and control one another's population this might be expected to cause an additive or antagonistic impact, because the two invaders have negative effects on the same group of native taxa. Alternatively a synergistic impact could occur if the two similar invaders reach higher densities than a single species would (Jackson, 2015). A recent meta-analysis found that antagonistic impacts were the most common when two similar invaders (e.g. two species of invasive crayfish) occurred together (in 4 out of 6 cases; Jackson, 2015).

In contrast, invasive species with distinct niches, or those which occupy different trophic levels, may have (1) opposing impacts which cancel each other out, causing an antagonistic impact; (2) similar impacts from different directions, causing a synergistic impact or (3) unrelated impacts in different parts of the food web, resulting in additive effects overall (Jackson, 2015; Jackson and Britton, 2014). Jackson's meta-analysis (2015; $n = 19$) found that functionally distinct invaders in freshwaters (e.g. an omnivorous crayfish and predatory fish) had additive effects in 74% of cases (i.e. scenario 3) and antagonistic effects in the remaining 26%.

3.5 Multiple Stressors

Many freshwater ecosystems are subject to multiple anthropogenic stressors (e.g. climate change, nutrient pollution, habitat modification; Fig. 2) which interact to affect food webs (Byers, 2002; Didham et al., 2007). These two networks of interactions (1) among stressors, such as global warming and species invasions, and (2) among species in a food web are interlinked and complex. Interactions between invasions and anthropogenic stressors can cause impacts that are not additive and therefore difficult to predict from their individual effects (i.e. ecological surprises; Jackson et al., 2016b). Further, an invasive species as a single stressor may have opposing effects within different parts of the food web (i.e. different energy pathways or trophic levels), and multiple stressors may act in opposite ways to each other (Jackson et al., 2016b).

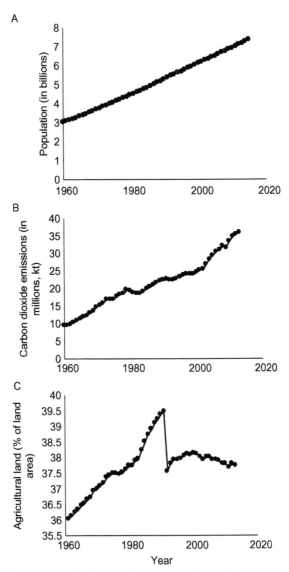

Fig. 2 World development indicators. Human population size (A), CO_2 emissions (B), and farmed land area (C) are all increasing from year to year across the planet. This raises the potential of multiple stressors, such as warming and pollution, occurring simultaneously in one ecosystem. *Data taken from the World Bank, 2016. World development indicators. Retrieved from http://data.worldbank.org/indicator. October 2016.*

Anthropogenic stressors may also promote species invasions, or favour invaders over native species, exacerbating their impacts on food webs (Byers, 2002). For instance, climate change is expected to facilitate some invaders such that they expand their range (Rahel and Olden, 2008). Another major context driven by climate change is the availability of oxygen in freshwaters, and Laverty et al. (2015) showed that invaders would be less affected in their predation strength than natives under decreased oxygen concentrations. It is, therefore, important to consider the impacts of other stressors when quantifying how invaders alter trophic links, especially with accelerating rates of global change (e.g. Fig. 2).

4. WHAT ARE THE AVAILABLE METHODS TO QUANTIFY THE IMPACTS OF INVADERS ON TROPHIC LINKS?

The impact of invasive species on trophic links can be measured from individuals to whole networks (i.e. food webs) in both experimental and natural settings (Fig. 3). Laboratory and field experiments (e.g. mesocosms; Fig. 3B) benefit from a high level of control such that the impacts of invasion can be separated from other variables. However, experiments lack the realism of field studies in natural settings, and therefore, a combination of the two is often the best approach when deducing invader impacts. Here, we discuss the available methods to quantify the impacts of invaders on trophic links, starting with experiments which focus on individuals before moving up to higher levels of biological organisation and complexity.

4.1 Functional Responses

Invasive species commonly use resources more rapidly and with greater efficiency than native counterparts (Chapple et al., 2012; Morrison and Hay, 2011; Strayer et al., 1999), and as a result, these resources often experience severe herbivory or predation pressure and subsequent declines (Roy et al., 2012; Salo et al., 2007). This per capita effect, important in the overall determination of invasive species impact (Parker et al., 1999) is measureable by the functional response (i.e. the rate of consumption as a function of resource density). Furthermore, consumer impact is often density dependent (Abrams, 2000; Jackson et al., 2015) and it follows that the impact of invasive species may vary accordingly to resource density. Therefore, approaches that permit consideration of the density-dependence, such as functional responses, of such interactions have been suggested as an informative measure of the elusive per capita effect of predators (Vucic-Pestic et al., 2010)

Fig. 3 Experimental setup. It is possible to quantify the impacts of invasive species on trophic links and/or food webs in controlled experimental conditions which can involve individual invaders (e.g. (A) functional response experiments at the South African Institute of Aquatic Biodiversity, photo by MA), simplified communities of a small number of species (e.g. community modules) or more complex communities in outdoor mesocosms (e.g. (B) mesocosm facilities at Queen Mary University of London, UK, photo by JG) or in enclosures in the field (e.g. (C) in an estuary in Eastern Cape, South Africa, photo by RW).

and have been shown to be particularly useful in the invasion context (e.g. Dick et al., 2014).

Functional responses, as first described by Holling (1959), are generally considered to take one of three forms that encompasses the linear Type I response, the Type II hyperbolic response that follows a saturating curve, and the sigmoidal Type III response, with a belated realisation by invasion ecologists that comparing invader and native functional responses could predict ecological impact of the former (Fig. 4). The form that such responses take is considered important for understanding consumer impact and community dynamics because response type contributes differently to stability and persistence of the resource (Eggleston et al., 1992; Hassell, 1978; Ward et al., 2008). This may be of particular importance in an

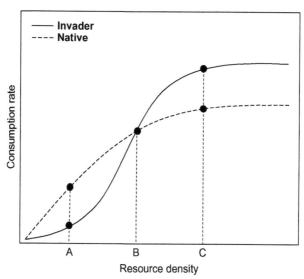

Fig. 4 Functional responses in invasion biology. Point estimation of consumption rates can be poor predictors of consumer–resource dynamics. As shown, consumption of the invader can be higher (point A), equal (B) or lower (C) than that of the native, depending on the resource density at which it is measured. This potentially leads to erroneous conclusions of interactions across a range of resource availabilities. Many other potential combinations of invader/native comparisons exist that are much more informative of snapshot designs where resource density is fixed (see Dick et al., 2014).

invader-native context and the alterations that occur to trophic links (Box 1). Furthermore, the use of functional responses considers a full spectrum of resource availability, therefore avoiding 'snapshot' conclusions of per capita effects drawn from single resource densities, which can sometimes be misleading (Dick et al., 2014). For example, while an invader may have a lower maximum feeding rate than a comparable native, the invader may have a more destabilising Type II interaction with shared prey and the native a more stabilising Type III interaction, which cannot be revealed in snapshot designs (Dick et al., 2014).

In freshwater systems, the application of functional responses has been used widely as a method to measure and quantify impacts of invasive species across a range of trophic and taxonomic groups (Alexander et al., 2014; Dick et al., 2013a; Haddaway et al., 2012; Médoc et al., 2015). The association between heightened responses (i.e. faster feeding rates) of invaders compared to natives and degree of ecological impact in the field has been well demonstrated (e.g. Alexander et al., 2014; Dick et al., 2013a; MacNeil et al., 2003), with studies considering impacts towards a range of prey species, including generic representative prey (e.g. tadpoles as fish prey; Alexander et al., 2014), in addition to ecologically relevant prey for which field impacts are known (e.g. Bovy et al., 2015; Dick et al., 2013b; Haddaway et al., 2012).

Invader feeding rates may vary with other stressors, such as temperature (Rall et al., 2010, 2012), and therefore functional response experiments provide ideal opportunities to quantify the combined impacts of multiple stressors (Iacarella et al., 2015b; Laverty et al., 2015; O'Gorman, 2014). For example, habitat simplification may result in greater functional responses, and therefore impact, in invasive fish species (Alexander et al., 2015). Similarly, changes from stabilising Type III responses towards destabilising Type II responses as a result of decreases in predator free space may result in prey becoming more vulnerable to predation by invaders (Barrios-O'Neill et al., 2015). Functional response investigations can also be used to quantify interaction effects between invasive and native predator species (Barrios-O'Neill et al., 2014) or among invasive predator species (Wasserman et al., 2016b) using a multiple predator experimental designs.

The main focus of functional response studies to date has been for mobile predators and the consumption of prey species. However, the approach has recently been applied to investigating impacts of herbivores toward plant resources. For example, the invasive and ecologically damaging golden apple snail has faster feeding rates on primary plant sources compared to non-invasive and native snails (Xu et al., 2016).

4.2 Microcosms and Mesocosms

As a progression from functional responses, which usually focus on examining individuals and their interaction with increasing densities of one or a few prey species, microcosm and mesocosm experiments can involve more complex multi-species communities. For instance, simplified food webs of strongly interacting species, which are also known as community modules, provide a convenient unit in the study of trophic links that balance realism with tractability (Hatcher et al., 2006; Holt, 1997). By focussing on strong interactions among a relatively small number of species in defined configurations, basic features of empirical systems can be described and key driving processes and measures of impact may be identified (Holt and Hochberg, 2001; Paterson et al., 2015). Thus, functional response analyses within experimental community modules may provide excellent explanatory and predictive power (e.g. Paterson et al., 2015). However, in many ecosystems, numerous weak trophic links are common and act as major stabilising forces and, therefore, simple community modules with strong links may be unrealistic.

In the context of biological invasions, the arrival of an invasive species may cause fundamental changes to a community module (Polis et al., 1997). As nonnative species become an increasing component of natural communities, modules may offer a greater insight into how invaders influence food webs both through direct and indirect processes (Gilman et al., 2010; Box 1). For example, the presence of higher order predators may in fact increase the interaction strength and thus impacts of invasive mesopredators (Barrios-O'Neill et al., 2015) and the effects of parasite loading of intermediate predators within a community module can either increase or decrease predatory impacts (Paterson et al., 2015). Such species interactions within modules can be further influenced by abiotic conditions, including warming as a result of global climate change, which can alter the strength of novel multitrophic interactions (Miller et al., 2014). Similarly the relative roles of top-down and bottom-up forces within a food web may vary as a result of the identity of the participants (i.e. native versus invasive) affecting how impacts are propagated through a system. For example, the presence of invasive round gobies (*N. melanostomus*) results in stronger top-down effects leading to greater reductions in the abundance of grazers with a corresponding increase in benthic algae compared to when native logperch (*Percina caprodes*) are present (Pagnucco et al., 2016).

Community module experiments are usually conducted in highly controlled conditions (i.e. microcosms) with exact starting abundance and no scope for colonisation. They are also relatively new to both freshwater ecology and invasion biology and, in contrast, mesocosm experiments have been used by freshwater ecologists to measure the impact of stressors for decades (Stewart et al., 2013). Mesocosms are typically subject to lower experimental control than microcosms because they are often outside and are seeded with homogenised samples, usually where exact starting abundances are unknown. However, they provide an opportunity to quantify invader impacts in a more complex and natural environment. Mesocosm experiments can therefore be employed to study invaders and their connections within food webs in freshwater ecosystems (e.g. Jackson et al., 2013, 2014; Pagnucco et al., 2016; Rudnick and Resh, 2005). They have been used to quantify the impacts of invaders on native species through competition and predator–prey links, as well as the impacts of invasions on trophic cascades and trophic subsidies (Shurin et al., 2012), and interactions between invasive species (Jackson et al., 2014; Johnson et al., 2009).

Experimental approaches using microcosms, mesocosms or enclosure/exclosure designs allow the preferred niche and trophic position of invaders to be quantified using a variety of methods, including direct observations, stomach content and stable isotope analyses. Further, the diet of native species can be compared when they occur in allopatry and sympatry with a single invasive species, or multiple invaders. Aside from the usual caveats and criticisms levelled at the use of mesocosms (e.g. lack of realism), there is the added consideration from diet tracer perspective techniques (Sections 4.4 and 4.5), such as stable isotopes, as to whether the experiments are of sufficient duration to allow the test organism to at least partially equilibrate with new dietary sources (e.g. Jackson et al., 2016c; Rudnick and Resh, 2005). However, a recent experiment demonstrated that trophic interactions between native and invasive species in mesocosm experiments can be scaled up to the real world (Tran et al., 2015).

4.3 Direct Observations of Diet

When using both experimental and field approaches to quantify the impacts of invaders on trophic links, direct observations of diet (i.e. behavioural observation, gut/stomach content analysis) are frequently used to give a 'snapshot' estimate of where a species fits within a food web (see chapter "Invasions Toolkit: Current Methods for Tracking the Spread and Impact of Invasive Species" by Kamenvova et al.). For example, gut content analysis of *N. melanostomus* and *Ponticola kessleri*, both invasive gobiid fishes in the Danube River demonstrated differences in their diet between seasons (Brandner et al., 2013b). In freshwater studies, behavioural observations are used less frequently and usually under experimental conditions due to the difficulty of more natural underwater observations. For instance, Karraker and Dudgeon (2014) showed that invasive apple snails (*Pomacea canaliculata*) consume the eggs of native amphibians in China. In some highly resolved food webs, where the ecology of many species is well known, literature inference of trophic links are also used (e.g. Layer et al., 2010). However, this method is less reliable when considering the impact of an invader because their diet can vary considerably both within, and between, their native and invasive ranges (Jackson and Britton, 2014; Tillberg et al., 2007).

Long-term field studies incorporating diet observations provide valuable information on how an invaders impact on food webs can vary over time (e.g. Jackson et al., 2012; Winfield et al., 2012), expanding 'snapshot' estimates to more useful time series. For instance, in Lake Windermere, UK, the gut contents of pike *Esox lucius* was investigated over >30 years, indicating a shift towards nonnative roach *Rutilus rutilus* as the latter become more abundant (Winfield et al., 2012). Long-term observations of predator diet from gut content have also been used in Lake Naivasha, Kenya to elucidate shifts in dominance in the ever changing invasive species pool (Britton et al., 2010; Jackson et al., 2012). Additionally, long-term studies are often combined with tracer techniques (particularly stable isotopes) to demonstrate shifts in food web structure before and after invasion (Jackson et al., 2012; Maguire and Grey, 2006).

4.4 Stable Isotopes

There are a variety of natural and label tracer techniques at the disposal of ecologists to assess how invasive species may alter trophic relationships. None of the approaches described below can be described as a panacea; each has limitations which must be acknowledged, but they have all contributed to the study of invasion ecology.

The measurement of natural abundance stable carbon and nitrogen isotope ratios is a well-established tool to qualitatively and quantitatively describe how organisms as individuals, populations or communities interact via diet (Cucherousset et al., 2012; Grey, 2006; Peterson and Fry, 1987), provided that dietary sources have distinct stable isotope values. It is a particularly useful technique to assess the longer term diet and what is actually assimilated by organisms as compared to gut content analyses, and for predators that often have empty guts when caught (Grey, 2016; Grey et al., 2002). For example, stable isotopes have been used to determine resource use and trophic position, identify trophic redundancy in native and invasive species that are assumed functionally equivalent (Ercoli et al., 2014; Jackson et al., 2014; Rudnick and Resh, 2005), identify specific traits that help explain the success of invaders (Grey and Jackson, 2012; Jackson et al., 2016c; Olsson et al., 2009) and examine the wider food web consequences resulting from establishment of an invader (Vander Zanden et al., 1999) or multiple invaders over time (Jackson et al., 2012).

Provided that the parameters of interest such as trophic position, food chain length, size and position of the isotopic 'niche', and relative contributions from putative dietary sources have been adequately described isotopically prior to the introduction of invaders, then the magnitude and direction of any shift in those parameters resulting from invader perturbations can be determined (Grey et al., 2004; Schmidt et al., 2007). If natural abundance stable isotopes cannot offer the resolution required to separate basal resources or identify the assimilation of specific resources within a consumer tissues, then it may be feasible to artificially enrich or 'label' the resource with the heavier isotope and hence trace it more easily (Aberle et al., 2005). Alternatively, compound-specific stable isotope analyses of amino acids are becoming more prevalent, especially when it is difficult to collect or assign basal resources using bulk analyses (Chikaraishi et al., 2009).

A common criticism of many studies purporting to assess the impacts of invasive species is that the temporal element is rarely quantified; whether a study was conducted in the acute or chronic phase of invasion (or encompassed both) is likely to result in different findings (Strayer et al., 2006). Hence, a particularly appealing aspect of the stable isotope approach is that there may be archived samples spanning the complete invasion process from which to retrospectively detect changes in long-term data series. These can then be rigorously tested against other known environmental parameters to robustly assign any changes to invasion (e.g. Feuchtmayr and Grey, 2003). For example, Maguire and Grey (2006) used formalin-preserved lake water samples to detect that zooplankton became more reliant upon allochthonous resources from the period when invasive zebra mussels, an accomplished filter feeder and hence strong competitor for algal resources, became established in a lake, thereby indicating a subtle shift in ecosystem functioning.

4.5 Fatty Acids

The fatty acid composition or profile of an organism can also be a useful and efficient natural tracer of trophic interactions in much the same way as stable isotopes. Fatty acids are derived from diet and some can be synthesised so resource availability and trophic position govern the resultant profile. As with any natural tracer, signal attenuation with each trophic transfer may limit the approach when assessing consumers of high trophic position. Fatty acids are particularly useful in plankton studies to establish the importance of various microbial groups which are difficult to assess using conventional means, provided that the resource of interest contains a diagnostic biomarker (or specific combination thereof) that cannot be synthesised and hence must be derived from diet (Arts, 1998). Good examples are the invasive marine green alga *Codium fragile* which could be traced through the diets of two consecutive consumers under experimental conditions (Kelly et al., 2009), or the use of fatty acid profiles of mysid consumers in lakes with or without the invader *Bythotrephes longimanus*; mysids from lakes with the invader exhibited fatty acid profiles consistent with feeding upon algae and competing with Cladocera, indicating a shift to a lower trophic position (Nordin et al., 2008).

Fatty acid analyses can bring a slightly different emphasis to the study of invaded food webs because of their association as indicators of diet quality; the factors that affect the relative nutritional value of resources are still poorly understood, despite their potential to decouple trophic interactions (Kratina and Winder, 2015). Several authors have demonstrated that biotic invasions can alter the nutritional composition of zooplankton populations or communities: Fink (2013) demonstrated that the Ponto-Caspian invader *Limnomysis benedeni* contained high concentrations of essential polyunsaturated fatty acids which are important for fish nutrition and that, as a consequence, food webs might undergo trophic upgrading following invasion. Kratina and Winder (2015) reported temporal shifts in fatty acid classes of zooplankton of San Francisco Bay following establishment of invaders. This perspective clearly warrants further attention.

4.6 Molecular Tools

Development of molecular methodologies, especially high-throughput sequencing techniques, now provides cost-effective means of examining biodiversity at a scale and level of precision previously unavailable (Clare, 2014). Organisms will leave traces of DNA in their surrounding environment so that is possible to sample water to detect rare species or suspected invaders at the invasion front, and reconstruct past communities from sediment analyses (Bohmann et al., 2014; Secondi et al., 2016; Thomsen and Willerslev, 2014). Using this approach, the diet of an individual can also be attained from scat, regurgitates or stomach contents (Côté et al., 2013; Jo et al., 2016; Thomsen and Willerslev, 2014). The data can then be used to elucidate the diet or prey range of invaders (Ward and Ramón-Laca, 2013; Witczuk et al., 2013), or whether invaders may become novel prey for native predators (Gorokhova, 2006), and hence determine any unobserved trophic interaction. For example, molecular analyses of the stomach contents of invasive brown trout *S. trutta* in Tasmania, Australia, indicated that the species consumes 44 unique taxa, showing a greater accuracy than visual inspection alone (Jo et al., 2016). However, while molecular approaches may provide additional taxonomic resolution, there still may be clear advantages of using multiple analytical techniques, for example, to distinguish between different life stages of prey groups that may require subtly different foraging methods. Furthermore, molecular techniques do not accurately quantify the relative importance of each prey item to overall diet.

4.7 Modelling Approaches

Potential future invaders are often forecast by modelling techniques (e.g. Fletcher et al., 2016; Gallardo and Aldridge, 2013; Zambrano et al., 2006) and simulations can be used to predict impact (e.g. Cook et al., 2007; Leung et al., 2002; Pinnegar et al., 2014). For example, abiotic and biotic habitat characteristics were used to estimate invasive bass (*M. dolomieu*) occurrence and impact in the Great Lakes, North America (Vander Zanden et al., 2004). This approach found that only 6% of lakes were likely to be both invaded *and* impacted by bass, with impact predicted based on food web structure (Vander Zanden et al., 2004). Models suggest that invaders alter food web structure by decreasing species richness and the number of links per species (Galiana et al., 2014). Other studies, using simulated invasions, found that invaders can result in increased food web connectance but decreased modularity, as well as a reduction in the number of energy pathways from basal resources to top predators (Lurgi et al., 2014). Additionally, experimental results are often combined with modelling techniques to scale-up to real-world scenarios for a given invader. For instance, the functional response and satiation of invasive red swamp crayfish (*P. clarkii*), estimated under laboratory conditions, were used to calculate that one crayfish could consume 150–160 invasive zebra mussels (*D. polymorpha*) a month in the wild (Gonçalves et al., 2016).

4.8 Combining Multiple Methods

The Ponto-Caspian amphipod *D. villosus* (the 'killer shrimp'; Fig. 5A) is one of the most damaging invaders in Europe (Rewicz et al., 2014), and it is probably one of the few invasive species to date that has been studied using most of the approaches above. Interestingly, while functional response and feeding trial type tests tend to conclude that it is a top predator (Dodd et al., 2014; Fig. 5B), field studies using stable isotopes (J. Grey, unpublished data; Koester and Gergs, 2014; Fig. 5C) or fatty acids (Maazouzi et al., 2007, 2009) report that the dietary niche of *D. villosus* is similar to that of native amphipods with which they are sympatric (i.e. opportunistic and

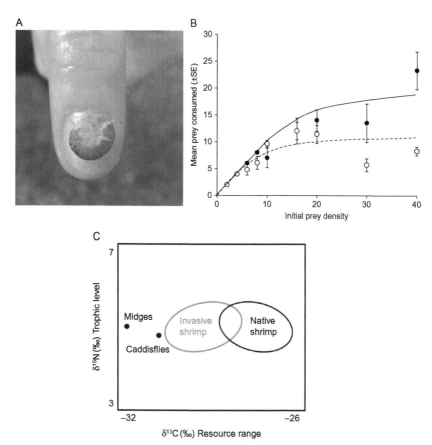

Fig. 5 Studies on the invasive *D. villosus* (A, photo by J.G.) using functional response experiments (B) and field surveys (C) indicate different degrees of predatory behaviour. (B) Comparison of functional response curves of *D. villosus* (*filled circles, solid line*) and native *Gammarus pulex* (*open circles, dashed line*) indicates heighted response towards the prey *Asellus aquaticus*. (C) In contrast, a carbon and nitrogen stable isotope biplot (where *ellipses* represent trophic niche) indicates a lower trophic level of *D. villosus* which is similar to native invertebrates, suggesting some herbivory, and a niche over-lapping with native shrimps. *(B) Redrawn from Dodd, J.A., Dick, J.T.A., Alexander, M.E., MacNeil, C., Dunn, A.M., Aldridge, D.C., 2014. Predicting the ecological impacts of a new freshwater invader: functional responses and prey selectivity of the 'killer shrimp',* Dikerogammarus villosus, *compared to the native* Gammarus pulex. *Freshw. Biol. 59, 337–352. (C) Redrawn from Koester, M., Gergs, R., 2014. No evidence for intraguild preda-tion of* Dikerogammarus villosus *(Sowinsky, 1894) at an invasion front in the Untere Lorze, Switzerland. Aquat. Invasions. 9, 489–497.*

omnivorous). Moreover, Koester and Gergs (2014) did not detect DNA from native gammarids in any of 62 individuals of *D. villosus* sampled from their study at an invasion front and implied that intraguild predation was not the key process driving the displacement of native amphipods. This suggests that a combination of natural field studies, experiments and different tracer techniques are needed to fully predict the impact of invaders on trophic links in freshwater ecosystems. However, the differences between experimental results and those from a more natural setting could simply reflect the lack of realism in many laboratory studies. As the number of complementary studies incorporating both stable isotopes and fatty acids has grown in recent years, we might expect a similar proliferation in combined approaches in the near future which will no doubt help to elucidate the more subtle interactions between invaders and natives mediated via diet.

5. CONCLUSIONS AND IMPLICATIONS

The impacts of invaders on trophic links can vary from palpable predator–prey links and indirect trophic cascades, to less obvious and far-reaching impacts which vary according to invader traits, abiotic factors, the receiving community and other environmental stressors. This complexity of interactions, including those within the food web and among other variables and stressors, often leads to 'ecological surprises'—unpredictable impacts that vary over space and time (Jackson et al., 2016b). Such ecological surprises makes it difficult to create a predictive framework for the impacts of invaders on trophic links and food webs, especially since observed impact can also vary according to study design and sampling technique (e.g. Fig. 5). We suggest a combination of controlled experiments, field surveys across the invasive range, and predictive modelling are needed to fully understand the impacts of a single invasive species.

Many invaded food webs in freshwater ecosystems are left considerably different from their preinvaded state, even following invader eradication. As such, food web resistance and resilience to invasion should be considered as distinct from each other. Resistance is the ability of the nodes and links in the food web to remain intact following invasion, while resilience refers to the ability of nodes, trophic links and, ultimately, food web structure to return to their preinvaded state. This recovery may occur following eradication, once the invasive population has naturally declined, or following the invasion front where populations tend to be larger and more voracious. For instance, in the Hudson River, North America, the food web has shifted to a state which is similar to that before it was invaded by the zebra mussel *D. polymorpha*, despite the invader still populating the ecosystem (Strayer et al., 2014).

Here we have shown how food webs can shift drastically following invasion, due to both novel and disrupted trophic links. However, despite the evidence described here for freshwater ecosystems, the respective roles of invaders and other (e.g. abiotic) environmental stressors remain are often confounded. We suggest that future research be directed towards disentangling the impacts of invaders and other stressors, including multiple invasions, with consideration given to how they vary over space and time. For instance, will the impact of an invader and a second stressor vary depending on which stressor comes first and/or the stage of invasion? Research should also be directed towards a more strategic approach that draws on a range of methodologies to fully elucidate invasion impacts.

REFERENCES

Aberle, N., Hillebrand, H., Grey, J., Wiltshire, K.H., 2005. Selectivity and competitive interactions between two benthic invertebrate grazers (*Asellus aquaticus* and *Potamopyrgus antipodarum*): an experimental study using 13C-and 15N-labelled diatoms. Freshw. Biol. 50, 369–379.

Abrams, P.A., 2000. The evolution of predator-prey interactions: theory and evidence. Annu. Rev. Ecol. Evol. Syst. 31, 79–105.

Adams, M.J., Pearl, C.A., Bury, R.B., 2003. Indirect facilitation of an anuran invasion by non-native fishes. Ecol. Lett. 6, 343–351.

Alexander, M.E., Dick, J.T.A., Weyl, O.L.F., Robinson, T.B., Richardson, D.M., 2014. Existing and emerging high impact invasive species are characterised by higher functional responses than natives. Biol. Lett. 10, 20130946.

Alexander, M.E., Kaiser, H., Weyl, O.L.F., Dick, J.T.A., 2015. Habitat simplification increases the impact of a freshwater invasive fish. Environ. Biol. Fishes 98, 477–486.

Alonso, F., Martínez, R., 2006. Shelter competition between two invasive crayfish species: a laboratory study. Bull. Fr. Pêche Piscic. 380–381, 1121–1132.

Arts, M.T., 1998. Ecophysiology of lipids in pelagic crustacean zooplankton communities. In: Molecular Approaches to the Study of the Ocean. Springer, Netherlands, pp. 329–341.

Atwood, T.B., Wiegner, T.N., Turner, J.P., MacKenzie, R.A., 2010. Potential effects of an invasive nitrogen-fixing tree on a Hawaiian stream food web. Pac. Sci. 64, 367–379.

Baiser, B., Russell, G.J., Lockwood, J.L., 2010. Connectance determines invasion success via trophic interactions in model food webs. Oikos 119, 1970–1976.

Baltz, D.M., Moyle, P.B., 1993. Invasion resistance to introduced species by a native assemblage of California stream fishes. Ecol. Appl. 3, 246–255.

Barrios-O'Neill, D., Dick, J.T.A., Emmerson, M., Ricciardi, A., MacIsaac, H., Alexander, M.E., Bovy, H., 2014. Fortune favours the bold: a higher predator reduces the impact of a native but not an invasive intermediate predator. J. Anim. Ecol. 83, 693–701.

Barrios-O'Neill, D., Dick, J.T.A., Emmerson, M.C., Ricciardi, A., MacIsaac, H.J., 2015. Predator-free space, functional responses and biological invasions. Funct. Ecol. 29, 377–384.

Baxter, C.V., Fausch, K.D., Murakami, M., Chapman, P.L., 2004. Fish invasion restructures stream and forest food webs by interrupting reciprocal prey subsidies. Ecology 85, 2656–2663. http://dx.doi.org/10.1890/04-138.

Belote, R., Jones, R., Hood, S., Wender, B., 2008. Diversity-invasibility across an experimental disturbance gradient in Appalachian forests. Ecology 89, 183–192.

Benjamin, J.R., Lepori, F., Baxter, C.V., Fausch, K.D., 2013. Can replacement of native by non-native trout alter stream-riparian food webs? Freshw. Biol. 58, 1694–1709.

Bergstrom, M.A., Mensinger, A.F., 2009. Interspecific resource competition between the invasive round goby and three native species: logperch, slimy sculpin, and spoonhead sculpin. Trans. Am. Fish. Soc. 138, 1009–1017.

Bobeldyk, A.M., Lamberti, G.A., 2010. Stream food web responses to a large omnivorous invader Orconectes rusticus (Decapoda, Cambaridae). Crustaceana 83, 641–657.

Bohmann, K., Evans, A., Gilbert, M.T.P., Carvalho, G.R., Creer, S., Knapp, M., Douglas, W.Y., de Bruyn, M., 2014. Environmental DNA for wildlife biology and biodiversity monitoring. Trends Ecol. Evol. 29, 358–367.

Boudreau, S.A., Yan, N.D., 2003. The differing crustacean zooplankton communities of Canadian shield lakes with and without the nonindigenous zooplanktivore Bythotrephes longimanus. Can. J. Fish. Aquat. Sci. 60, 1307–1313.

Bovy, H.C., Barrios-O'Neill, D., Emmerson, M.C., Aldridge, D.C., Dick, J.T.A., 2015. Predicting the predatory impacts of the "demon shrimp" Dikerogammarus haemobaphes, on native and previously introduced species. Biol. Invasions 17, 597–607.

Brandner, J., Cerwenka, A.F., Schliewen, U.K., Geist, J., 2013a. Bigger is better: characteristics of Round Gobies forming an invasion front in the Danube river. PLoS One 8, e73036.

Brandner, J., Auerswald, K., Cerwenka, A.F., Schliewen, U.K., Geist, J., 2013b. Comparative feeding ecology of invasive Ponto-Caspian gobies. Hydrobiologia 703, 113–131.

Brennan, A.C., Woodward, G., Seehausen, O., Muñoz-Fuentes, V., Moritz, C., Guelmami, A., Abbott, R.J., Edelaar, P., 2014. Hybridization due to changing species distributions: adding problems or solutions to conservation of biodiversity during global change? Evol. Ecol. Res. 16, 475–491.

Britton, J.R., Harper, D.M., Oyugi, D.O., Grey, J., 2010. The introduced Micropterus salmoides in an equatorial lake: a paradoxical loser in an invasion meltdown scenario? Biol. Invasions 12, 3439–3448.

Brown, G.P., Ujvari, B., Madsen, T., Shine, R., 2013. Invader impact clarifies the roles of top-down and bottom-up effects on tropical snake populations. Funct. Ecol. 27, 351–361.

Burlakova, L.E., Karatayev, A.Y., Padilla, D.K., Cartwright, L.D., Hollas, D.N., 2009. Wetland restoration and invasive species: apple snail (Pomacea insularum) feeding on native and invasive aquatic plants. Restoration Ecol. 17, 433–440.

Byers, J.E., 2002. Impact of non-indigenous species on natives enhanced by anthropogenic alteration of selection regimes. Oikos 97, 449–458.

Cameron, E.K., Vilà, M., Cabeza, M., 2016. Global meta-analysis of the impacts of terrestrial invertebrate invaders on species, communities and ecosystems. Glob. Ecol. Biogeogr.

Cano, L., Escarré, J., Vrieling, K., Sans, F., 2009. Palatability to a generalist herbivore, defence and growth of invasive and native Senecio species: testing the evolution of increased competitive ability hypothesis. Oecologia 159, 95–106.

Capinha, C., Leung, B., Anastacio, P., 2011. Predicting worldwide invasiveness for four major problematic decapods: an evaluation of using different calibration sets. Ecography 34, 448–459.

Carpenter, S.R., Kitchell, J.F., Hodgson, J.R., 1985. Cascading trophic interactions and lake productivity. BioScience 35, 634–639.

Chapple, D.G., Simmonds, S.M., Wong, B.B.N., 2012. Can behavioural and personality traits influence the success of unintentional species introductions? Trends Ecol. Evol. 27, 57–64.

Chikaraishi, Y., Ogawa, N.O., Kashiyama, Y., Takano, Y., Suga, H., Tomitani, A., Miyashita, H., Kitazato, H., Ohkouchi, N., 2009. Determination of aquatic food-web structure based on compound-specific nitrogen isotopic composition of amino acids. Limnol. Oceanogr. Methods 7, 740–750.

Clare, E.L., 2014. Molecular detection of trophic interactions: emerging trends, distinct advantages, significant considerations and conservation applications. Evol. Appl. 7, 1144–1157.

Clavero, M., Garcia-Berthou, E., 2005. Invasive species are a leading cause of animal extinctions. Trends Ecol. Evol. 20, 110.

Cohen, A.N., Carlton, J.T., 1998. Accelerating invasion rate in a highly invaded estuary. Science 279, 555–558.

Cohen, J.E., Jonsson, T., Carpenter, S.R., 2003. Ecological community description using the food web, species abundance, and body size. Proc. Natl. Acad. Sci. 100, 1781–1786.

Cook, D.C., Thomas, M.B., Cunningham, S.A., Anderson, D.L., De Barro, P.J., 2007. Predicting the economic impact of an invasive species on an ecosystem service. Ecol. Appl. 17, 1832–1840.

Correia, A.M., 2001. Seasonal and interspecific evaluation of predation by mammals and birds on the introduced red swamp crayfish *Procambarus clarkii* (Crustacea, Cambaridae) in a freshwater marsh (Portugal). J. Zool. 255, 533–541.

Côté, I.M., Green, S.J., Morris Jr., J.A., Akins, J.L., Steinke, D., 2013. Diet richness of invasive Indo-Pacific lionfish revealed by DNA barcoding. Mar. Ecol. Prog. Ser. 472, 249–256.

Covich, A.P., 2010. Winning the biodiversity arms race among freshwater gastropods: competition and coexistence through shell variability and predator avoidance. Hydrobiologia 653, 191–215.

Cox, J.G., Lima, S.L., 2006. Naiveté and an aquatic terrestrial dichotomy in the effects of introduced predators. Trends Ecol. Evol. 21, 674–680.

Crain, C.M., Bertness, M.D., 2006. Ecosystem engineering across environmental gradients: implications for conservation and management. BioScience 56, 211–218.

Crowder, D.W., Snyder, W.E., 2010. Eating their way to the top? Mechanisms underlying the success of invasive insect generalist predators. Biological Invasions 12, 2857–2876.

Cucherousset, J., Olden, J.D., 2011. Ecological impacts of non-native freshwater fishes. Fisheries 36 (5), 215–230.

Cucherousset, J., Boulêtreau, S., Martino, A., Roussel, J.M., Santoul, F., 2012. Using stable isotope analyses to determine the ecological effects of non-native fishes. Fish. Manag. Ecol. 19, 111–119.

D'antonio, C., Dudley, T., 1995. Biological invasions as agents of change on islands versus mainlands. In: Islands. Springer, Berlin Heidelberg, pp. 103–121.

David, P., Thébault, E., Anneville, O., Duyck, P.-F., Chapuis, E., Loeuille, N., 2017. Impacts of invasive species on food webs: a review of empirical data. Adv. Ecol. Res. 56, 1–60.

Dick, J.T.A., 2008. Role of behaviour in biological invasions and species distributions; lessons from interactions between the invasive *Gammarus pulex* and the native *G. duebeni* (Crustacea: Amphipoda). Contrib. Zool. 77, 91–98.

Dick, J.T.A., Platvoet, D., 1996. Intraguild predation and species exclusions in amphipods: the interaction of behaviour, physiology and environment. Freshw. Biol. 36, 375–383.

Dick, J.T.A., Gallagher, K., Avlijas, S., Clarke, H.C., Lewis, S.E., Leung, S., Minchin, D., Caffrey, J., Alexander, M.E., Maguire, C., Harrod, C., Reid, N., Haddaway, N.R., Farnsworth, K.D., Penk, M., Ricciardi, A., 2013a. Ecological impacts of an invasive predator explained and predicted by comparative functional responses. Biol. Invasions 15, 837–846.

Dick, J.T.A., MacNeil, C., Alexander, M., Dodd, J., Ricciardi, A., 2013b. Predators vs. alien: differential biotic resistance to an invasive species by two resident predators. NeoBiota 19, 1.

Dick, J.T.A., Alexander, M.E., Jeschke, J.M., Ricciardi, A., MacIsaac, H.J., Robinson, T.B., Kumschick, S., Weyl, O.L.F., Dunn, A.M., Hatcher, M.J., Paterson, R.A., Farnsworth, K.D., Richardson, D.M., 2014. Advancing impact prediction and hypothesis testing in invasion ecology using a comparative functional response approach. Biol. Invasions 16, 735–753.

Didham, R.K., Tylianakis, K.M., Gemmell, N.J., Rand, T.A., Ewers, R.M., 2007. The interactive effects of habitat loss and species invasion on native species decline. Trends Ecol. Evol. 22, 489–496.

Dodd, J.A., Dick, J.T.A., Alexander, M.E., MacNeil, C., Dunn, A.M., Aldridge, D.C., 2014. Predicting the ecological impacts of a new freshwater invader: functional responses and prey selectivity of the 'killer shrimp', Dikerogammarus villosus, compared to the native Gammarus pulex. Freshw. Biol. 59, 337–352.

Dunne, J.A., Williams, R.J., Martinez, N.D., 2002. Food-web structure and network theory: the role of connectance and size. Proc. Natl. Acad. Sci. U.S.A. 99, 12917–12922.

Eggleston, D.B., Lipcius, R.N., Hines, A.H., 1992. Density-dependent predation by blue crabs upon infaunal clam species with contrasting distribution and abundance patterns. Mar. Ecol. Prog. Ser. 85, 55–68.

Ercoli, F., Ruokonen, T.J., Hämäläinen, H., Jones, R.I., 2014. Does the introduced signal crayfish occupy an equivalent trophic niche to the lost native noble crayfish in boreal lakes? Biol. Invasions 16, 2025–2036.

Feuchtmayr, H., Grey, J., 2003. Effect of preparation and preservation procedures on carbon and nitrogen stable isotope determinations from zooplankton. Rapid Commun. Mass Spectrom. 17, 2605–2610.

Findlay, C.S., Bert, D.G., Zheng, L.G., 2000. Effect of introduced piscivores on native minnow communities in Adirondack lakes. Can. J. Fish. Aquat. Sci. 57, 570–580.

Fink, P., 2013. Invasion of quality: high amounts of essential fatty acids in the invasive Ponto-Caspian mysid Limnomysis benedeni. J. Plankton Res. 35 (4), 907–913.

Fletcher, D.H., Gillingham, P.K., Britton, J.R., Blanchet, S., Gozlan, R.E., 2016. Predicting global invasion risks: a management tool to prevent future introductions. Sci. Rep. 6, 26316.

Galiana, N., Lurgi, M., Montoya, J.M., López, B.C., 2014. Invasions cause biodiversity loss and community simplification in vertebrate food webs. Oikos 123, 721–728.

Gallardo, B., Aldridge, D.C., 2013. Priority setting for invasive species management: risk assessment of Ponto-Caspian invasive species into Great Britain. Ecol. Appl. 23, 352–364.

Garcia, M.L., Protogino, L.C., 2005. Invasive freshwater molluscs are consumed by native fishes in South America. J. Appl. Ichthyol. 21, 34–38.

Gergs, R., Koester, M., Schulz, R.S., Schulz, R., 2014. Potential alteration of cross-ecosystem resource subsidies by an invasive aquatic macroinvertebrate: implications for the terrestrial food web. Freshw. Biol. 59, 2645–2655.

Gherardi, F., Acquistapace, P., 2007. Invasive crayfish in Europe: the impact of Procambarus clarkii on the littoral community of a Mediterranean lake. Freshw. Biol. 52, 1249–1259.

Gilman, S.E., Urban, M.C., Tewksbury, J., Gilchrist, G.W., Holt, R.D., 2010. A framework for community interactions under climate change. Trends Ecol. Evol. 25, 325–331.

Gonçalves, V., Gherardi, F., Rebelo, R., 2016. Modelling the predation effects of invasive crayfish, Procambarus clarkii (Girard, 1852), on invasive zebra mussel, Dreissena polymorpha (Pallas, 1771), under laboratory conditions. Ital. J. Zool. 83, 59–67.

Gonzalez, A., Lambert, A., Ricciardi, A., 2008. When does ecosystem engineering cause invasion and species replacement? Oikos 117, 1247–1257.

Gorokhova, E., 2006. Molecular identification of the invasive cladoceran *Cercopagis pengoi* (Cladocera: Onychopoda) in stomachs of predators. Limnol. Oceanogr. Methods 4, 1–6.

Goudswaard, K., Witte, F., Katunzi, E., 2008. The invasion of an introduced predator, Nile perch (*Lates niloticus*) in Lake Victoria (East Africa): chronology and causes. Environ. Biol. Fishes 81, 127–139.

Grey, J., 2006. The use of stable isotope analyses in freshwater ecology: current awareness. Pol. J. Ecol. 54, 563–584.

Grey, J., 2016. The incredible lightness of being methane-fuelled: stable isotopes reveal alternative energy pathways in aquatic ecosystems and beyond. Front. Ecol. Evol. 4, 8.

Grey, J., Jackson, M.C., 2012. "Leaves and eats shoots": direct terrestrial feeding can supplement invasive red swamp crayfish in times of need. PLoS One 7, e425575.

Grey, J., Thackeray, S.J., Jones, R.I., Shine, A., 2002. Ferox Trout (*Salmo trutta*) as 'Russian dolls': complementary gut content and stable isotope analyses of the Loch Ness foodweb. Freshw. Biol. 47, 1235–1243.

Grey, J., Waldron, S., Hutchinson, R., 2004. The utility of carbon and nitrogen isotope analyses to trace contributions from fish farms to the receiving communities of freshwater lakes: a pilot study in Esthwaite water, UK. Hydrobiologia 524, 253–262.

Haddaway, N.R., Wilcox, R.H., Heptonstall, R.E., Griffiths, H.M., Mortimer, R.J., Christmas, M., Dunn, A.M., 2012. Predatory functional response and prey choice identify predation differences between native/invasive and parasitised/unparasitised crayfish. PLoS One 7, e32229.

Hassell, M.P., 1978. Functional responses. In: Hassell, M.P. (Ed.), The Dynamics of Arthropod Predator-Prey Systems. Princeton University Press, Princeton, pp. 28–49.

Hatcher, M., Dick, J.T.A., Dunn, A., 2006. How parasites affect interactions between competitors and predators. Ecol. Lett. 9, 1253–1271.

Heath, M.R., Speirs, D.C., Steele, J.H., 2014. Understanding patterns and processes in models of trophic cascades. Ecol. Lett. 17, 101–114.

Hernández, L., Laundré, J.W., 2005. Foraging in the "landscape of fear" and its implications for habitat use and diet quality of elk *Cervus elaphus* and bison *Bison bison*. Wildl. Biol. 11, 215–220.

Hill, A.M., Lodge, D.M., 1999. Replacement of resident crayfishes by an exotic crayfish: the roles of competition and predation. Ecol. Appl. 9, 678–690.

Hladyz, S., Gessner, M.O., Giller, P.S., Pozo, J., Woodward, G., 2009. Resource quality and stoichiometric constraints on stream ecosystem functioning. Freshw. Biol. 54, 957–970.

Hladyz, S., Abjornsson, K., Giller, P.S., Woodward, G., 2011. Impacts of an aggressive riparian invader on community structure and ecosystem functioning in stream food webs. J. Appl. Ecol. 48, 443–452.

Holling, C.S., 1959. Some characteristics of simple types of predation and parasitism. Can. J. Entomol. 91, 38–398.

Holt, R.D., 1997. Community modules. In: Gange, A.C., Brown, V.K. (Eds.), Multitrophic Interactions in Terrestrial Systems. Black-Well Science, Oxford, pp. 333–350.

Holt, R.D., Hochberg, M.E., 2001. Indirect interactions, community modules, and biological control: a theoretical perspective. In: Waijnberg, E., Scott, J.K., Quimby, P.C. (Eds.), Evaluation of Indirect Ecological Effects of Biological Control. CABI International; Wallingford, UK, pp. 13–37.

Hulme, P.E., 2009. Trade, transport and trouble: managing invasive species pathways in an era of globalization. J. Appl. Ecol. 46, 10–18.

Iacarella, J.C., Ricciardi, A., 2015. Dissolved ions mediate body mass gain and predatory response of an invasive fish. Biol. Invasions 17, 3237–3246.

Iacarella, J.C., Dick, J.T.A., Alexander, M.E., Ricciardi, A., 2015a. Ecological impacts of invasive alien species along temperature gradients: testing the role of environmental matching. Ecol. Appl. 25, 706–716.

Iacarella, J.C., Dick, J.T.A., Ricciardi, A., 2015b. A spatio-temporal contrast of the predatory impact of an invasive freshwater crustacean. Divers. Distrib. 21, 803–812.

Inger, R., McDonald, R.A., Rogowski, D., Jackson, A.L., Parnell, A., Preston, S.J., Harrod, C., Goodwin, C., Griffiths, D., Dick, J.T.A., Elwood, R.W., Newton, J., Bearhop, S., 2010. Do non-native invasive fish support elevated lamprey populations? J. Appl. Ecol. 47, 121–129.

Jackson, M.C., 2015. Interactions among multiple invasive animals. Ecology 96, 2035–2041.

Jackson, M.C., Britton, J.R., 2014. Divergence in the trophic niche of sympatric freshwater invaders. Biol. Invasions 16, 1095–1103.

Jackson, M.C., Grey, J., 2013. Accelerating rates of freshwater invasions in the catchment of the river Thames. Biol. Invasions 15, 945–951.

Jackson, M.C., Jackson, A.L., Britton, J.R., Donohue, I., Harper, D., Grey, J., 2012. Population-level metrics of trophic structure based on stable isotopes and their application to invasion ecology. PLoS One 7, e31757.

Jackson, M.C., Allen, R., Pegg, J., Britton, J.R., 2013. Do trophic subsidies affect the outcome of introductions of a non-native freshwater fish? Freshw. Biol. 58, 2144–2153.

Jackson, M.C., Jones, T., Milligan, M., Sheath, D., Taylor, J., Ellis, A., England, J., Grey, J., 2014. Niche differentiation among invasive crayfish and their impacts on ecosystem structure and functioning. Freshw. Biol. 59, 1123–1135.

Jackson, M.C., Ruiz-Navarro, A., Britton, J.R., 2015. Population density modifies the ecological impacts of invasive species. Oikos 124, 880–887.

Jackson, M.C., Woodford, D.J., Bellingan, T.A., Weyl, O.L.F., Potgieter, M.J., Rivers-Moore, N.A., Ellender, B.R., Fourie, H.E., Chimimba, C.T., 2016a. Trophic overlap between fish and riparian spiders: potential impacts of an invasive fish on terrestrial consumers. Ecol. Evol. 6, 1745–1752.

Jackson, M.C., Loewen, C.J., Vinebrooke, R.D., Chimimba, C.T., 2016b. Net effects of multiple stressors in freshwater ecosystems: a meta-analysis. Glob. Chang. Biol. 22, 180–189.

Jackson, M.C., Grey, J., Miller, K., Britton, J.R., Donohue, I., 2016c. Dietary niche constriction when invaders meet natives: evidence from freshwater decapods. J. Anim. Ecol. 85, 1098–1107.

James, J., Thomas, J.R., Ellis, A., Young, K.A., England, J., Cable, J., 2016. Over-invasion in a freshwater ecosystem: newly introduced virile crayfish (Orconectes virilis) outcompete established invasive signal crayfish (Pacifastacus leniusculus). Mar. Freshwater Behav. Physiol. 49, 9–18.

Jo, H., Ventura, M., Vidal, N., Gim, J.S., Buchaca, T., Barmuta, L.A., Jeppesen, E., Joo, G.J., 2016. Discovering hidden biodiversity: the use of complementary monitoring of fish diet based on DNA barcoding in freshwater ecosystems. Ecol. Evol. 6, 219–232.

Johnson, B.M., Martinez, P.J., Hawkins, J.A., Bestgen, K.R., 2008. Ranking predatory threats by non-native fishes in the Yampa river, Colorado via bioenergetics modelling. N. Am. J. Fish. Manage. 28, 1941–1953.

Johnson, P., Olden, J., Solomon, C., Vander Zanden, M., 2009. Interactions among invaders: community and ecosystem effects of multiple invasive species in an experimental aquatic system. Oecologia 159, 161–170.

Karraker, N.E., Dudgeon, D., 2014. Invasive apple snails (Pomacea canaliculata) are predators of amphibians in South China. Biol. Invasions 16, 1785–1789.

Kaufman, L., 1992. Catastrophic change in species-rich freshwater ecosystems. Bioscience 42, 846–858.

Keane, R., Crawley, M., 2002. Exotic plant invasions and the enemy release hypothesis. Trends Ecol. Evol. 17, 164–170.

Kelleher, B., Bergers, P.J.M., Van Den Brink, F.W.B., Giller, P.S., Van Der Velde, G., Bij De Vaate, A., 1998. Effects of exotic amphipod invasions on fish diet in the lower Rhine. Arch. Hydrobiol. 143, 363–382.

Kelly, D., Hawes, I., 2005. Effects of invasive macrophytes on littoral-zone productivity and foodweb dynamics in a New Zealand high-country lake. J. North Am. Benthol. Soc. 24, 300–320.

Kelly, J.R., Scheibling, R.E., Iverson, S.J., 2009. Fatty acids tracers for native and invasive macroalgae in an experimental food web. Mar. Ecol. Prog. Ser. 391, 53–63.

Kestrup, Å.M., Ricciardi, A., 2009. Environmental heterogeneity limits the local dominance of an invasive freshwater crustacean. Biol. Invasions 11, 2095–2105.

Kestrup, Å.M., Dick, J.T., Ricciardi, A., 2011. Interactions between invasive and native crustaceans: differential functional responses of intraguild predators towards juvenile hetero-specifics. Biol. Invasions 13, 731–737.

King, R.B., Ray, J.M., Stanford, K.M., 2006. Gorging on gobies: beneficial effects of alien prey on a threatened vertebrate. Can. J. Zool. 84, 108–115.

Koester, M., Gergs, R., 2014. No evidence for intraguild predation of *Dikerogammarus villosus* (Sowinsky, 1894) at an invasion front in the Untere Lorze, Switzerland. Aquat. Invasions 9, 489–497.

Kolar, C.S., Lodge, D.M., 2001. Progress in invasion biology: predicting invaders. Trends Ecol. Evol. 16, 199–204.

Kratina, P., Winder, M., 2015. Biotic invasions can alter nutritional composition of zooplankton communities. Oikos 124, 1337–1345.

Larrañaga, A., Basaguren, A., Pozo, J., 2014. Resource quality controls detritivore consumption, growth, survival and body condition recovery of reproducing females. Mar. Freshw. Res. 65, 910–917.

Laverty, C., Dick, J.T.A., Alexander, M.E., Lucy, F.E., 2015. Differential ecological impacts of invader and native predatory freshwater amphipods under environmental change are revealed by comparative functional responses. Biol. Invasions 17, 1761–1770.

Layer, K., Riede, J.O., Hildrew, A.G., Woodward, G., 2010. Food web structure and stability in 20 streams across a wide ph gradient. Adv. Ecol. Res. 42, 265–299.

Layer, K., Hildrew, A.G., Jenkins, G.B., Riede, J.O., Rossiter, S.J., Townsend, C.R., Woodward, G., 2011. Long-term dynamics of a well-characterised food web: four decades of acidification and recovery in the Broadstone Stream model system. Adv. Ecol. Res. 44, 69–117.

Lepak, J.M., Kraft, C.E., Weidel, B.C., 2006. Rapid food web recovery in response to removal of an introduced apex predator. Can. J. Fish. Aquat. Sci. 63, 569–575.

Leung, B., Lodge, D.M., Finnoff, D., Shogren, J.F., Lewis, M.A., Lamberti, G., 2002. An ounce of prevention or a pound of cure: bioeconomic risk analysis of invasive species. Proc. R. Soc. Lond. B Biol. Sci. 269, 2407–2413.

Levine, J.M., Adler, P.B., Yelenik, S.G., 2004. A meta-analysis of biotic resistance to plant invasions. Ecol. Lett. 10, 975–989.

Li, Y., Ke, Z., Wang, S., Smith, G.R., Liu, X., 2011. An exotic species is the favorite prey of a native enemy. PLoS One 6, e24299.

Lurgi, M., Galiana, N., López, B.C., Joppa, L.N., Montoya, J.M., 2014. Network complexity and species traits mediate the effects of biological invasions on dynamic food webs. Front. Ecol. Evol. 2, 36.

Maazouzi, C., Masson, G., Izquierdo, M.S., Pihan, J.C., 2007. Fatty acid composition of the amphipod Dikerogammarus villosus: feeding strategies and trophic links. Comp. Biochem. Physiol. A Mol. Integr. Physiol. 147, 868–875.

Maazouzi, C., Piscart, C., Pihan, J.C., Masson, G., 2009. Effect of habitat-related resources on fatty acid composition and body weight of the invasive Dikerogammarus villosus in an artificial reservoir. Fundam. Appl. Limnol. 175, 327–338.

MacNeil, C., Dick, J.T.A., Hatcher, M.J., Terry, R.S., Smith, J.E., Dunn, A.M., 2003. Parasite-mediated predation between native and invasive amphipods. Proc. R. Soc. Lond. B Biol. Sci. 270, 1309–1314.

MacNeil, C., Dick, J.T.A., Johnson, M.P., Hatcher, M.J., Dunn, A.M., 2004. A species invasion mediated through habitat structure, intraguild predation, and parasitism. Limnol. Oceanogr. 49, 1848–1856.

Maezono, Y., Miyashita, T., 2003. Community-level impacts induced by introduced largemouth bass and bluegill in farm ponds in Japan. Biol. Conserv. 109, 111–121.

Maguire, C.M., Grey, J., 2006. Determination of zooplankton dietary shift following a zebra mussel invasion, as indicated by stable isotope analysis. Freshw. Biol. 51, 1310–1319.

Martínez, A., Larrañaga, A., Pérez, J., Basaguren, A., Pozo, J., 2013. Leaf-litter quality effects on stream ecosystem functioning: a comparison among five species. Fundam. Appl. Limnol. 183, 239–248.

May, R.M., 1973. Stability and Complexity in Model Ecosystems, vol. 6. Princeton University Press; Woodstock, UK.

Mayer, C.M., Rudstam, L.G., Mills, E.L., Cardiff, S.G., Bloom, C.A., 2001. Zebra mussels (Dreissena polymorpha), habitat alteration, and yellow perch (Perca flavescens) foraging: system-wide effects and behavioural mechanisms. Can. J. Fish. Aquat. Sci. 58, 2459–2467.

McCann, K.S., 2000. The diversity-stability debate. Nature 405, 228–233.

McIntosh, A.R., McHugh, P.A., Dunn, N.R., Goodman, J.M., Howard, S.W., Jellyman, P.G., O'Brien, L.K., Nyström, P., Woodford, D.J., 2010. The impact of trout on galaxiid fishes in New Zealand. N. Z. J. Ecol. 34, 195.

Médoc, V., Albert, H., Spataro, T., 2015. Functional response comparisons among freshwater amphipods: ratio-dependence and higher predation for Gammarus pulex compared to the non-natives Dikerogammarus villosus and Echinogammarus berilloni. Biol. Invasions 17, 3625–3637.

Michelan, T.S., Silveira, M.J., Petsch, D.K., Pinha, G.D., Thomaz, S.M., 2014. The invasive aquatic macrophyte Hydrilla verticillata facilitates the establishment of the invasive mussel Limnoperna fortunei in Neotropical reservoirs. J. Limnol. 73, 598–602.

Miller, L.P., Matassa, C.M., Trussell, G.C., 2014. Climate change enhances the negative effects of predation risk on an intermediate consumer. Glob. Chang. Biol. 20, 3834–3844.

Mollot, G., Pantel, J.H., Romanuk, T.N., 2017. The effects of invasive species on the decline in species richness: a global meta-analysis. Adv. Ecol. Res. 56, 61–83.

Molloy, D.P., Powell, J., Ambrose, P., 1994. Short term reduction of adult zebra mussels (Dreissena polymorpha) in the Hudson river near Catskill, New York: an effect of juvenile blue crab (Callinectes sapidus) predation? J. Shellfish Res. 13, 367–371.

Morrison, W.E., Hay, M.E., 2011. Feeding and growth of native, invasive and non-invasive alien apple snails (Ampullariidae) in the United States: invasives eat more and grow more. Biol. Invasions 13, 945–955.

Moyle, P.B., Light, T., 1996. Biological invasions of fresh water: empirical rules and assembly theory. Biol. Conserv. 78, 149–161.

Njiru, M., Okeyo-Owuor, J.B., Muchiri, M., Cowx, I.G., 2004. Shifts in the food of Nile tilapia, Oreochromis niloticus in Lake Victoria, Kenya. Afr. J. Ecol. 42, 163–170.

Noonburg, E.G., Byers, J.E., 2005. More harm than good: when invader vulnerability to predators enhances impact on native species. Ecology 86, 2555–2560.

Nordin, L.J., Arts, M.T., Johannsson, O.E., Taylor, W.D., 2008. An evaluation of the diet of Mysis relicta using gut contents and fatty acid profiles in lakes with and without the invader Bythotrephes longimanus (Onychopoda, Cercopagidae). Aquat. Ecol. 42, 421–436.

O'Gorman, E.J., 2014. Integrating comparative functional response experiments into global change research. J. Anim. Ecol. 83, 525–527.

O'Gorman, E.J., Emmerson, M.C., 2011. Body mass–abundance relationships are robust to cascading effects in marine food webs. Oikos 120, 520–528.

Olsson, K., Stenroth, P., Nyström, P., Granéli, W., 2009. Invasions and niche width: does niche width of an introduced crayfish differ from a native crayfish? Freshw. Biol. 54, 1731–1740.

Pagnucco, K.S., Remmal, Y., Ricciardi, A., 2016. An invasive benthic fish magnifies trophic cascades and alters pelagic communities in an experimental freshwater system. Freshw. Sci. 35, 654–665.

Parker, J., Hay, M., 2005. Biotic resistance to plant invasions? Native herbivores prefer non-native plants. Ecol. Lett. 89, 959–967.

Parker, I.M., Simberloff, D., Lonsdale, W.M., Goodell, K., Wonham, M., Kareiva, P.M., Williamson, M.H., Von Holle, B., Moyle, P.B., Byers, J.E., Goldwasser, L., 1999. Impact: toward a framework for understanding the ecological effects of invaders. Biol. Invasions 1, 3–19.

Parker, J.D., Caudill, C.C., Hay, M.E., 2007. Beaver herbivory on aquatic plants. Oecologia 151, 616–625.

Paterson, R.A., Dick, J.T., Pritchard, D.W., Ennis, M., Hatcher, M.J., Dunn, A.M., 2015. Predicting invasive species impacts: a community module functional response approach reveals context dependencies. J. Anim. Ecol. 84, 453–463.

Pearl, C.A., Adams, M.J., Schuytema, G.S., Nebeker, A.V., 2003. Behavioral responses of anuran larvae to chemical cues of native and introduced predators in the Pacific Northwestern United States. J. Herpetol. 37, 572–576.

Pejchar, L., Mooney, H.A., 2009. Invasive species, ecosystem services and human well-being. Trends Ecol. Evol. 24, 497–504.

Petchey, O.L., Brose, U., Rall, B.C., 2010. Predicting the effects of temperature on food web connectance. Philos. Trans. R. Soc. Lond. B Biol. Sci. 365, 2081–2091.

Peterson, B.J., Fry, B., 1987. Stable isotopes in ecosystem studies. Annu. Rev. Ecol. Syst. 18, 293–320.

Phillips, B.L., 2009. The evolution of growth rates on an expanding range edge. Biol. Lett. 5, 802–804.

Phillips, B., Brown, G.P., Shine, R., 2003. Assessing the potential impact of cane toads (Bufo marinus) on Australian snakes. Conserv. Biol. 17, 1738–1747.

Pinnegar, J.K., Tomczak, M.T., Link, J.S., 2014. How to determine the likely indirect food-web consequences of a newly introduced non-native species: a worked example. Ecol. Model. 272, 379–387.

Polis, G.A., Myers, C., Holt, R.D., 1989. The ecology and evolution of intraguild predation: potential competitors that eat each other. Annu. Rev. Ecol. Evol. Syst. 20, 297–330.

Polis, G.A., Anderson, W.B., Holt, R.D., 1997. Toward an integration of landscape and food web ecology: the dynamics of spatially subsidized food webs. Annu. Rev. Ecol. Evol. Syst. 28, 289–316.

Polo-Cavia, N., López, P., Martín, J., 2011. Aggressive interactions during feeding between native and invasive freshwater turtles. Biol. Invasions 13, 1387–1396.

Prenter, J., MacNeil, C., Dick, J.T., Dunn, A.M., 2004. Roles of parasites in animal invasions. Trends Ecol. Evol. 19, 385–390.

Radke, R.J., Kahl, U., 2002. Effects of a filter-feeding fish [silver carp, Hypophthalmichthys molitrix] on phyto- and zooplankton in a mesotrophic reservoir: results from an enclosure experiment. Freshw. Biol. 47, 2337–2344.

Rahel, F.J., Olden, J.D., 2008. Assessing the effects of climate change on aquatic invasive species. Conserv. Biol. 22, 521–533.

Rall, B.C., Vucic-Pestic, O., Ehnes, R.B., Emmerson, M., Brose, U., 2010. Temperature, predator–prey interaction strength and population stability. Glob. Change Biol. 16, 2145–2157.

Rall, B.C., Brose, U., Hartvig, M., Kalinkat, G., Schwarzmüller, F., Vucic-Pestic, O., Petchey, O.L., 2012. Universal temperature and body-mass scaling of feeding rates. Philos. Trans. R. Soc. B 367, 2923–2934.

Raw, J.L., Miranda, N.A., Perissinotto, R., 2013. Chemical cues released by an alien invasive aquatic gastropod drive its invasion success. PLoS One 8, e64071.

Reid, A.J., Chapman, L.J., Ricciardi, A., 2013. Wetland edges as peak refugia from an introduced piscivore. Aquat. Conserv. 23, 646–655.

Reinhart, K., VandeVoort, R., 2006. Effect of native and exotic leaf litter on macroinvertebrate communities and decomposition in a western Montana stream. Divers. Distrib. 12, 776–781.

Rewicz, T., Grabowski, M., MacNeil, C., Bacela-Spychalska, K., 2014. The profile of a 'perfect' invader–the case of killer shrimp, Dikerogammarus villosus. Aquat. Invasions 9, 267–288.

Ricciardi, A., 2001. Facilitative interactions among aquatic invaders: is an "invasional meltdown" occurring in the Great Lakes? Can. J. Fish. Aquat. Sci. 58, 2513–2525.

Ricciardi, A., 2005. Facilitation and synergistic interactions among introduced aquatic species. In: Mooney, H.A., Mack, R.N., McNeely, J., Neville, L.E., Schei, P.J., Waage, J.K. (Eds.), Invasive Alien Species: A New Synthesis. Island Press, Washington, DC, pp. 162–178.

Ricciardi, A., 2006. Patterns of invasion of the Laurentian Great Lakes in relation to changes in vector activity. Divers. Distrib. 12, 425–433.

Ricciardi, A., Atkinson, S.K., 2004. Distinctiveness magnifies the impact of biological invaders in aquatic ecosystems. Ecol. Lett. 7, 781–981.

Ricciardi, A., Kipp, R., 2008. Predicting the number of ecologically harmful exotic species in an aquatic system. Divers. Distrib. 14, 374–380.

Ricciardi, A., MacIsaac, H.J., 2011. Impacts of biological invasions on freshwater ecosystems. In: Richardson, D.M. (Ed.), Fifty Years of Invasion Ecology: The Legacy of Charles Elton. Wiley-Blackwell, pp. 211–224.

Ricciardi, A., Hoopes, M.F., Marchetti, M.P., Lockwood, J.L., 2013. Progress toward understanding the ecological impacts of non-native species. Ecol. Monograph. 83, 263–282.

Robinson, J.V., Wellborn, G.A., 1988. Ecological resistance to the invasion of a freshwater clam, Corbicula fluminea: fish predation effects. Oecologia 77, 445–452.

Rodriguez, L.F., 2006. Can invasive species facilitate native species? Evidence of how, when, and why these impacts occur. Biol. Invasions 8, 927–939.

Roemer, G.W., Donlan, C.J., Courchamp, F., 2002. Golden eagles, feral pigs, and insular carnivores: how exotic species turn native predators into prey. Proc. Natl. Acad. Sci. U.S.A. 99, 791–796.

Roy, H.E., Adriaens, T., Isaac, N.J.B., Kenis, M., Onkelinx, T., San Martin, G., Brown, P.M.J., Hautier, L., Poland, R., Roy, D.B., Comont, R., Eschen, R., Frost, R., Zindel, R., Van Vlaenderen, J., Nedved, O., Ravn, H.P., Gregoire, J.C., de Biseau, J.C., Maes, D., 2012. Invasive alien predator causes rapid declines of native European ladybirds. Divers. Distrib. 18, 717–725.

Rudnick, D., Resh, V., 2005. Stable isotopes, mesocosms and gut content analysis demonstrate trophic differences in two invasive decapod crustacea. Freshw. Biol. 50, 1323–1336.

Ryan, M.E., Johnson, J.R., Fitzpatrick, B.M., 2009. Invasive hybrid tiger salamander genotypes impact native amphibians. Proc. Natl. Acad. Sci. U.S.A. 106, 11166–11171.

Salo, P., Korpimaki, E., Banks, P.B., Nordstrom, M., Dickman, C.R., 2007. Alien predators are more dangerous than native predators to prey populations. Proc R Soc Lond B. 274, 1237–1243.

Sampson, S.J., Chick, J.H., Pegg, M.A., 2009. Diet overlap among two Asian carp and three native fishes in backwater lakes on the Illinois and Mississippi rivers. Biol. Invasions 11, 483–496.

Sardiña, P., Beringer, J., Roche, D., Thompson, R.M., 2015. Temperature influences species interactions between a native and a globally invasive freshwater snail. Freshwater Sci. 34, 933–941.

Schmidt, S.N., Olden, J.D., Solomon, C.T., Vander Zanden, M.J., 2007. Quantitative approaches to the analysis of stable isotope food web data. Ecology 88, 2793–2802.

Schultz, R., Dibble, E., 2011. Effects of invasive macrophytes on freshwater fish and macroinvertebrate communities: the role of invasive plant traits. Hydrobiologia 684, 1–14.

Secondi, J., Dejean, T., Valentini, A., Audebaud, B., Miaud, C., 2016. Detection of a global aquatic invasive amphibian, *Xenopus laevis*, using environmental DNA. Amphib. Reptil. 37, 131–136.

Shine, R., 2010. The ecological impact of invasive cane toads (*Bufo marinus*) in Australia. Q. Rev. Biol. 85, 253–291.

Shurin, J.B., Clasen, J.L., Greig, H.S., Kratina, P., Thompson, P.L., 2012. Warming shifts top-down and bottom-up control of pond food web structure and function. Philos. Trans. R. Soc. Lond. B Biol. Sci. 367, 3008–3017.

Simberloff, D., Von Holle, B., 1999. Positive interactions of nonindigenous species: invasional meltdown? Biol. Invasions 1, 21–32.

Simberloff, D., Martin, J.L., Genovesi, P., Maris, V., Wardle, D.A., Aronson, J., Courchamp, F., Galil, B., García-Berthou, E., Pascal, M., Pyšek, P., 2013. Impacts of biological invasions: what's what and the way forward. Trends Ecol. Evol. 28, 8–66.

Simon, K.S., Townsend, C.R., 2003. Impacts of freshwater invaders at different levels of ecological organisation, with emphasis on salmonids and ecosystem consequences. Freshw. Biol. 48, 982–994.

Smetacek, V., Assmy, P., Henjes, J., 2004. The role of grazing in structuring Southern Ocean pelagic ecosystems and biogeochemical cycles. Antarct. Sci. 16, 541–558.

Smith, G.R., Boyd, A., Dayer, C.B., Winter, K.E., 2007. Behavioral responses of American toad and bullfrog tadpoles to the presence of cues from the invasive fish, *Gambusia affinis*. Biol. Invasions 10, 743–748.

Spencer, R.J., Van Dyke, J.U., Thompson, M.B., 2016. The 'Ethological Trap': functional and numerical responses of highly efficient invasive predators driving prey extinctions. Ecol. Appl. 26, 1969–1983.

Stewart, R.I., Dossena, M., Bohan, D.A., Jeppesen, E., Kordas, R.L., Ledger, M.E., Meerhoff, M., Moss, B., Mulder, C., Shurin, J.B., Suttle, B., 2013. Mesocosm experiments as a tool for ecological climate-change research. Adv. Ecol. Res. 48, 71–181.

Strayer, D.L., Caraco, N.F., Cole, J.J., Findlay, S., Pace, M.L., 1999. Transformation of freshwater ecosystems by bivalves: a case study of zebra mussels in the Hudson River. BioScience 49, 19–27.

Strayer, D.L., Lutz, C., Malcom, H.M., Munger, K., Shaw, W.H., 2003. Invertebrate communities associated with a native (*Vallisneria americana*) and an alien (*Trapa natans*) macrophyte in a large river. Freshw. Biol. 48, 1938–1949.

Strayer, D.L., Eviner, V.T., Jeschke, J.M., Pace, M.L., 2006. Understanding the long-term effects of species invasions. Trends Ecol. Evol. 1, 645–651.

Strayer, D.L., Hattala, K.A., Kahnle, A.W., Adams, R.D., 2014. Has the Hudson river fish community recovered from the zebra mussel invasion along with its forage base? Can. J. Fish. Aquat. Sci. 71, 1146–1157.

Swan, C., Healey, B., Richardson, D., 2008. The role of native riparian tree species in decomposition of invasive tree of heaven (*Ailanthus altissima*) leaf litter in an urban stream. Ecoscience 15, 27–35.

Taniguchi, Y., Fausch, K.D., Nakano, S., 2002. Size-structured interactions between native and introduced species: can intraguild predation facilitate invasion by stream salmonids? Biol. Invasions 4, 223–233.

Thomsen, P.F., Willerslev, E., 2014. Environmental DNA – an emerging tool in conservation for monitoring past and present biodiversity. Biol. Conserv. 183, 4–18.

Tillberg, C.V., Holway, D.A., LeBrun, E.G., Suarez, A.V., 2007. Trophic ecology of invasive Argentine ants in their native and introduced ranges. Proc. Natl. Acad. Sci. U.S.A. 104, 20856–20861.

Tran, T.N.Q., Jackson, M.C., Sheath, D., Verreycken, H., Britton, J.R., 2015. Patterns of trophic niche divergence between invasive and native fishes in wild communities are predictable from mesocosm studies. J. Anim. Ecol. 84, 1071–1080.

Trumpickas, J., Mandrak, N.E., Ricciardi, A., 2011. Nearshore fish assemblages associated with introduced predatory fishes in lakes. Aquat. Conserv. 21, 338–347.

Turesson, H., Brönmark, C., 2007. Predator–prey encounter rates in freshwater piscivores: effects of prey density and water transparency. Oecologia 153, 281–290.

Valley, R., Bremigan, M., 2002. Effects of selective removal of Eurasian watermilfoil on age-0 largemouth bass piscivory and growth in southern Michigan lakes. J. Aquat. Plant Manag. 40, 79–87.

van Riel, M.C., van der Velde, G., Rajagopal, S., Marguillier, S., Dehairs, F., de Vaate, A., 2006. Trophic relationships in the Rhine food web during invasion and after establishment of the Ponto-Caspian invader *Dikerogammarus villosus*. Hydrobiologia 565, 39–58.

Vander Zanden, M.J., Casselman, J.M., Rasmussen, J.B., 1999. Stable isotope evidence for the food web consequences of species invasions in lakes. Nature 401, 464–467.

Vander Zanden, M.J., Olden, J.D., Thorne, J.H., Mandrak, N.E., 2004. Predicting occurrences and impacts of smallmouth bass introductions in north temperate lakes. Ecol. Appl. 14, 132–148.

Vucic-Pestic, O., Rall, B.C., Kalinkat, G., Brose, U., 2010. Allometric functional response model: body masses constrain interaction strengths. J. Anim. Ecol. 79, 249–256.

Vye, S.R., Emmerson, M.C., Arenas, F., Dick, J.T., O'Connor, N.E., 2015. Stressor intensity determines antagonistic interactions between species invasion and multiple stressor effects on ecosystem functioning. Oikos 124, 1005–1012.

Ward, D.F., Ramón-Laca, A., 2013. Molecular identification of the prey range of the invasive Asian paper wasp. Ecol. Evol. 3, 4408–4414.

Ward, D.M., Nislow, K.H., Folt, C.L., 2008. Predators reverse the direction of density dependence for juvenile salmon mortality. Oecologia 156, 515–522.

Wasserman, R.J., Alexander, M.E., Weyl, O.L.F., Barrios-O'Neill, D., Froneman, P.W., Dalu, T., 2016a. Emergent effects of structural complexity and temperature on predator–prey interactions. Ecosphere 7, e01239.

Wasserman, R.J., Alexander, M.E., Dalu, T., Ellender, E.R., Kaiser, H., Weyl, O.L.F., 2016b. Using functional responses to quantify interaction effects among predators. Funct. Ecol. http://dx.doi.org/10.1111/1365-2435.12682.

Westhoff, J.T., Rabeni, C.F., 2013. Resource selection and space use of a native and an invasive crayfish: evidence for competitive exclusion? Freshwater Sci. 32, 1383–1397.

Williams, R.J., Berlow, E.L., Dunne, J.A., Barabási, A.L., Martinez, N.D., 2002. Two degrees of separation in complex food webs. Proc. Natl. Acad. Sci. U.S.A. 99, 12913–12916.

Winfield, I.J., Fletcher, J.M., James, J.B., 2012. Long-term changes in the diet of pike (*Esox lucius*), the top aquatic predator in a changing Windermere. Freshw. Biol. 57, 373–383.

Witczuk, J., Pagacz, S., Mills, L.S., 2013. Disproportionate predation on endemic marmots by invasive coyotes. J. Mammal. 94, 702–713.

Witte, F., Goldschmidt, T., Wanink, J., Oijen, M., Goudswaard, K., Witte-Maas, E., Bouton, N., 1992. The destruction of an endemic species flock: quantitative data on the decline of the haplochromine cichlids of Lake Victoria. Environ. Biol. Fishes 34, 1–28.

Wollrab, S., Diehl, S., Roos, A.M., 2012. Simple rules describe bottom-up and top-down control in food webs with alternative energy pathways. Ecol. Lett. 15, 935–946.

Woodward, G., Hildrew, A.G., 2001. Invasion of a stream food web by a new top predator. J. Anim. Ecol. 70, 273–288.

Xu, M., Mu, X., Dick, J.T., Fang, M., Gu, D., Luo, D., Zhang, J., Luo, J., Hu, Y., 2016. Comparative functional responses predict the invasiveness and ecological impacts of alien herbivorous snails. PLoS One 11, e0147017.

Young, K.A., Stephenson, J., Terreau, A., Thailly, A., Gajardo, G., Garcia de Leaniz, C., 2009. The diversity of juvenile exotic salmonids does not affect their competitive impact on a native galaxiid. Biol. Invasions 11, 1955–1961.

Zambrano, L., Martínez-Meyer, E., Menezes, N., Peterson, A.T., 2006. Invasive potential of common carp (*Cyprinus carpio*) and Nile tilapia (*Oreochromis niloticus*) in American freshwater systems. Can. J. Fish. Aquat. Sci. 63, 1903–1910.

Zavaleta, E., Hobbs, R., Mooney, H., 2001. Viewing invasive species removal in a whole-ecosystem context. Trends Ecol. Evol. 16, 454–459.

CHAPTER THREE

Importance of Microorganisms to Macroorganisms Invasions: Is the Essential Invisible to the Eye? (The Little Prince, A. de Saint-Exupéry, 1943)

L. Amsellem*,[1], C. Brouat[†], O. Duron[‡], S.S. Porter[§], A. Vilcinskas[¶,||], B. Facon[#,**]

*Unité "Evolution, Ecologie, Paléontologie", UMR CNRS 8198, Université de Lille, Villeneuve d'Ascq, France
[†]IRD, CBGP (UMR INRA/IRD/CIRAD/Montpellier SupAgro), Campus International de Baillarguet, Montferrier-sur-Lez, France
[‡]Laboratoire MIVEGEC, CNRS (UMR 5290), Université de Montpellier, IRD (UR 224), Montpellier, France
[§]School of Biological Sciences, Washington State University, Vancouver, WA, United States
[¶]Fraunhofer Institute for Molecular Biology and Applied Ecology, Giessen, Germany
[||]Institute for Insect Biotechnology, Justus-Liebig-University of Giessen, Giessen, Germany
[#]INRA, CBGP (UMR INRA/IRD/CIRAD/Montpellier SupAgro), Campus International de Baillarguet, Montferrier-sur-Lez Cedex, France
[**]INRA, UMR PVBMT, Saint-Pierre, France
[1]Corresponding author: e-mail address: laurent.amsellem@univ-lille1.fr

Contents

Advances in Ecological Research, Volume 57
ISSN 0065-2504
http://dx.doi.org/10.1016/bs.aecr.2016.10.005

Abstract

Microorganisms comprise the majority of earth's biodiversity and are integral to biosphere processes. Biological invasions are no exception to this trend. The success of introduced macroorganisms can be deeply influenced by diverse microorganisms (bacteria, virus, fungus and protozoa) occupying the whole range of species interaction outcomes, from parasitism to obligate mutualism. This large range of interactions, often coupled with complex historical and introduction events, can result in a wide variety of ecological dynamics. In this chapter, we review different situations in which microorganisms affect biological invasions. First, we consider outcomes of microorganism loss during the introduction of alien species. Second, we discuss positive effects of microorganisms on the invasiveness of their exotic hosts. Third, we examine the influence of microorganisms hosted by native species on the success of introduced species. Finally, in an applied perspective, we envisage how microorganisms can be used (i) to better decipher invasion processes and (ii) as biological control agents.

Life would not long remain possible in the absence of microbes

Louis Pasteur

1. INTRODUCTION

Throughout history, humans have changed their environment by introducing plants and animals into newly colonized habitats. The anthropogenic range expansion of organisms can be extended beyond domesticated plants and animals, to include the release of ornamental plants and animals as game, and the introduction of agents for biological pest control. More generally, human activities can contribute to unintentionally spread other exotic species (Banks et al., 2015; Blackburn & Ewen, 2016; Wilson et al., 2009). Thousands of species have been transported by humans from their native range to new habitats. Nonindigenous species that expand their range in their newly introduced habitats are considered invasive if they deeply modify the structure and population dynamics of the recipient community. For example, invasive species can lead to a decrease in recipient community species diversity, blocking of successional stages, changes in food webs and flows of matter, and thus can have a negative impact on the area of introduction (Simberloff et al., 2013). Research in invasion biology seeks to understand why some species become successful invaders, while others do not, even if they are closely phylogenetically related.

Invasive populations are generally founded by relatively few individuals that are likely infected with only a subset of all possible microorganisms from the native range source population (Blackburn et al., 2015;

Colautti et al., 2004). In addition, abiotic (e.g. climate) and biotic (e.g. occurrence of alternative hosts) differences between source and arrival environments can disrupt the life cycles of some microorganisms, leading to further microorganism loss in the arrival ranges (MacLeod et al., 2010). Furthermore, introduced macroorganisms also have to cope with increasing numbers of novel parasites during their range expansion. The loss of either antagonist or mutualist species of microbial symbionts can have strong ecological and evolutionary consequences for host organisms and is at the core of prominent hypotheses to explain invasion success or failure.

To keep the discussion simple, hereafter we will use terms such as "microorganism" and "parasite" not in their strict biological signification, but more for their ecological significance. We intend "microorganism" to include all uni- or multicellular microscopic organisms such as fungi, bacteria and protozoan, and also viruses. Similarly, "parasite" will encompass viral, bacterial or fungal pathogens.

Microbial symbionts cointroduced with invasive species can play critical roles in the colonization of new habitats. In fact, recent evidence suggests that microorganisms can have broader ranging impacts on community interactions during biological invasion processes than previously acknowledged (Brown et al., 2014; Rizzo et al., 2016; van Elsas et al., 2012). Aside from viral, bacterial or fungal pathogens, more than 40% of all animals display a parasitic lifestyle (including macroparasites), providing at least a crude estimate of their emerging role in invasion biology (Dobson et al., 2008). Pathogens do not walk alone, since other passengers, such as commensals and mutualistic symbionts, are also common. Most remarkable examples are found in terrestrial arthropods: recent surveys estimated that >40% of insect species harbour maternally inherited endosymbionts, such as *Wolbachia* which is among the most abundant intracellular bacteria so far discovered, infecting thousands of insect species (Duron et al., 2008; Zug and Hammerstein, 2012). In some cases, these endosymbionts have evolved towards mutualism by performing key metabolic functions required to support normal host development: these endosymbionts have become obligate, meaning that both partners entirely depend on each other for survival and achieve their life cycle (Moran et al., 2008). Furthermore, in most cases, there are secondary (e.g. facultative) symbionts as well, which can exert a variety of effects: increasing host survival and reproductive success, conferring advantages under certain environmental conditions, facilitating resource acquisition, protecting against natural enemies, interfering with the replication and transmission of parasites or subtly manipulating host reproduction (Bonfante and Genre, 2010; Brownlie et al., 2009; Dion et al., 2011; Moran et al., 2008; Oliver et al., 2010; Simon et al., 2011). As arthropods vary in the numbers and types of harboured endosymbionts, this provides heritable and functionally important variations within host populations—which in turn will influence their invasiveness (Ferrari and Vavre, 2011; Oliver et al., 2010).

A number of hypotheses link detrimental microbial symbionts to ecological outcomes in biological invasions. First, the Enemy Release Hypothesis (ERH) posits that invasive species can be more competitive in invaded habitats because they have left a significant fraction of their natural enemies in their native range (Mitchell and Power, 2003).

Second, the Novel Weapons Theory posits that invasive species carry traits that provide increased competitive ability, such as chemical weapons that provide a selective advantage over naive competitors (Callaway and Ridenour, 2004). This theory can be expanded beyond chemical factors that benefit the invader to include coinvading parasites. Indeed, parasites tolerated by the invasive carrier, but which can harm native competitors in newly colonized habitats, can be considered as biological weapons (Strauss et al., 2012; Vilcinskas, 2015). This phenomenon described as "spillover" (Power and Mitchell, 2004) is based on the expectation of a lower virulence of parasites in hosts that have coevolved with them, but a higher virulence in new phylogenetically closed hosts because of a lack of evolved immunological resistance (Schmid-Hempel, 2011).

In the same vein, the Biotic Resistance Hypothesis claims that the success of some exotic invaders can be dampened due to novel parasite invaders encountered in newly colonized habitats. These parasites can display higher virulence on invaders than on native species because they lack a coevolutionary history with the invader (Schmid-Hempel, 2011). However, invasive species can also function as novel reservoirs for native pathogens normally hosted by native species closely related to invaders. Indeed, they can act as vectors of infection towards native species in the area of introduction, amplifying the detrimental effect of invasion and increasing native pathogen loads. This may be explained for instance by immune traits of the invader population that allow it to overcome native parasites within the colonized areas (Lee and Klasing, 2004). This phenomenon of apparent competition mediated by native pathogens is known as "spill-back" (Kelly et al., 2009) and is related to the "Accumulation of Native Pathogens Hypothesis" (Eppinga et al., 2006; Strauss et al., 2012). In our approach to deciphering the roles played by microorganisms in the invasive success of their hosts in newly colonized habitats, this chapter will be structured around two major points. First, we will review literature studies focused on microorganisms in order to evaluate current evidence for the hypotheses we have presented. Second, we will explore the potential to use microorganisms as practical tools to manage invasions either by (i) using cointroduced species to retrace invasion histories and routes or by (ii) using microorganisms present in the native range as biocontrol agents to manage invasive species.

2. IMPACT OF MICROORGANISM LOSSES ON BIOLOGICAL INVASIONS

2.1 The Enemy Release Hypothesis

The ERH (Fig. 1) predicts that in newly colonized areas, invasive species will escape a portion of the pathogens that are present in their native range. As a consequence, this release from enemies will have negative effects for native species in the recipient community (Colautti et al., 2004; Keane and Crawley, 2002; Torchin et al., 2003). Numerous reviews have explored

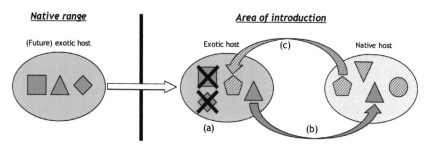

Fig. 1 Schematic representation of some hypotheses linking invasion success or failure and the presence of microbial passengers (parasitic or mutualistic symbionts). Concordant processes apply to both types of symbionts, but with possible contrasting outcomes (*symbols* represent introduced [*solid*] and native [*dashed*] hosts, with their respective original symbionts): (a) enemy (parasite) release will have positive impacts on exotic host fitness, while mutualistic symbiont loss may have negative impacts on exotic host fitness; (b) coinvading pathogens can have negative impacts on host invasion, though exotic pathogen spillover to native hosts can have positive impacts on host invasion. Exotic mutualistic symbiont coinvasion could have positive impacts on host invasion, and mutualistic symbiont spillover to native hosts can accentuate these positive impacts on host invasion if exotic symbionts are less beneficial than native ones to native hosts; (c) acquisition of novel pathogens can have negative impacts on invasion, though the net effect of these pathogens can be positive on exotic hosts due to spillback. Acquisition of novel mutualistic symbionts can have positive impacts on invasion if increasing exotic host fitness.

the complexity of the ERH (Colautti et al., 2004; Heger and Jeschke, 2014; Jeschke et al., 2012; Prior et al., 2015). A key condition of the ERH is that "enemy loss" occurs. This pattern is expected because introduced populations are likely to harbour only a subset of the parasites of their parental population(s) (Torchin et al., 2003), either because of sampling effect (introduced hosts were by chance not infected) or because of the absence of conditions required for parasite persistence in the introduction area (e.g. absence of a host species or a small host population size; MacLeod et al., 2010). Parasite loss may be accentuated further if the process of introduction favours healthy individuals or particular individual stages (e.g. juveniles) that are free of parasitic infections (Perkins et al., 2008). Evaluating enemy loss has been the subject of numerous studies focusing on microorganisms, as shown later (see also Blackburn and Ewen, 2016). Another key condition of the ERH is that enemy loss has positive consequences on the performance of the introduced host (Colautti et al., 2004; Prior et al., 2015), allowing the host to reallocate resources from immunity to invasiveness-related traits, such as rapid reproduction, dispersal ability or the ability to cope with challenging environmental factors. However, we will see later that this condition of the ERH often remains untested.

2.1.1 Is There Any Loss of Pathogenic Microorganisms by Invasive Species?

Several empirical studies on microorganisms support the idea that successfully invading hosts frequently lose microbial enemies as they colonize new areas. Introduced plants commonly leave behind many viruses and fungal pathogens from their native ranges (Mitchell and Power, 2003; Rúa et al., 2011; van Kleunen and Fischer, 2009). In insects, the loss of *Wolbachia*—an intracellular bacterium manipulating host reproduction—is commonly observed, such as in invasive populations of ants (Reuter et al., 2005; Rey et al., 2013; Schoemaker et al., 2000; Tsutsui et al., 2003; Yang et al., 2010) and *Citrus* thrips (Nguyen et al., 2016). In crustaceans, the absence of microsporidian parasites in British invasive populations of the amphipod *Dikerogammarus villosus* compared to continental European ones suggested enemy loss (Arundell et al., 2015). In vertebrate hosts, the prevalence and diversity of avian malaria infections were shown to be lower in introduced than in native populations of the house sparrow *Passer domesticus* (Marzal et al., 2011), the barred owl *Strix varia* (Lewicki et al., 2015) and the New Zealand bell-bird *Anthornis melanura* (Baillie et al., 2012). A comparison on seven nonnative freshwater fishes introduced into England

and Wales also showed evidence of enemy release from their native parasite fauna, which was only partially present in their region of introduction (Sheath et al., 2015).

However, the loss of microorganism enemies is not ubiquitous across successful invaders. For example, the total number of microbial taxa was found to be similar between native and invaded ranges of the common wasp *Vespula vulgaris*, though regionally distinct pathogen communities could indicate that some pathogens are missing in the area of introduction (Lester et al., 2015). Moreover, wasps are highly susceptible to honeybee pathogens and can have rapidly acquired some of them during their spread (Lester et al., 2015). No evidence for enemy loss was found for microsporidian parasites in invasive populations of the amphipod *Crangonyx pseudogracilis* (Slothouber Galbreath et al., 2010), or for haemosporidian infections in the range expanding populations of the house sparrow *P. domesticus* in Kenya (Coon and Martin, 2014). In the latter case, one possible explanation was that the important period of invasion relative to this pattern was outdated (Coon and Martin, 2014).

Understanding how enemy pressure varies over the time course of an invasion and how this relates to invasion dynamics is important in the context of the ERH (Colautti et al., 2004). Microorganism loss can persist for a long time in invasive populations. For instance, populations of the alfalfa weevil *Hypera postica*, an invasive pest of various leguminous crops, were still free of their original endosymbiotic *Wolbachia* 20 years after their introduction into Japan, though they were occasionally infected by an exotic new strain from an independent origin (1.43% infection prevalence in Japan, compared to 100% in the European source area; Iwase et al., 2015). Several lines of evidence suggest that microorganism loss can be a transitory situation (Gendron et al., 2012; Hawkes, 2007). First, new host–microorganism associations can form over time in the invaded range, leading microorganisms to be progressively recruited by introduced hosts. For example, the diversity and frequency of local endophytes were shown to increase with time since introduction of the weed *Ageratina adenophora* in China (Mei et al., 2014). Similarly, an experimental study on 12 introduced plants species in New Zealand showed that negative soil feedbacks increase with time since their establishment, suggesting an accumulating effect of below-ground native microorganism enemies over time (Diez et al., 2010). The rapid accumulation of enemies from the local fauna can therefore result in similar enemy loads in invasive and native host species within the invaded range. The rapid acquisition of local trypanosome and microsporidian pathogens from native hosts can underlie high parasite loads in invasive populations of the bumble-bee *Bombus hypnorum* (Jones and Brown, 2014). Furthermore, the pace of enemy accumulation in invasive species can depend on their phylogenetic novelty in a recipient community because more phylogenetically novel introduced species experience a stronger escape from local pathogens (Parker et al., 2015). In addition, microorganism loss can be transient if coinvading microorganism enemies experience a lagged population growth in the invaded range, e.g., resulting from an Allee effect (Liebhold and Bascompte, 2003). This is the case for two specific pathogens of the invasive forest defoliator *Lymantria dispar*, the entomophthoralean fungus *Entomophaga maimaiga* and a baculovirus, which were shown to catch up with their host in its introduction area in only a few years (Hajek and Tobin, 2011), and for a microsporidian parasite of the freshwater amphipod, *D. villosus* (Wattier et al., 2007). It is possible that the empty niche space freed up by a transient loss of microorganism enemies in invasive populations can facilitate the emergence of novel native or invasive pathogen communities: this idea was suggested by the surprisingly diverse assemblage of undescribed foliar pathogens across the invasive range of the grass *Microstegium vimineum* one century after its introduction (Stricker et al., 2016).

2.1.2 Is There Any Effect of Pathogenic Microorganism Loss on Introduced Hosts?

Enemy loss does not necessarily mean enemy release (Colautti et al., 2004; Prior et al., 2015). For many invasions, there is little convincing evidence that enemy loss is a determinant of invasion success in natural populations (Blackburn and Ewen, 2016). For instance, enemy release does not appear to be the main driver of the invasion success of the barred owl in North America, as only rare lineages of their *Haemoproteus* enemies were lost in invasive populations (Lewicki et al., 2015). However, there are also cases supporting a role for enemy loss in invasion success. For example, invasive house sparrows exhibit a loss of virulent avian malaria lineages (Marzal et al., 2011). Partial support for effective enemy release has also been reported by correlative studies at intra- or interspecific levels. Invasive plants are often less damaged by pathogenic microbes in their invaded range as compared to in their native range, and they often suffer lower impacts of pathogen in the invaded ranges than do native species (reviewed in Mitchell et al., 2006). For example, plant species introduced to the United States from Europe are infected by 84% fewer fungi and 24% fewer virus species in the naturalized range, and the greater the release from pathogens, the more invasive these species tend to be (Mitchell and Power, 2003). The latter pattern was, however, shown to be reversed for 140 North American plants that have spread in Europe (van Kleunen and Fisher, 2009). As stated in this more recent study, these contrasting results remain difficult to explain. A greater postfire reduction in pathogen damage for native grassland plants, as compared to invasive ones, is consistent with the ERH, as fire eliminates litter and thus eggs and spores of native parasites (Roy et al., 2014). Also, the diversity of avian malaria lineages was shown to be negatively related to body condition during winter in reintroduced populations of the New Zealand passerine *A. melanura* (Baillie et al., 2012).

Experimental studies are more convincing concerning effective enemy release. The only experimental study found in animals suggested that the escape from the protozoan gut parasite *Ascogregarina taiwanensis* can explain the competitive advantage of the mosquito *Aedes albopictus* in its introduction range (Aliabadi and Juliano, 2002). Otherwise, experimental studies are still largely restricted to plants since plant ecologists have long recognized the insufficiency of observational evidence to unequivocally assess the ERH (Keane and Crawley, 2002). Common garden experiments on taxonomically paired plants have shown that introduced plant species were subject to less negative soil microbial feedbacks compared to native ones (Agrawal et al., 2005; van Grunsven et al., 2007). Cross-inoculations have more precisely demonstrated a more negative effect of local soil microbiota on plant growth for native plants than for invasive ones (Callaway and Ridenour, 2004; Gundale et al., 2014; Maron et al., 2014). However, some other experimental studies show contrasting results concerning the ERH. As an illustration, fungicide treatments had a greater negative impact on fungal pathogens affecting the survival of seeds of native compared to exotic herbaceous plants (Dostal, 2010). Also, pesticide treatments had no effect on the survival of the neotropical shrub *Clidemia hirta* in its native or introduced range (deWalt et al., 2004). Similar results were obtained with inoculation/sterilization experiments on the invasive forb *Centaurea solstitialis* (Andonian et al., 2012) or with plant–soil feedbacks measurements on the leguminous tree *Robinia pseudoacacia* (Callaway et al., 2011) or the forb *Solidago gigantea* (Maron et al., 2015). Soil feedbacks were, however, found to be variable in time and/or space, suggesting a biogeographical mosaic of interaction strengths related to invasion success (Agrawal et al., 2005; Andonian et al., 2012; deWalt et al., 2004; Maron et al., 2014).

Fungal endophytes can allow invasive plants to escape enemies (Keane and Crawley, 2002) if they deter herbivores in the invaded range or if the loss of conditionally mutualistic endophytes functions as an enemy release allowing greater spread of the host plant in the exotic range. Fungal endophytes of plants are diverse (Saikkonen et al., 2006) and can span the continuum from having positive to negative direct effects on their hosts depending on the interaction and the ecological context (Johnson et al., 1997). In its exotic range where it is invasive, the grass *Brachypodium sylvaticum* has lost the fungal endophyte that is ubiquitous in its native area (Vandegrift et al., 2015). Endophyte-free genotypes in the introduced range not only display enhanced growth and competitive ability relative to symbiotic clones but also suffer increased herbivory pressure. These observations suggest that the net effect of this endophyte can be that of a specialist enemy, and its loss can thus facilitate invasion (Vandegrift et al., 2015).

2.1.3 Other Invasion Hypotheses Linked With Enemy Release

The ERH is strongly connected with other invasion hypotheses (Jeschke, 2014). Based on an optimal defence theory, the Evolution of Increased Competitive Ability (EICA) Hypothesis states that escape from parasites should favour introduced species by means of a change in resource allocation along an implicit allocation trade-off. If introduced host populations are free from enemies, then selection should favour the reallocation of some resources (usually invested in defences and immunity) towards life-history traits more directly related to invasion success, such as faster maturity, higher reproductive effort or dispersal ability (Blossey and Nötzold, 1995). Indeed, the release of avian malaria pathogens has been invoked to explain the dampening of inflammatory responses in the invasive house sparrow (Martin et al., 2014). In animals, testing this hypothesis has, however, proven to be more challenging than initially thought (Cornet et al., 2016). Soil feedback experiments on the invasive plant *S. gigantea* have shown that introduced genotypes are still resistant to the negative effects of their original soil biota, thus suggesting the occurrence of generalist pathogens in their invasive range (Maron et al., 2015).

The Resource Availability Hypothesis states that plant invasion is facilitated by high resource availability (Davis et al., 2000). Reviewing studies on 243 European plant species naturalized in North America, Blumenthal et al. (2009) showed that species adapted to high resource availability experience stronger release of pathogenic fungi compared to species requiring fewer resources, suggesting that enemy release and increases in resource availability would thus act synergistically to favour exotic species (Blumenthal et al., 2009).

2.2 Changes in Mutualist Assemblages

Invasive species do not only lose their pathogens and parasites, but loss of mutualistic symbionts can also occur. Similar mechanisms operate for pathogens, parasites and mutualistic symbionts: invasive populations are founded by few individuals, and only a subset of their microorganisms, including mutualistic symbionts, are introduced in the invasive range. The consequences of this process are, however, radically different depending on the nature of the relation between symbionts and their hosts: while enemy loss will benefit the invasive species, mutualist loss will have negative consequences on host fitness. Worthy of note is that the risk of mutualistic symbiont loss is high for secondary (e.g. facultative) mutualistic symbionts but not for primary (obligatory) mutualistic symbionts: invasive populations of pea aphids or whiteflies always carry their primary symbionts since these microorganisms are strictly required for host survival (Gueguen et al., 2010; Henry et al., 2013). By contrast, facultative mutualistic symbionts can be present at low frequency in the native range and can thus be lost during invasion process.

No clear example of loss of mutualistic symbionts is documented to our knowledge in invasive insect species. Obviously, any organism losing its beneficial symbionts would have lower chance to become invasive, and this may explain why this process is rare and thus underdocumented. Interestingly, possible cases may be found with *Wolbachia*, which is commonly lost in invasive populations of various ant species as stated earlier (Reuter et al., 2005; Rey et al., 2013; Schoemaker et al., 2000; Tsutsui et al., 2003; Yang et al., 2010). This symbiont is commonly viewed as a reproductive parasite of terrestrial arthropods, such as ants, e.g., manipulating the reproduction of their host species towards the production or survival of infected female hosts. However, it is now clear that *Wolbachia* is not simply a reproductive parasite since it has recently emerged as a conditional mutualist conferring advantages under certain environmental conditions in many insect species. For instance, *Wolbachia* increases fecundity of fruit flies reared on iron-restricted or -overloaded diets and can thus confer a direct fitness benefit during periods of nutritional stress (Brownlie et al., 2009; Fellous and Salvaudon, 2009). *Wolbachia* can also protect their hosts against attack by natural enemies. *Wolbachia* infection interferes with the replication and transmission of a wide range of pathogens and parasites (including RNA viruses, bacteria, protozoa and nematodes) and protects its host from parasite-induced mortality (Hedges et al., 2008; Moreira et al., 2009; Zélé et al., 2012). These properties suggest that multiple potencies (e.g. reproductive manipulations and conditional mutualism) may be a global feature of *Wolbachia*. In this context, loss of *Wolbachia* in invasive ant populations can be beneficial on the one hand because of the loss of reproductive manipulation (see Section 2.1), but negative on the other hand because of the loss of beneficial defensive effects. In addition, if the loss of *Wolbachia* is incomplete, this can create infection polymorphism that may hamper invasion: mating between infected males and uninfected females can be infertile due to cytoplasmic incompatibility and then limit population growth (Engelstadter and Hurst, 2009). Overall, this underlines how the effect of *Wolbachia* infection on an insect host is more complex than previously considered with multiple (and contrasted) impacts on host fitness.

3. EFFECTS OF MICROORGANISMS HOSTED BY THE ALIEN SPECIES

A typical feature of invasive species is their rapid population growth and spread after establishing a bridgehead in new locations. There, they encounter local species competing for the same resources and which are well adapted to the local environment. Parasitic microorganisms have been postulated to play a key role in determining the outcome of biological invasions (Hatcher et al., 2006). Thus, in this context, pathogens (especially generalist) can be viewed as biological weapons hosted by invasive species that can potentially infect and kill native competitors. There are many examples of invasive species benefiting from their ability to carry pathogens or parasites, therefore conferring a selective advantage during the invasion process (Figs. 1 and 2). Indeed, disease can then "spill over" from the nonindigenous host species to infect susceptible native competitors, increasing the probability of a successful invasion.

3.1 Spillover

De Castro and Bolker (2005) have collected a number of examples showing that cointroduced parasites can result in extinction of native species (Figs. 1 and 2). Increasing evidence is available that cointroduced parasites can become invasive too, and their spread often indirectly drives the decline of native host species. There are widely known spillover examples described in the literature. As a first illustration, we can cite the parapox virus, which has been cointroduced with the grey squirrel (*Sciurus carolinensis*) into the United Kingdom, and which helps this invasive species outcompete the

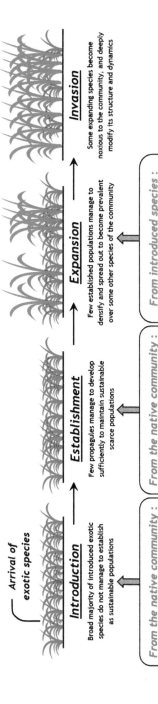

Arrival of exotic species

Introduction

Broad majority of introduced exotic species do not manage to establish as sustainable populations

Establishment

Few propagules manage to develop sufficiently to maintain sustainable scarce populations

Expansion

Few established populations manage to densify and spread out to become prevalent over some other species of the community

Invasion

Some expanding species become noxious to the community, and deeply modify its structure and dynamics

Introduction

From the native community :

(−) *Biotic resistance* : soil microbiota does not correspond to the needs, preventing or slowing down establishment.

(−) *Native pathogens* : the native pathogen cortege causes a huge parasitic load, preventing or slowing down establishment.

(+) *Favourization* : soil microbiota is convenient, enhancing or accelerating establishment.

From introduced species :

(+) *Spillover* : favourization of the introduced parasites towards the native community, enhancing or accelerating establishment.

Establishment

From the native community :

(−) *Native pathogens* : the native pathogen cortege prevents or slows down the expansion.

(+) *Favourization* : soil microbiota enhances or favours the expansion.

From introduced species :

(+) *Spillover* : favourization of the established parasites towards the native community, enhancing or accelerating expansion.

(+) *Spill-back* : established species serve as a reservoir of native pathogens causing surinfection to the native community, enhancing or accelerating expansion.

Expansion

From introduced species :

(+) *Spillover* : favourization of the expanding parasites towards the native community, enhancing or accelerating invasion.

(+) *Spill-back* : expanding species serve as a reservoir of native pathogens to definitely establish their prevalence in the native community, and thus becoming invasive.

Fig. 2 Where microorganisms may play a beneficial or detrimental role during the different phases of a biological invasion process. (−) and (+) symbols represent, respectively, favouring or preventing alien species from reaching the next phase of the invasion process.

native red squirrels (*Sciurus vulgaris*). Both squirrels compete for the same resources and the virus helps the invading grey squirrels by causing high mortality in the native species, while the invasive one acts as a pathogen reservoir (Darby et al., 2014). Another intriguing example is the decline of the noble crayfish (*Astacus astacus*) in Europe, which is caused by the fungal parasite *Aphanomyces astaci* that has been cointroduced with the signal crayfish *Pacifastacus leniusculus* from North America (Capinha et al., 2013; Söderhäll and Cerenius, 1999). In this context, we can also highlight the role of microsporidia closely related to *Nosema thompsoni*, which have recently been discovered in the invasive ladybird *Harmonia axyridis* (Vilcinskas et al., 2013a). These parasites do not harm the invasive carrier but do harm native ladybird species, which can become infected by microsporidia upon feeding on *H. axyridis* eggs or larvae. This predatory behaviour facilitates microsporidia to cross the species barrier among ladybirds (Vilcinskas et al., 2015). Cointroduced parasites must not necessarily kill native competitors to benefit the invader. It is sufficient if they harm the fitness of the native competitors to tip the balance in favour of the invasive carrier in the long run. The spillover of cointroduced parasites from invasive to native species represents an intriguing explanation for the success of many invaders but is often not the only factor and has to be evaluated in the context of other mechanisms (Vilcinskas, 2015). The role of native parasites, which are harmful for native competitors but are more or less tolerated by the invasive species, has been neglected in the past because of their limited effects on their original host (Strauss et al., 2012).

3.2 Vertically Inherited Symbionts

Aside from pathogens, alien species can also carry mutualistic symbiotic organisms that can influence the invasive properties of their hosts. In insects, such symbiotic organisms are common and often maternally inherited, through egg cytoplasm (Moran et al., 2008; Wernegreen, 2012). They typically use specific adaptive strategies to spread and persist within insect populations (Ferrari and Vavre, 2011; Moran et al., 2008; Oliver et al., 2010). Invasive insects feeding on plants commonly benefit from associations with facultative symbionts: for instance, aphids are subject to a range of ecological stressors, including specific natural enemies such as entomopathogenic fungi and parasitoid wasps, heat and changes in plant hosts. Facultative symbionts can compensate for limited adaptive capacities of their hosts to counteract these stressors (Moran et al., 2008; Oliver et al.,

2010, 2014). Infection frequencies by facultative symbionts are typically dynamic, varying across temporal and spatial gradients and ecological associations. Hence, the pea aphid *Acyrthosiphon pisum* harbours a facultative symbiont, *Serratia symbiotica*, that confers heat tolerance: aphid clones infected by this bacterium display substantially improved survival during heat stress (Chen et al., 2000; Montllor et al., 2002). As a result, *S. symbiotica* prevalence and distributions are most influenced by climate: infection frequency significantly increases during warmer seasons in natural populations (Russel and Moran, 2006) and also reaches higher values in arid region like California's Central Valley than in temperate regions (Chen and Purcell, 1997; Henry et al., 2013). Therefore, the facultative association with a symbiont can facilitate the persistence of aphid populations in previously unsuitable habitats and enhance their invasive properties.

Interestingly, facultative symbionts can also mediate colonization of new plant species by invasive insect species (Moran et al., 2008; Oliver et al., 2010, 2014). For example, pea aphids comprise several genetically distinct races that are specialized for feeding on a particular plant species and that host a distinct community of facultative symbionts, which can directly assist them in exploiting a particular food plant (Henry et al., 2013; Tsuchida et al., 2004). The distribution of some of these facultative symbionts, such as *Regiella insecticola*, is structured by the host plant of the aphids they infect, a pattern suggesting a fitness benefit to the aphid on a particular host plant (Henry et al., 2013). For instance, the successful transfer of *R. insecticola* from pea aphid to another aphid species that does not normally carry this symbiont enabled it to survive and reproduce on otherwise unsuitable clover (Tsuchida et al., 2004). This is a strong support for a direct effect of symbionts on host plant utilization, and thus a role for symbionts in determining aphid ecological niche.

In recent years, a great deal of attention has been devoted to defensive symbionts that protect insects against attack by natural enemies (Jaenike et al., 2010; Oliver et al., 2008, 2010, 2014). The presence of these symbionts is particularly relevant for alien insect species because their range expansion may result in novel antagonistic interactions with enemies absent in their native ranges. Indeed, infection of the pea aphid by the facultative symbiont *Hamiltonella defensa* provides resistance to parasitoid wasps by killing the developing wasp larvae or embryos in the aphid haemocoel (Oliver et al., 2003, 2005). Through reducing the mortality risk after wasp oviposition, the pea aphid thus relies heavily on *H. defensa* for its protection. Studies of caged populations have shown that *H. defensa* can rapidly invade in the presence of the parasitoid but is not favoured and decline in populations without this stress, a pattern suggesting that there are costs and benefits associated with this symbiont (Oliver et al., 2008; Polin et al., 2014). As a hypothesis, defensive symbionts can significantly enhance the capacity of alien insect species to colonize new habitats filled by native enemies. Although this pattern has not been demonstrated, it is worth noting that many invasive insect species carry defensive symbionts like *H. defensa* (Moran et al., 2005; Russell et al., 2003; Zchori-Fein and Brown, 2002), suggesting an adaptive role for these symbionts in invasion processes.

Perhaps the most remarkable recent observation is the speed at which symbiont-associated adaptation can occur, as shown in populations of an invasive whitefly, *Bemisia tabaci*, in the southwestern United States. Fast evolution of these populations was observed after the spread of a new facultative symbiont, *Rickettsia* sp. nr. *bellii*, that spread from less than 1% of individuals infected to 97% in only 6 years and dramatically increased whitefly performance (Himler et al., 2011). *Rickettsia*-infected whiteflies produce offspring at nearly twice the rate of individuals lacking the infection, and a higher proportion of the offspring survived to adulthood. *Rickettsia* infection also causes a strong female bias, favouring its diffusion through an increased production of the transmitting sex and thus also enhancing the intrinsic growth rate of its host species (Himler et al., 2011). More recently, *Rickettsia* was further shown to decrease whitefly mortality rate due to *Pseudomonas syringae*, a common environmental bacterium, some strains of which are pathogenic to insects (Hendry et al., 2014). The simultaneous expression of these distinct effects (e.g. enhancing fecundity and survival, reproductive manipulations and defensive symbiosis) by *Rickettsia* demonstrates the profound and potentially rapid effects of heritable symbionts. They should be increasingly recognized as major players in the ecology of animals and will certainly be a key feature of agricultural entomology in years to come. An interesting ecological context relating this with biological invasions is when herbivorous insects introduced to control weed populations fail to establish owing to attack from natural enemies (Julien and Griffiths, 1998), and could benefit from the presence of defensive symbionts. Facultative symbionts that influence host plant range or thermal tolerance can also influence the pest status of herbivorous insects, thus either aiding or hampering biological control efforts.

Microbes hosted by alien plants can affect host invasiveness by acting as biological weapons, either allowing their host to escape enemies or increasing its competitive abilities (Johnson et al., 2013; Mitchell et al., 2006; Pringle et al., 2009; Reinhart and Callaway, 2006; Traveset and

Richardson, 2014). In vertically inherited mutualisms in plant species, symbionts are transmitted through the seed from mother to progeny, with infrequent losses or gains of symbiosis over evolutionary time, and thus these symbionts typically coinvade with their invasive host plants by "hitch-hiking". Vertically transmitted symbionts are common in plants and play important roles in plant evolution and ecology (Friesen et al., 2011). Vertically transmitted endophytes occur in a quarter of all grass species as well as a diversity of other plant forms, in which they are systemic and can confer diverse benefits (Lymbery et al., 2014; Mitchell et al., 2006; Rudgers et al., 2004; Strauss et al., 2012) such as enhancing resistance to herbivory due to mycotoxin and alkaloid production (Rudgers et al., 2004; Saikkonen et al., 2006). For example, the agriculturally important, but also invasive, grasses tall fescue (*Lolium arundinaceum*) and perennial ryegrass (*Lolium perenne*) experience reduced herbivory when infected with *Neotyphodium* endophytes, as compared to clones from which the endophyte has been excluded (Saikkonen et al., 2006). Fungal endophytes are diverse and can span the continuum from having positive to negative direct effects on their hosts depending of the context (Johnson et al., 1997). In the exotic range of the invasive grass, *B. sylvaticum*, the grass has lost its fungal endophyte that is ubiquitous in the native range (Vandegrift et al., 2015). Endophyte-free genotypes in the exotic range display increased growth and competitive ability relative to symbiotic clones, despite also showing increased herbivory pressure, suggesting that the net effect of this endophyte can be that of a specialist enemy and that the loss of this symbiotic enemy may facilitate invasion (Vandegrift et al., 2015). Fungal endophyte impacts are not exclusively mediated through herbivory.

3.3 Horizontally Inherited Symbionts

Environmentally acquired symbionts are taken up from the environment rather than passed from mother to offspring. These symbionts can either coinvade alongside their hosts or hosts can switch to associate themselves with different symbionts in the invaded range. As plants and mutualistic microorganisms coinvade across heterogeneous landscapes, the costs and benefits of mutualism can shift and coinvading partners can evolve specialized ecotypes that maximize mutualism benefits in a particular habitat type (Porter et al., 2011). Therefore, the coinvasive spread of critical microbial mutualists can drive invasion dynamics of invasive plants with specialized symbiotic requirements. Understanding how the interaction between invaders and their symbiotic microbes evolves during biological invasions is an exciting frontier—during invasion, environmental context can shift rapidly, and symbionts could experience major changes in selection that intensify or reduce cooperation or antagonism (Nunez et al., 2009; Porter and Simms, 2014; Porter et al., 2011; Pringle et al., 2009; Schwartz et al., 2006), and rapid evolutionary responses in symbionts could play an important role in the invasion dynamics of their hosts.

Symbiotic microorganisms in alien species can function as novel biological weapons that facilitate invasion if they provide a selective advantage for invasive species over native ones (Lymbery et al., 2014; Strauss et al., 2012), this process being characterized as "spill-back" (see Section 4.1). When pathogens that are boosted by invasive plants have stronger negative effects on native than on invasive plants, pathogens shared by invaders and natives can promote invasion (Vilcinskas, 2015). For example, the destructive tropical weed *Chromolaena odorata* accumulates *Fusarium* fungal pathogens, which suppress native plant competitors (Mangla and Callaway, 2008); the plant *Vincetoxicum rossicum* hosts soil fungal pathogens that can suppress the growth of native plants (Day et al., 2016); and cheatgrass (*Bromus tectorum*) accumulates the fungal seed pathogen, *Pyrenophora semeniperda*, which then infects the seeds of native plants (Beckstead et al., 2010). More subtly, the spillover of even mutualistic symbionts from invasive plants to native ones can lead to native plant decline if the symbionts are less beneficial to native plants. For example, there are exotic nitrogen-fixing *Bradyrhizobium* bacterial symbionts that are coinvading with the invasive plant *Acacia longifolia*. These exotic symbionts have become dominant within symbiont populations on co-occurring native species (Rodriguez-Echeverria, 2010), likely due to their amplification in exotic invasive hosts. However, these exotic rhizobia confer less benefit to native plants than do native rhizobia, and thus this symbiont community shift could contribute to native species decline (Rodriguez-Echeverria et al., 2012). For example, environmentally acquired *Alternaria* endophytes can directly increase the competitive ability of *Centaurea stoebe*, a forb that is a problematic invader in grasslands, and this increased competitive ability is not mediated indirectly via reduced herbivore impacts on the host (Aschehoug et al., 2012).

A lack of appropriate microbial mutualists can be a major constraint to establishment if alien plant species with specialized microbial mutualist requirements are introduced into a novel area devoid of these symbionts (Parker, 2001; Pringle et al., 2009; Schwartz et al., 2006). For example, repeated introductions of exotic pine trees to novel continents can be unsuccessful until compatible exotic mycorrhizae are cointroduced (Nunez et al., 2009; Pringle et al., 2009; Schwartz et al., 2006). Similarly, the successful introduction of exotic leguminous forbs to novel areas for pasture development has depended upon cointroduction of compatible exotic rhizobium bacteria (Sullivan et al., 1995). Suboptimal densities of microbial mutualists in the invaded range can slow invasive spread until sufficient densities have been accumulated in the soil (Nunez et al., 2009; Parker et al., 2006). It is important to notice that invasion dynamics linked to environmentally acquired symbionts is slightly different to the one with vertically inherited symbionts. In the former, the main constraint to establishment is that the new area is "devoid of symbionts", whereas the loss of vertically transmitted symbionts can constrain invasion through a failure to transmit the symbiont to the new portion of the range.

3.4 The Role of Immunity

Immune systems of native and invasive hosts determine their susceptibility to cointroduced microorganisms. Lee and Klasing (2004) postulated that invasive species should have a superior immune system in comparison with noninvasive related species because in the area of introduction invaders encounter parasites with which they have not coevolved. On the other hand, investing into a more efficient immune system can trade off against other invasiveness-related traits such as fecundity or the ability to cope with enemies or competitors. The eco-immunology of invasive species has therefore emerged as a novel field in invasion biology (White and Perkins, 2012). The successful defence against spillover of parasites from native species is solely mediated by a powerful immune system. This hypothesis has been supported by findings from the invasive ladybird *H. axyridis*, which is protected by a bilayered defence system encompassing chemical defences based on an antimicrobial alkaloid (called harmonine) and inducible antimicrobial peptides (Schmidtberg et al., 2013). Harmonine has been shown to display antibacterial and antiparasitic activities and has been postulated as a constitutive defence compound providing protection in newly colonized habitats (Röhrich et al., 2012). Strikingly, a next-generation sequencing-based analysis of the immunity-related transcriptome of *H. axyridis* resulted in the discovery of more than 50 putative genes encoding antimicrobial peptides. No other animal is known thus far to possess such a high number of antimicrobial peptides (Vilcinskas et al., 2013b), and the antimicrobial peptide repertoire of native ladybird species is significantly lower. Contrary to the ERH, these findings support the idea that invasive species have a superior immune system when compared with closely related noninvasive species, at least until shortly after their arrival in their area of introduction. Interestingly, injection of bacteria into *H. axyridis* solely reflects a trade-off between fitness-related costs associated with the simultaneous decline of harmonine along with the induction of antimicrobial peptides' synthesis (Schmidtberg et al., 2013): the fitness penalty for the constitutive synthesis of harmonine is inferior to that of the induced synthesis of a broad spectrum of antimicrobial peptides.

The postulated role of immunity in invasion biology can be expanded beyond the protection against pathogens encountered in newly colonized habitats to include a role in keeping cointroduced microbes, which function like bioweapons, under control. In the case of *H. axyridis*, it has been postulated that harmonine is involved in keeping the growth of its associated microsporidia under control (Vilcinskas et al., 2015).

4. EFFECTS OF MICROORGANISMS HOSTED BY NATIVE SPECIES

4.1 Spill-Back

In the context of introduced species, alien host species phylogenetically close to native ones increase the diversity of potential hosts that local parasites can infect in the community. Exotic and native species sharing the same cortege of parasites in the area of introduction can act as reciprocal reservoirs, cross-exchanging their parasites and directly infecting each other (Hudson and Greenman, 1998) (Fig. 1). Among hypotheses based on parasite transmission between native and introduced species, the Spill-Back Hypothesis states that if exotic hosts are appropriate and competent for native parasites, they may consequently act as new hosts for these native pathogens and thus be considered as reservoirs from which native hosts can also be infected (Daszak et al., 2000; Dunn and Hatcher, 2015; Kelly et al., 2009; Prenter et al., 2004). This can profoundly alter the epidemiology of native parasites and directly involve them in processes of apparent competition between local and introduced host species with, respectively, differential vulnerabilities, by affecting their dynamics, densities and life-history traits (Dunn et al., 2012). As a feedback, the introduction of a competent host can increase the abundance of the infectious stages of a local parasite in the invaded habitat (Figs. 1 and 2), including intensification of parasitism in local native populations (Kelly et al., 2009).

Parasites from the area of introduction represent a major part of the parasite cortege of invasive species (Kelly et al., 2009; Mastitsky et al., 2010). The "Naive Host Syndrome theory" states that local parasites will have higher pathogenic effects on exotic hosts with which they did not coevolve (Mastitsky et al., 2010). Moreover, under the ERH, the immune capacity of invasive species should evolve towards attenuated responses (at least for costly pathways) and should in turn increase chances for parasite acquisition to occur. In the same vein, it seems also likely that a newly introduced host species can allow or facilitate the proliferation of parasites previously cointroduced with another invasive species (Kelehear et al., 2013); such a process may illustrate the Invasional Meltdown Hypothesis (Ricciardi et al., 2013; Simberloff and Von Holle, 1999).

Until recently, parasite spill-back has rarely been considered in the literature (Kelly et al., 2009; Roy and Lawson Handley, 2012), and one reason to explain this lack of empirical data is a bias in considering a realistic context of host and parasite communities. For example, most studies related to spill-back in rodents (Lopez-Darias et al., 2008; Smith and Carpenter, 2006; Torchin et al., 2003) or birds (MacLeod et al., 2010) involve two host species (one native and one invasive) and one parasite group or even only one parasite species. Host communities and associated parasites are rarely—if ever—considered in a more complex perspective, including both native and invasive species (Johnson et al., 2008). However, some convincing examples of spill-back processes involving microorganisms have been highlighted in the last years. For examples, in plants, Californian serotypes of barley and cereal yellow dwarf viruses appear to have been essential to the widespread invasion of Mediterranean grass species (Malmstrom et al., 2005). Flory and Clay (2013) reviewed some other potential examples of spill-back involving soil or seed pathogenic fungi in native host plants. In animals, community studies at biogeographical and local scales have convincingly shown that the invasion of brine shrimps (*Artemia franciscana*) in France was facilitated by native *Microsporidia* (Rode et al., 2013). Also, the spill-back of the native parasite *Myxosporea* is strongly suspected to explain a main part of the invasion success of the cane toad *Rhinella marina* (formerly known as *Bufo marinus*) to the detriment of native frogs in Australia (Brown et al., 2015; Hartigan et al., 2011). In birds, house sparrows (*P. domesticus*) have been shown to increase the prevalence of Buggy Creek virus in native cliff swallows (*Petrochelidon pyrrhonota*), by increasing numbers and densities of hosts and vectors in mixed colonies as a result of their sedentary behaviour (O'Brien et al., 2011).

However, acquisition of native parasites by alien species may have outcomes different than a spill-back process and may contrarily result in a final benefit to native species. For example, invading hosts can be insufficiently competent to ensure the transmission of newly acquired native parasites, therefore representing a dead end for them. Consequently, their transmission and abundance can thus decrease through a "dilution effect" in the area of introduction, finally benefiting native hosts (Johnson et al., 2013). Another scenario is to consider that alien host species are too sensitive to these newly acquired native parasites, resulting in a considerable competitive advantage for native host species. As an illustration, the coexistence of native and invasive amphipods in benthic communities of St. Lawrence River is facilitated by a parasitic oomycete (the invasive character of which is still unknown), which is highly more virulent in the invasive host (Kestrup et al., 2011).

4.2 Invasive Species Interactions With Beneficial Native Symbionts

In plants, invaders routinely associate with soil mutualists that are naturally hosted by native species, and they can play diverse roles in invasion dynamics (Johnson et al., 2013; Mitchell et al., 2006; Pringle et al., 2009; Reinhart and Callaway, 2006; Richardson et al., 2007; Traveset and Richardson, 2014). The microbial mutualists of native plants can be inhibited by invasive plants, according to the "Degraded Mutualism Hypothesis" (Fedriani et al., 2012; Shah et al., 2009). This theory states that the disruption of the mutualistic relationship between microbes and native plants is caused by an invasive species and can promote invasive spread (Brouwer et al., 2015; Johnson et al., 2013; Richardson et al., 2007; Traveset and Richardson, 2014). This can occur when native plants are more dependent on microbial mutualists than invasive ones (Shah et al., 2009). For example, invasion by species with low mycorrhizal dependence can reduce overall mycorrhizal density in the soil, resulting in decreased performance of native plants and facilitating further invasion (Meinhardt and Gehring, 2012; Vogelsang and Bever, 2009). This dynamics could be accentuated over the course of an invasion because, during range expansion, invasive plants species can be under selection to further reduce their dependence on microbial mutualists (Seifert et al., 2009). Native mutualisms can also be degraded if invasive species directly inhibit the microbial mutualists of natives (Cipollini et al., 2012). For example, allelopathic exudates from invasive garlic mustard, *Alliaria petiolata*, disrupt the mutualism between native plants and both native mycorrhizae

(Wolfe et al., 2008) and native rhizobium bacteria (Portales-Reyes et al., 2015), which can inhibit the growth of native plants (Callaway et al., 2008; Hale et al., 2016).

Alternatively, native plants can host mutualistic symbionts that benefit generalist invader species, according to the "Enhanced Novel Mutualism Hypothesis" (Reinhart and Callaway, 2006; Richardson et al., 2007). This theory states that microbial mutualists present in native plants are also beneficial to invasive plants species. For example, some invaders like the spotted knapweed, *Centaurea maculosa*, are symbiotic generalists that can associate with arbuscular mycorrhizal fungi hosted by native plants and thus gain access to nutrient pools that contribute to invasive spread (Callaway and Ridenour, 2004; Harner et al., 2010). Similarly, invasive legumes that are symbiotic generalists, such as *Cytisus scoparius*, can benefit from the nitrogen fixation provided by novel *Rhizobium* symbiont partners that normally associate with native legumes (Horn et al., 2014). An interesting area for future research is the extent to which horizontal transfer of genes or genomic regions from native microbes facilitates the invasion of exotic microbial symbionts and their associated host species. Such genomic chimerism (e.g. the transfer of exotic chromosomal symbiotic genes from exotic rhizobia to native soil microbes) was critical to the naturalization of *Lotus corniculatus* in New Zealand (Sullivan et al., 1995, 2002) and appears to play a role in the invasion of other legumes such as *C. scoparius* (Horn et al., 2014), *R. pseudoacacia* (Wei et al., 2009) and *Biserrula pelecinus* (Nandasena et al., 2006).

Furthermore, the intergenerational feedbacks occurring when a plant species cultivates a microbial community that positively or negatively affects the growth of that species (Bever et al., 2012; Klironomos, 2002) can play an important role in invasion dynamics. Native plants often experience negative plant–soil microbe feedbacks, in that they accumulate soil microbial symbiont communities that have net negative impacts on the growth of the next generation of this species (Klironomos, 2002). However, plants successful at colonizing novel areas, such as *C. maculosa* (Reinhart and Callaway, 2006), can instead experience positive plant–soil microbe feedbacks in the invaded range, which can help invasion (Callaway et al., 2004; Zhang et al., 2010). Over time, positive plant–soil microbe feedbacks for invaders can accelerate the decline of the native community and lead to invasional meltdown (Rodriguez-Echeverria, 2010; Simberloff and Von Holle, 1999), although a meta-analysis suggests that the role of feedbacks in plant invasions is complex and context-dependent (Suding et al., 2013).

The impact of native symbionts on invasion can be quite different in insects, in which research has revealed effects of horizontal transmission of beneficial symbionts that are usually maternally inherited. Here, maternally inherited symbionts originally hosted by native insect species can be acquired through horizontal transfer by alien insect species and then favour their invasion. While maternal inheritance ensures the maintenance of these symbionts within host species, occasional horizontal transmission between insect species also occurs: phylogenetic evidence indicates how the wide distribution of some facultative symbionts is largely due to their ability to shift from one insect species to another (Baldo et al., 2008; Cordaux et al., 2001; Jousselin et al., 2013; Vavre et al., 1999). For invasive species, their expanding ranges create new opportunities for horizontal acquisition of new symbionts from native species. In this context, horizontal transfers can serve as an immediate and powerful mechanism of rapid adaptation resulting in the instantaneous acquisition by an invasive species of ecologically important traits, such as heat tolerance or parasite defence. This process can greatly facilitate the adaption of an alien species to a new habitat (Henry et al., 2013; Moran et al., 2008; Oliver et al., 2010), as in whitefly with *Rickettsia* (Himler et al., 2011).

5. ANTHROPIC USE OF MICROORGANISMS
5.1 As Tools to Retrace Invasion Histories

A prerequisite for the understanding of biological invasions lies in a clear picture of the geographic pathways followed by propagules from their source to the introduced area. Indeed, a precise knowledge of the invasion routes is fundamental for (i) planning relevant strategies to prevent or control an invasion and (ii) testing meaningful hypotheses concerning the ecological and evolutionary factors driving invasive success (Estoup and Guillemaud, 2010). Initially, the reconstruction of invasion pathways hinged on historical introduction records of species invasions. However, this kind of survey data can be incomplete or even misleading (Fischer et al., 2015). More recently, population genetics studies using mitochondrial and microsatellite markers have allowed obtaining a much more in-depth understanding of the history of the invasion process including the origin and genetic composition of the invading populations but also inference about some demographic and genetic parameters (Estoup and Guillemaud, 2010). This genetic approach has pointed out some invasion stories with complex and counterintuitive introduction events (see for instance Gaudeul et al., 2011; Leduc et al., 2015;

Lombaert et al., 2014). However, the reconstruction of invasion pathways may be unresolved for invasive organisms displaying pauperized genetic diversity and low population structure, mainly due to severe bottlenecks during introduction events or peculiar modes of reproduction such as parthenogenesis (Amsellem et al., 2000; Zhang et al., 2014). In this context, it has been suggested that microorganisms may serve as an alternative tool to better reconstruct invasion histories of genetically uniform hosts (Nieberding and Olivieri, 2007; Wirth et al., 2005). Indeed, when both actors (hosts and their associated microorganisms) are closely associated, microorganisms may offer a higher resolution picture of the invasion history of their host.

A first option is to focus on the comparison of microorganism communities in host populations from native and nonnative areas. If these communities are composed of strict host-specific lineages, the presumption is that their composition may reflect the origin area of the invasive host. This approach is illustrated by the studies of associated bacterial communities to unravel the geographical sources of invasive marine green algae (Aires et al., 2013; Meusnier et al., 2001). For instance, using cloned bacterial 16S rDNA gene sequences from the associated microflora of *Caulerpa taxifolia* collected from different areas, Meusnier et al. (2001) detected an analogous composition of the bacterial flora shared between the Australian and Mediterranean sites. They were thus able to conclude that the Australian sample was the original biogeographical source population of the accidental introduction into the Mediterranean Sea. By characterizing the bacterial communities associated to native and nonnative populations of *Caulerpa racemosa* through pyrosequencing, Aires et al. (2013) also showed that bacterial community may be an effective tracer of the origin of invasion. An alternative method is to focus on a single microorganism species and use fast-evolving regions of DNA in the microorganisms as a magnifying lens to decipher introduction pathways of the invasive host (Wilfert and Jiggins, 2014). Indeed, studying the genetic structure of microorganisms may allow to detect recent demographic processes in host populations, when microorganisms have lower effective population size (N_e) and stronger population structure than their hosts (Criscione and Blouin, 2005; Holmes, 2004; Nieberding and Olivieri, 2007). Endosymbiotic bacteria in insects constitute an excellent illustration of this approach. For instance, the routes of invasion of the Russian wheat aphid *Diuraphis noxia*, one of the world's most invasive cereal (wheat and barley) crop pests, have been unresolved due to the limited polymorphism of the available markers (Liu et al., 2010; Zhang et al., 2014). By targeting genomic and plasmid DNA of its obligate endosymbiont

Buchnera aphidicola, Zhang et al. (2014) succeeded in retracing the recent worldwide invasion of this damaging pest species. *Wolbachia* infections were also shown to confirm the common invasion history of several invasive populations of the little fire ant *Wasmannia auropunctata* (Rey et al., 2013). For invasive plants, rhizobial symbionts may be as helpful. For instance, the phylogenetic analysis of several DNA regions of the *Bradyrhizobium japonicum* symbionts has allowed confirming the relationship between native populations from Australia and invasive populations from South Africa of its invasive host plant *Acacia pycnantha* (Ndlovu et al., 2013). Parasitic microorganisms may also mirror the invasion routes of their hosts. In the case of *Drosophila melanogaster*, its vertically transmitted sigma virus DMelSV has recently been used as a tool to infer the global invasion pattern of its host (Wilfert and Jiggins, 2014). The phylogenetic study of several genes from this parasitic virus has highlighted that its host invaded Europe from Africa and that the North American fruit fly population stemmed from a single immigration from Europe in the late 19th century. To our knowledge, such an approach has never been conducted on fungal pathogens, while numerous phylogeographic studies indicate host tracking and thus congruent expansion histories between these microorganisms and their host plants (Gladieux et al., 2008; Stukenbrock et al., 2007; Vercken et al., 2010).

Despite the image of success provided by these different examples, the use of microorganisms as a tracer of the invasion pathways of their hosts requires some precautions. First, it is important to keep in mind that numerous microorganisms are lost by their host through drift or selection during the invasion process (see Section 2 for more details). Microorganism populations or communities in invaded areas may thus be a pale reflection of those found in the native area, restricting their use as a reliable tracer of host invasion history. For both approaches (microorganism community and genome of a single microorganism species), microorganisms need to share a common evolutionary history with their host (Nieberding and Olivieri, 2007). Several characteristics can influence this intimacy (Whiteman and Parker, 2005) such as the transmission mode (vertical vs horizontal) or the type of biotic relationship (mutualistic endosymbiosis, commensalism, parasitism, etc.). For instance, it is widely assumed that microorganisms with vertical transmission should more realistically reflect the invasion history of their hosts (Johnson and Clayton, 2004). In the same vein, concerning parasitic microorganisms, it is much better to use a very specific parasite with a direct life cycle (Nieberding and Olivieri, 2007; Wirth et al., 2005). Several studies have illustrated that microorganisms are not always adequate tracers of the invasion history of their host. For instance, a characterization of the bacterial microbiota of *A. albopictus* by 16S rDNA metabarcoding showed no correlation between microbiota and the genetic structure (estimated on neutral genetic markers) of the host populations in invaded and native areas (Minard et al., 2015). This study provides support to the suggestion that the gut microbiota of animals would be rather a reflection of local environmental conditions than of host genotype (Boissière et al., 2012; Linnenbrink et al., 2013). Genetic differentiation has been shown to be higher in hosts than in parasites having complex cycles, free-living stages and hermaphroditic reproduction (Mazé-Guilmo et al., 2016). As another example, a population genetic study of the horizontally transmitted fungal pathogen *Microbotryum lychnidis-dioicae* of the invasive plant *Silene latifolia* confirmed contrasted invasion histories between the host and its specific parasite (Fontaine et al., 2013).

Concerning the monospecific approach (e.g. genome of a microorganism as a magnifying lens), additional recommendations have to be followed. An invasion occurs over a relatively short timescale and thus, to be informative, microorganisms have to accumulate more mutations in this short time period than their hosts. Selection may also result in artificial groupings of populations with different genealogies (Wirth et al., 2005) and thus alter the efficiency of microorganism to retrace the invasion routes. Therefore, asexual microorganisms should be avoided due to nonindependence of loci and potential selective sweeps (Rosenberg and Nordborg, 2002). Finally, genes used to reconstruct the invasion history have to be chosen in such a way as to prevent problems linked to horizontal gene transfer that could disturb the historical pattern (Lerat et al., 2005).

In conclusion, only few microorganisms have been used, up to now, as tracers for the invasion histories of their hosts. But the already existing case studies seem promising and, bearing in mind some caveats, such approaches could be helpful to precisely retrace the invasion history of an increasing number of species.

5.2 As Tools to Manage Invasive Species

Invasive species can induce detrimental consequences in both natural and human-altered habitats, including erosion of biodiversity, decrease of farm and forest productivity and impact on public health (Lodge, 1993; Murphy and Romanuk, 2014; Pimentel et al., 2001). In many situations, classic control methods (such as culling, trapping or chemical agents) have, in different

ways and for different reasons, either failed to provide an adequate control or
are too costly/detrimental for the environment and/or the health to be sustainable (Cleaveland et al., 1999; Di Giallonardo and Holmes, 2015). Biological control (e.g. the use of any organism, its genes or its products to cause
negative impact on pest populations) has thus been considered as an alternative for managing invasive species (Thomas and Reid, 2007; Zavaleta et al.,
2001). The idea behind this strategy is that release from enemies (see
Section 2.1) is commonly recognized as a major determinant of the successful invasion of an exotic species (Keane and Crawley, 2002). Thus, adding a
specific enemy may reverse the trend and allows to reduce—if not
suppress—invasive populations. Besides predator and parasitoid macroorganisms, a wide diversity of microorganisms can also serve as biocontrol
agents against a range of invasive species in a wide diversity of circumstances
(Lacey et al., 2001; Miller et al., 1983). Here, we first outline the different
approaches that use microorganisms as control agents based on several
emblematic examples. We then discuss the benefits and limitations of microorganisms as biocontrol agents. Finally, we highlight future developments to
improve the role of microorganisms as biocontrol agents in invasive species
management.

As early as the 19th century, some biocontrol programmes involving
microbial control agents (mainly fungi) were carried out against invasive pest
insects. However, the use of pathogenic microorganisms really started with
the discovery of the bacterium *Bacillus thuringiensis* (*Bt*) at the beginning of
the 20th century (Lacey et al., 2001). The mode of action of this bacterium
against insects relies on several insecticidal proteins known as Cry, Cyt, Vip
and Sip proteins (Palma et al., 2014). Since its first uses by farmers as a pesticide in 1920, *Bt* has become the main microorganism used in biological
control and thousands of strains are now available (Hajek and Tobin,
2010; Miller et al., 1983; Mnif and Ghribi, 2015). Besides traditional formu-
lations (such as suspensions), additional means of application have been
developed to produce and deliver the toxins to the target insects such as
other plant-colonizing bacteria and the insertion of *Bt* toxin genes into transgenic crops (Romeis et al., 2006; Sanahuja et al., 2011; Stock et al., 1990).
Following the development of this leading microbial agent, the use of
microorganisms as biocontrol agents has been diversified so that nowadays
any type of pathogenic microorganisms (e.g. viruses, bacteria, fungi and protists) can be successfully used for eradicating or controlling a range of invasive species (e.g. arthropods, plants, nematodes, fungi and vertebrates) in
diverse ecosystems worldwide (Hajek et al., 2007; Lacey et al., 2001;

Meyer and Foudrigniez, 2011; Nakaew et al., 2015). It is noteworthy that genetic engineering techniques are now used to increase the efficiency of microorganisms used as biocontrol agents (Georgievska et al., 2010; Giddings, 1998; Godfray, 1995; Zwart et al., 2009).

In recent years, endosymbiont bacteria (especially those belonging to the genus *Wolbachia*) have also been recognized as powerful tools for biocontrol programmes (Vavre and Charlat, 2012; Zindel et al., 2011). First, *Wolbachia* is well known for manipulating its hosts' reproduction through different mechanisms such as feminization of genetic males and parthenogenesis induction or cytoplasmic incompatibility (Vavre and Charlat, 2012). This attribute can be directly used to suppress invasive populations through the mass release of cytoplasmic incompatibility—*Wolbachia* infects males in a similar way to the "Sterile Insect Technique" (Atyame et al., 2011; Zabalou et al., 2004). Reproductive manipulations by *Wolbachia* may also affect the rearing and establishment of macroorganisms used as biocontrol agents (Zindel et al., 2011). In the majority of parasitoid biocontrol agents for instance, only females are effective by laying their eggs in their hosts. Thus, parthenogenetic reproduction induced by *Wolbachia* may facilitate mass production and provide a large number of beneficial insects at the appropriate time (Zindel et al., 2011). In addition to the reproductive manipulation of their hosts, endosymbionts bacteria can affect a variety of traits such as resistance to host enemies, lifespan of their hosts or their vectorial capacity (Graham et al., 2012; Hoffmann et al., 2011; Vavre and Charlat, 2012). As it has been shown with *Wolbachia*, endosymbiont bacteria can be used to decrease the transmission of pathogens (such as dengue) by invasive insects either directly by reducing pathogen load in insect vectors or indirectly by reducing host lifespan (Hedges et al., 2008; Hoffman and Turelli, 2013; Hughes et al., 2011; McMeniman et al., 2009; Moreira et al., 2009). Recent findings show that *Wolbachia* can also be used to directly regulate invasive populations of insects through the increase of susceptibility of their hosts to viral infection (Graham et al., 2012).

In the case of vertebrates, biocontrol strategies using pathogenic microorganisms (virus, bacteria, fungi and protozoans) have given mixed results (Saunders et al., 2010). Although some biocontrol programmes are currently under investigation (McColl et al., 2014; Oliveira and Hilker, 2010), only three successful viral biocontrols have been identified to date: (i) the release of a parvovirus to eliminate cats on Marion island, and the well-documented cases of (ii) *Myxoma* and (iii) rabbit haemorrhagic disease viruses, which have been released to control invasive rabbits around the world (Saunders et al.,

2010). Following the *Myxoma* virus release, rabbit populations were suppressed over a decade, followed by a recovery of rabbit populations reflecting a combination of evolving virus attenuation and rabbit resistance (Saunders et al., 2010). Privately released as biocontrol in France, the virus spreads rapidly through European rabbit populations, with a number of negative ecological and economic consequences (Di Giallonardo and Holmes, 2015). Viral-vectored immune contraception was another use of microorganisms (e.g. species–specific viruses) envisaged to control mammals (Courchamp and Cornell, 2000). It gave promising results on the house mouse but required significant research efforts to implement (Campbell et al., 2015).

As any biocontrol agents, microorganisms may allow to reduce or eliminate chemical use and their adverse consequences such as pollution (Bourguet and Guillemaud, 2016) and development of resistance (Rex Consortium, 2013; but see Asser-Kaiser et al., 2007). As compared to chemical pesticides, microorganisms used as biocontrol agents are mostly host-specific (Hajek and Tobin, 2010; Lacey et al., 2001). The absence of evidence that voluntary release of insect pathogens has resulted in long-term negative effects on nontarget species or pathogen establishment supports this assertion (Cory and Myers, 2000). Their pretty narrow host range may allow using microorganisms in combination with other biocontrol agents, such as predators or parasitoids (Miller et al., 1983). Besides, the production and storage of microorganisms as biocontrol agents are very often easier and cheaper than those of macroorganisms (Lacey et al., 2001).

There are also limits to the use of microorganisms as biocontrol agents. Historically, it was thought that resistance to microorganisms should be more difficult to evolve than resistance to chemical pesticides (Miller et al., 1983), but this precept is nowadays challenged. For instance, there are an increasing number of field-evolved resistance reports to *Bt* crops in several major pest species (Cory and Franklin, 2012; Tabashnik et al., 2013). Also in the case of Australian rabbits, resistance to *Myxoma* evolved in only a decade, suggesting the need to work precisely on the relationships between resistance, transmission and virulence (Kerr et al., 2015). Disadvantages of biocontrol strategies also include low environmental persistence, relatively slow action and ethical and environmental issues associated with the release of lethal pathogens (Bravo et al., 2011; Messing and Wright, 2006). In vertebrates, there is a high fear of inadvertent infection of other species (including humans) because barriers preventing successful host jumps are still largely unknown (Di Giallonardo and Holmes, 2015). Also, total eradication of the host is rarely achieved by a pathogen, and more intensive techniques are necessary to remove pest individual below the density corresponding to the threshold allowing pathogen establishment (McColl et al., 2014).

The use of microorganisms as cheap and safe tools to manage invasive species is a clearly aimed objective. Achieving this goal still requires fundamental and applied studies combining genetic, molecular, theoretical and ecological approaches (Douglas, 2007; Thomas and Reid, 2007; Vinale et al., 2008). More specifically, the following advances are pivotal: first, we need to strengthen our understanding of the interactions not only between microorganisms and their hosts but also with the whole environment from which these partners belong, and interactions among microorganisms within the host. This could consecutively improve the speed of action and the environmental persistence but also allows a better management of potential pest resistance and enhanced efficiency of production and application. Second, genetic engineering, noticeably with the new development of the CRISPR–Cas9 technology (which allows to make precise and targeted changes to the genome of living cells; Makarova et al., 2002), can offer great improvements in all these topics (Champer et al., 2016; Sander and Joung, 2014). However, we are only at the beginning of the use of this innovative technology and obtaining a better understanding of its potential unintended consequences is of paramount importance (Webber et al., 2015).

6. CONCLUSIONS

Among the many potential biotic and abiotic factors that can interact to determine the invasiveness of a species in a newly invaded community, empirical studies have historically focused on macroscaled processes, such as population dynamics (mainly of the introduced species, but also of those belonging to the native community; Huntley et al., 2010; Ibanez et al., 2014; Pearson and Dawson, 2003), or changes of the invasive species between its native range and area of introduction from a genetic (Amsellem et al., 2000; Cristescu, 2015; Fitzpatrick et al., 2012) or a phenotypic point of view (Amsellem et al., 2001, 2002). It is now assumed that these macroscopic patterns often represent the visible consequences of underlying mechanisms occurring at (much) smaller scales. Hence, proximal inductors caused by biological activities and ecological roles of both native and cointroduced microorganisms are now totally doubtless about determinism of biological invasion (as for beneficial or detrimental consequences). Microorganism-based effects (Fig. 2) are now widely considered because they are recognized as potential major determinants of biological invasions processes (Brown et al., 2014; Rizzo et al., 2016; van Elsas et al., 2012) and give the largest overview and the most inclusive understanding of the biological and ecological processes during the different invasion stages.

Most of the various case studies about microorganisms influencing the invasion process of their introduced host species depicted here are focused on a short timescale, whereas the long-term balance of positive and negative interactions has received less attention (Agrawal et al., 2005; Gilbert and Parker, 2010). However, both host–microorganism interactions and biological invasions are prone to induce rapid and drastic changes over time, and long-term outcomes are rather hard to predict (Altizer et al., 2003; Kelehear et al., 2012; Lambrinos, 2004; Thompson, 1998; Weeks et al., 2007). Pathogen accumulation constitutes an outstanding example to illustrate how the dynamic relationships between microorganisms and their invasive hosts may induce potentially different and unexpected eco-evolutionary trajectories. While introduced species may initially benefit from the release of their pathogens (see Section 2), several case studies, mainly focused on plant species, suggest that pathogen accumulation could be widespread (Diez et al., 2010; Mitchell et al., 2010; Phillips et al., 2010; Stricker et al., 2016; Strong and Levin, 1975). Such pathogen accumulation can be driven by several ecological and evolutionary mechanisms, such as an increase in density of the invasive host species, cointroductions of coevolved microorganisms with the invasive candidate species, adaptation of native pathogens to invasive species and the evolution of reduced defences from the invasive species (EICA Hypothesis—Flory and Clay, 2013). For now, potential consequences of pathogen accumulation in the area of introduction are poorly known and may only be hypothesized according to several outcomes. First, pathogen accumulation can decrease the density and extent of invasive species and, as a consequence, it facilitates recovery of native species over the long-term (Stricker et al., 2016). Second, invasive species could respond to pathogen accumulation through phenotypic plasticity or tolerance and thus not be truly affected through their population dynamics (Flory and Clay, 2013). Finally, pathogen accumulation may also strengthen the superiority of invasive species over native ones (spillover and spill-back—Kelly et al., 2009) via increasing impacts of diseases on co-occurring native hosts (see Sections 3.1 and 4.1). These three alternative outcomes emphasize the urgent need for a deeper understanding of the long-term dynamics of invasive species and their microorganisms (Dietz and Edwards, 2006; Hawkes, 2007). Such an approach requires a multidisciplinary point of view, including experimental manipulations, long-term surveys of invasions of known ages, genomic studies of host–microorganism interactions as well as theoretical modelling of these interactions within an explicit invasion context (Flory and Clay, 2013; Hawkes, 2007). These complementary approaches would allow assessing the emergence of general patterns and would thus enhance our knowledge about the outcomes of biological invasions (Flory and Clay, 2013; White and Perkins, 2012).

Beyond the putative significant role played by microorganisms during biological invasions, they can also have significant involvements in many related biological and ecological phenomena linked to biological invasions. For example, microorganisms may be considered in the context of conservation biology (i) to stop invasion processes (via their use through biological control) or (ii) to consider risks of new emerging diseases for native species (including humans) during or shortly after biological invasions processes (Bellard et al., 2016; Blackburn and Ewen, 2016; Santini et al., 2013). Moreover, we draw attention to conservation actions with potentially antagonistic effects, putatively leading to important and irreversible damages. These dramatic outcomes may be caused during reintroduction of individuals on purpose either (i) to reintroduce some species in areas where they are not found anymore or (ii) to genetically strengthen populations involved in a vortex of extinction. In both cases, overlooking the possible effects of cointroduced microorganisms would be a mistake that could lead to adverse consequences (Seddon et al., 2014; Walker et al., 2008).

ACKNOWLEDGEMENTS

The authors thank the editors of the issue and the CESAB COREIDS project for the opportunity to present this review. This study was supported by the GDR 3647 "Invasions biologiques" and a grant from the ERA-Net BiodivERsA, with the national funders ANR (France), DFG (Germany) and BELSPO (Belgium), as part of the 2012–2013 BiodivERsA call for research proposals. S.S.P. acknowledges support from DEB-1355216 from the National Science Foundation.

REFERENCES

Agrawal, A.A., Kotanen, P.M., Mitchell, C.E., Power, A.G., Godsoe, W., Klironomos, J., 2005. Enemy release? An experiment with congeneric plant pairs and diverse above- and below-ground enemies. Ecology 86, 2979–2989.

Aires, T., Serrão, E.A., Kendrick, G., Duarte, C.M., Arnaud-Haond, S., 2013. Invasion is a community affair: clandestine followers in the bacterial community associated to green algae, *Caulerpa racemosa*, track the invasion source. PLoS One 8. e68429.

Aliabadi, B.W., Juliano, S.A., 2002. Escape from gregarine parasites affects the competitive interactions of an invasive mosquito. Biol. Invasions 4, 283–297.

Altizer, S., Harvell, C.D., Friedle, E., 2003. Rapid evolutionary dynamics and disease threats to biodiversity. Trends Ecol. Evol. 18, 589–596.

Amsellem, L., Noyer, J.-L., Le Bourgeois, T., Hossaert-McKey, M., 2000. Comparison of genetic diversity of the invasive weed *Rubus alceifolius* Poir. (Rosaceae) in its native range and in areas of introduction, using amplified fragment length polymorphism (AFLP) markers. Mol. Ecol. 9, 443–455.

Amsellem, L., Noyer, J.-L., Hossaert-McKey, M., 2001. Evidences of a switch in the reproductive biology of *Rubus alceifolius* (Rosaceae) towards apomixis, between its native range and its area of introduction. Am. J. Bot. 88, 2243–2251.

Amsellem, L., Noyer, J.-L., Pailler, T., Hossaert-McKey, M., 2002. Characterisation of pseudogamous apospory in the reproductive biology of the invasive weed *Rubus alceifolius* (Rosaceae), in its area of introduction. Acta Bot. Gallica 149, 217–224.

Andonian, K., Hierro, J.L., Khetsuriani, L., Becerra, P.I., Janoyan, G., Villareal, D., Cavieres, L.A., Fox, L.R., Callaway, R.M., 2012. Geographic mosaics of plant–soil microbe interactions in a global plant invasion. J. Biogeogr. 39, 600–608.

Arundell, K., Dunn, A., Alexander, J., Shearman, R., Archer, N., Ironside, J.E., 2015. Enemy release and genetic founder effects in invasive killer shrimp populations of Great Britain. Biol. Invasions 17, 1439–1451.

Aschehoug, E.T., Metlen, K.L., Callaway, R.M., Newcombe, G., 2012. Fungal endophytes directly increase the competitive effects of an invasive forb. Ecology 93, 3–8.

Asser-Kaiser, S., Fritsch, E., Undorf-Spahn, K., Kienzle, J., Eberle, K.E., Gund, N.A., Reineke, A., Zebitz, C.P.W., Heckel, D.G., Huber, J., Jehle, J.A., 2007. Rapid emergence of baculovirus resistance in codling moth due to dominant, sex-linked inheritance. Science 317, 1916–1918.

Atyame, C.M., Pasteur, N., Dumas, E., Tortosa, P., Tantely, M.L., Pocquet, N., Licciardi, S., Bheecarry, A., Zumbo, B., Weill, M., Duron, O., 2011. Cytoplasmic incompatibility as a means of controlling *Culex pipiens quinquefasciatus* mosquito in the islands of the South-Western Indian Ocean. PLoS Negl. Trop. Dis. 5. e1440.

Baillie, S.M., Gudex-Cross, D., Barraclough, R.K., Blanchard, W., Brunton, D.H., 2012. Patterns in avian malaria at founder and source populations of an endemic New Zealand passerine. Parasitol. Res. 111, 2077–2089.

Baldo, L., Ayoub, N.A., Hayashi, C.Y., Russell, J.A., Stahlhut, J.K., Werren, J.H., 2008. Insight into the routes of *Wolbachia* invasion: high levels of horizontal transfer in the spider genus *Agelenopsis* revealed by *Wolbachia* strain and mitochondrial DNA diversity. Mol. Ecol. 17, 557–569.

Banks, N.C., Paini, D.R., Bayliss, K.L., Hodda, M., 2015. The role of global trade and transport network topology in the human-mediated dispersal of alien species. Ecol. Lett. 18, 188–199.

Beckstead, J., Meyer, S.E., Connolly, B.M., Huck, M.B., Street, L.E., 2010. Cheatgrass facilitates spillover of a seed bank pathogen onto native grass species. J. Ecol. 98, 168–177.

Bellard, C., Genovesi, P., Jeschke, J.M., 2016. Global patterns in threats to vertebrates by biological invasions. Proc. R. Soc. Lond. B Biol. Sci. 283, 20152454.

Bever, J.D., Platt, T.G., Morton, E.R., 2012. Microbial population and community dynamics on plant roots and their feedbacks on plant communities. Annu. Rev. Microbiol. 66, 265–283.

Blackburn, T.M., Ewen, J.G., 2016. Parasites as drivers and passengers of human-mediated biological invasions. Ecohealth, 1–13. http://dx.doi.org/10.1007/s10393-015-1092-6.

Blackburn, T.M., Lockwood, J.L., Cassey, P., 2015. The influence of numbers on invasion success. Mol. Ecol. 24, 1942–1953.

Blossey, B., Nötzold, R., 1995. Evolution of increased competitive ability in invasive nonindigenous plants—a hypothesis. J. Ecol. 83, 887–889.

Blumenthal, D., Mitchell, C.E., Pysek, P., Jarosik, V., 2009. Synergy between pathogen release and resource availability in plant invasion. Proc. Natl. Acad. Sci. U.S.A. 106, 7899–7904.

Boissière, A., Tchioffo, M.T., Bachar, D., Abate, L., Marie, A., Nsango, S.E., Shahbazkia, H.R., Awono-Ambene, P.H., Levashina, E.A., Christen, R., Morlais, I., 2012. Midgut microbiota of the malaria mosquito vector *Anopheles gambiae* and interactions with *Plasmodium falciparum* infection. PLoS Pathog. 8. e1002742.

Bonfante, P., Genre, A., 2010. Mechanisms underlying beneficial plant–fungus interactions in mycorrhizal symbiosis. Nat. Commun. 1, 48.

Bourguet, D., Guillemaud, T., 2016. The hidden and external costs of pesticide use. Sustain. Agric. Rev. 19, 35–120.

Bravo, A., Likitvivatanavong, S., Gill, S.S., Soberon, M., 2011. *Bacillus thuringiensis*: a story of a successful bioinsecticide. Insect Biochem. Mol. Biol. 41, 423–431.

Brouwer, N.L., Hale, A.N., Kalisz, S., 2015. Mutualism-disrupting allelopathic invader drives carbon stress and vital rate decline in a forest perennial herb. AoB Plants 7, plv014.

Brown, A., Huynh, L.Y., Bolender, C.M., Nelson, K.G., McCutcheon, J.P., 2014. Population genomics of a symbiont in the early stages of a pest invasion. Mol. Ecol. 23, 1516–1530.

Brown, G.P., Phillips, B.L., Dubey, S., Shine, R., 2015. Invader immunology: invasion history alters immune system function in cane toads (*Rhinella marina*) in tropical Australia. Ecol. Lett. 18, 57–65.

Brownlie, J.C., Cass, B.N., Riegler, M., Witsenburg, J.J., Iturbe-Ormaetxe, I., McGraw, E.A., O'Neill, S.L., 2009. Evidence for metabolic provisioning by a common invertebrate endosymbiont, *Wolbachia pipientis*, during periods of nutritional stress. PLoS Pathog. 5. e1000368.

Callaway, R.M., Ridenour, W.M., 2004. Novel weapons: invasive success and the evolution of increased competitive ability. Front. Ecol. Environ. 2, 436–443.

Callaway, R.M., Thelen, G.C., Rodriguez, A., Holben, W.E., 2004. Soil biota and exotic plant invasion. Nature 427, 731–733.

Callaway, R.M., Cipollini, D., Barto, K., Thelen, G.C., Hallett, S.G., Prati, D., Stinson, K., Klironomos, J., 2008. Novel weapons: invasive plant suppresses fungal mutualists in America but not in its native Europe. Ecology 89, 1043–1055.

Callaway, R.M., Bedmar, E.J., Reinhart, K.O., Gomez Silvan, C., Kliromonos, J., 2011. Effects of soil biota from different ranges on *Robinia* invasion: acquiring mutualists and escaping pathogens. Ecology 92, 1027–1035.

Campbell, K.J., Beek, J., Eason, C.T., Glen, A.S., Godwin, J., Gould, F., Holmes, N.D., Howald, G.R., Madden, F.M., Ponder, J.B., Threadgill, D.W., Wegmann, A.S., Baxter, G.S., 2015. The next generation of rodent eradications: innovative technologies and tools to improve species specificity and increase their feasibility on islands. Biol. Conserv. 185, 47–58.

Capinha, C., Brotons, L., Anastácio, P., 2013. Geographical variability in propagule pressure and climatic suitability explain the European distribution of two highly invasive crayfish. J. Biogeogr. 40, 548–558.

Champer, J., Buchman, A., Akbari, O.S., 2016. Cheating evolution: engineering gene drives to manipulate the fate of wild populations. Nat. Rev. Genet. 17, 146–159.

Chen, D.Q., Purcell, A.H., 1997. Occurrence and transmission of facultative endosymbionts in aphids. Curr. Microbiol. 34, 220–225.

Chen, D.Q., Montllor, C.B., Purcell, A.H., 2000. Fitness effects of two facultative endosymbiotic bacteria on the pea aphid, *Acyrthosiphon pisum*, and the blue alfalfa aphid, *A. kondoi*. Entomol. Exp. Appl. 95, 315–323.

Cipollini, D., Rigsby, C.M., Barto, E.K., 2012. Microbes as targets and mediators of allelopathy in plants. J. Chem. Ecol. 38, 714–727.

Cleaveland, S., Thirgood, S., Laurenson, K., 1999. Pathogens as allies in island conservation? Trends Ecol. Evol. 14, 83–84.

Colautti, R.I., Ricciardi, A., Grigorovich, I.A., MacIsaac, H.J., 2004. Is invasion success explained by the enemy release hypothesis? Ecol. Lett. 7, 721–733.

Coon, C.A.C., Martin, L.B., 2014. Patterns of haemosporidian prevalence along a range expansion in introduced Kenyan house sparrows *Passer domesticus*. J. Avian. Biol. 45, 34–42.

Cordaux, R., Michel-Salzat, A., Bouchon, D., 2001. *Wolbachia* infection in crustaceans: novel hosts and potential routes for horizontal transmission. J. Evol. Biol. 14, 237–243.

Cornet, S., Brouat, C., Diagne, C.A., Charbonnel, N., 2016. Eco-immunology and bioinvasion: revisiting the evolution of increased competitive ability hypotheses. Evol. Appl. 9, 952–962.

Cory, J.S., Franklin, M.T., 2012. Evolution and the microbial control of insects. Evol. Appl. 5, 455–469.

Cory, J.S., Myers, J.H., 2000. Direct and indirect ecological effects of biological control. Trends Ecol. Evol. 15, 137–139.

Courchamp, F., Cornell, S.J., 2000. Virus-vectored immunocontraception to control feral cats on islands: a mathematical model. J. Appl. Ecol. 37, 903–913.

Criscione, C.D., Blouin, M.S., 2005. Effective sizes of macroparasite populations: a conceptual model. Trends Parasitol. 21, 212–217.

Cristescu, M.E., 2015. Genetic reconstructions of invasion history. Mol. Ecol. 24, 2212–2225.

Darby, A.C., McInnes, C.J., Kjær, K.H., Wood, A.R., Hughes, M., Martensen, P.M., Radford, A.D., Hall, N., Chantrey, J., 2014. Novel host-related virulence factors are encoded by squirrelpox virus, the main causative agent of epidemic disease in red squirrels in the UK. PLoS One 9. e96439.

Daszak, P., Cunningham, A.A., Hyatt, A.D., 2000. Emerging infectious diseases of wildlife—threats to biodiversity and human health. Science 287, 443–449.

Davis, M.A., Grime, J.P., Thompson, K., 2000. Fluctuating resources in plant communities: a general theory of invasibility. J. Ecol. 88, 528–534.

Day, N.J., Dunfield, K.E., Antunes, P.M., 2016. Fungi from a non-native invasive plant increase its growth but have different growth effects on native plants. Biol. Invasions 18, 231–243.

de Castro, F., Bolker, B., 2005. Mechanisms of disease-induced extinction. Ecol. Lett. 8, 117–126.

deWalt, S.J., Denslow, J.S., Ickes, K., 2004. Natural-enemy release facilitates habitat expansion of the invasive tropical shrub *Clidemia hirta*. Ecology 85, 471–483.

Di Giallonardo, F., Holmes, E.C., 2015. Exploring host-pathogen interactions through biological control. PLoS Pathog. 11. e1004865.

Dietz, H., Edwards, P.J., 2006. Recognition that causal processes change during plant invasion helps explain conflicts in evidence. Ecology 87, 1359–1367.

Diez, J.M., Dickie, I., Edwards, G., Hulme, P.E., Sullivan, J.J., Duncan, R.P., 2010. Negative soil feedbacks accumulate over time for non-native plant species. Ecol. Lett. 13, 803–809.

Dion, E., Polin, S.E., Simon, J.-C., Outreman, Y., 2011. Symbiont infection affects aphid defensive behaviours. Biol. Lett. 7, 743–746.

Dobson, A., Lafferty, K.D., Kuris, A.M., Hechinger, R.F., Jetz, W., 2008. Homage to Linnaeus: how many parasites? How many hosts? Proc. Natl. Acad. Sci. U.S.A. 105, 11482–11489.

Dostal, P., 2010. Post-dispersal seed mortality of exotic and native species: effects of fungal pathogens and seed predators. Basic Appl. Ecol. 11, 676–684.

Douglas, A.E., 2007. Symbiotic microorganisms: untapped resources for insect pest control. Trends Biotechnol. 25, 338–342.

Dunn, A.M., Hatcher, M.J., 2015. Parasites and biological invasions: parallels, interactions, and control. Trends Parasitol. 31, 189–199.

Dunn, A.M., Torchin, M.E., Hatcher, M.J., Kotanen, P.M., Blumenthal, D.M., Byers, J.E., Coon, C.A.C., Frankel, V.M., Holt, R.D., Hufbauer, R.A., Kanarek, A.R., Schierenbeck, K.A., Wolfe, L.M., Perkins, S.E., 2012. Indirect effects of parasites in invasions. Funct. Ecol. 26, 1262–1274.

Duron, O., Bouchon, D., Boutin, S., Bellamy, L., Zhou, L., Engelstadter, J., Hurst, G.D.D., 2008. The diversity of reproductive parasites among arthropods: *Wolbachia* do not walk alone. BMC Biol. 6, 27.

Engelstadter, J., Hurst, G.D., 2009. The ecology and evolution of microbes that manipulate host reproduction. Annu. Rev. Ecol. Evol. Syst. 40, 127–149.

Eppinga, M.B., Rietkerk, M., Dekker, S.C., De Ruiter, P.C., Van der Putten, W.H., 2006. Accumulation of local pathogens: a new hypothesis to explain exotic plant invasions. Oikos 114, 168–176.

Estoup, A., Guillemaud, T., 2010. Reconstructing routes of invasion using genetic data: why, how and so what? Mol. Ecol. 19, 4113–4130.

Fedriani, J.M., Zywiec, M., Delibes, M., 2012. Thieves or mutualists? Pulp feeders enhance endozoochore local recruitment. Ecology 93, 575–587.

Fellous, S., Salvaudon, L., 2009. How can your parasites become your allies? Trends Parasitol. 25, 62–66.

Ferrari, J., Vavre, F., 2011. Bacterial symbionts in insects or the story of communities affecting communities. Philos. Trans. R. Soc. Lond. B Biol. Sci. 366, 1389–1400.

Fischer, M.L., Hochkirch, A., Heddergott, M., Schulze, C., Anheyer-Behmenburg, H.E., Lang, J., Michler, F.U., Hohmann, U., Ansorge, H., Hoffmann, L., Klein, R., Frantz, A.C., 2015. Historical invasion records can be misleading: genetic evidence for multiple introductions of invasive raccoons (*Procyon lotor*) in Germany. PLoS One 10. e0125441.

Fitzpatrick, B.M., Fordyce, J.A., Niemiller, M.L., Reynolds, R.G., 2012. What can DNA tell us about biological invasions? Biol. Invasions 14, 245–253.

Flory, S.L., Clay, K., 2013. Pathogen accumulation and long-term dynamics of plant invasions. J. Ecol. 101, 607–613.

Fontaine, M.C., Gladieux, P., Hood, M.E., Giraud, T., 2013. History of the invasion of the anther smut pathogen on *Silene latifolia* in North America. New Phytol. 198, 946–956.

Friesen, M.L., Porter, S.S., Stark, S.C., von Wettberg, E.J., Sachs, J.L., Martinez-Romero, E., 2011. Microbially mediated plant functional traits. Annu. Rev. Ecol. Evol. Syst. 42, 23–46.

Gaudeul, M., Giraud, T., Kiss, L., Shykoff, J.A., 2011. Nuclear and chloroplast microsatellites show multiple introductions in the worldwide invasion history of common ragweed, *Ambrosia artemisiifolia*. PLoS One 6. e17658.

Gendron, A.D., Marcogliese, D.J., Thomas, M., 2012. Invasive species are less parasitized than native competitors, but for how long? The case of the round goby in the Great Lakes-St. Lawrence Basin. Biol. Invasions 14, 367–384.

Georgievska, L., Hoover, K., van der Werf, W., Munoz, D., Caballero, P., Cory, J.S., Vlak, J.M., 2010. Effects of single and mixed infections with wild type and genetically modified *Helicoverpa armigera* nucleopolyhedrovirus on movement behaviour of cotton bollworm larvae. J. Invertebr. Pathol. 104, 44–50.

Giddings, G., 1998. The release of genetically engineered micro-organisms and viruses into the environment. New Phytol. 140, 173–184.

Gilbert, G.S., Parker, I.M., 2010. Rapid evolution in a plant-pathogen interaction and the consequences for introduced host species. Evol. Appl. 3, 144–156.

Gladieux, P., Zhang, X.G., Afoufa-Bastien, D., Sanhueza, R.M.V., Sbaghi, M., Le Cam, B., 2008. On the origin and spread of the scab disease of apple: out of Central Asia. PLoS One 3. e1455.

Godfray, H.C.J., 1995. Field experiments with genetically manipulated insect viruses: ecological issues. Trends Ecol. Evol. 10, 465–469.

Graham, R.I., Grzywacz, D., Mushobozi, W.L., Wilson, K., 2012. *Wolbachia* in a major African crop pest increases susceptibility to viral disease rather than protects. Ecol. Lett. 15, 993–1000.

Gueguen, G., Vavre, F., Gnankine, O., Peterschmitt, M., Charif, D., Chiel, E., Gottlieb, Y., Ghanim, M., Zchori-Fein, E., Fleury, F., 2010. Endosymbiont metacommunities, mtDNA diversity and the evolution of the *Bemisia tabaci* (Hemiptera: Aleyrodidae) species complex. Mol. Ecol. 19, 4365–4378.

Gundale, M.J., Kardol, P., Nilsson, M.-C., Nilsson, U., Lucas, R.W., Wardle, D.A., 2014. Interactions with soil biota shift from negative to positive when a tree species is moved outside its native range. New Phytol. 202, 415–421.

Hajek, A.E., Tobin, P.C., 2010. Micro-managing arthropod invasions: eradication and control of invasive arthropods with microbes. Biol. Invasions 12, 2895–2912.

Hajek, A.E., Tobin, P.C., 2011. Introduced pathogens follow the invasion front of a spreading alien host. J. Anim. Ecol. 80, 1217–1226.

Hajek, A.E., McManus, M.L., Delalibera Jr., I., 2007. A review of introductions of pathogens and nematodes for classical biological control of insects and mites. Biol. Control 41, 1–13.

Hale, A.N., Lapointe, L., Kalisz, S., 2016. Invader disruption of belowground plant mutualisms reduces carbon acquisition and alters allocation patterns in a native forest herb. New Phytol. 209, 542–549.

Harner, M.J., Mummey, D.L., Stanford, J.A., Rillig, M.C., 2010. Arbuscular mycorrhizal fungi enhance spotted knapweed growth across a riparian chronosequence. Biol. Invasions 12, 1481–1490.

Hartigan, A., Fiala, I., Dykova, I., Jirku, M., Okimoto, B., Rose, K., Phalen, D.N., Slapeta, J., 2011. A suspected parasite spill-back of two novel *Myxidium* spp. (Myxosporea) causing disease in Australian endemic frogs found in the invasive cane toad. PLoS One 6. e18871.

Hatcher, M.J., Dick, J.T., Dunn, A.M., 2006. How parasites affect interactions between competitors and predators. Ecol. Lett. 9, 1253–1271.

Hawkes, C.V., 2007. Are invaders moving targets? The generality and persistence of advantages in size, reproduction and enemy release in invasive plant species with time since introduction. Am. Nat. 170, 832–843.

Hedges, L.M., Brownlie, J.C., O'Neill, S.L., Johnson, K.N., 2008. *Wolbachia* and virus protection in insects. Science 322, 702.

Heger, T., Jeschke, J.M., 2014. The enemy release hypothesis as a hierarchy of hypotheses. Oikos 123, 741–750.

Hendry, T.A., Hunter, M.S., Baltrus, D.A., 2014. The facultative symbiont *Rickettsia* protects an invasive whitefly against entomopathogenic *Pseudomonas syringae* strains. Appl. Environ. Microbiol. 80, 7161–7168.

Henry, L.M., Peccoud, J., Simon, J.C., Hadfield, J.D., Maiden, M.J.C., Ferrari, J., Godfray, H.C., 2013. Horizontally transmitted symbionts and host colonization of ecological niches. Curr. Biol. 23, 1713–1717.

Himler, A.G., Adachi-Hagimori, T., Bergen, J.E., Kozuch, A., Kelly, S.E., Tabashnik, B.E., Chiel, E., Duckworth, V.E., Dennehy, T.J., Zchori-Fein, E., Hunter, M.S., 2011.

Rapid spread of a bacterial symbiont in an invasive whitefly is driven by fitness benefits and female bias. Science 332, 254–256.

Hoffmann, A.A., Turelli, M., 2013. Facilitating *Wolbachia* introductions into mosquito populations through insecticide-resistance selection. Proc. R. Soc. Lond. B Biol. Sci. 280, 20130371.

Hoffmann, A.A., Montgomery, B.L., Popovici, J., Iturbe-Ormaetxe, I., Johnson, P.H., Muzzi, F., Greenfield, M., Durkan, M., Leong, Y.S., Dong, Y., Cook, H., Axford, J., Callahan, A.G., Kenny, N., Omodei, C., McGraw, E.A., Ryan, P.A., Ritchie, S.A., Turelli, M., O'Neill, S.L., 2011. Successful establishment of *Wolbachia* in *Aedes* populations to suppress dengue transmission. Nature 476, 454–457.

Holmes, E.C., 2004. Phylogeography of human viruses. Mol. Ecol. 13, 745–756.

Horn, K., Parker, I.M., Malek, W., Rodriguez-Echeverria, S., Parker, M.A., 2014. Disparate origins of *Bradyrhizobium* symbionts for invasive populations of *Cytisus scoparius* (Leguminosae) in North America. FEMS Microbiol. Ecol. 89, 89–98.

Hudson, P., Greenman, J., 1998. Competition mediated by parasites: biological and theoretical progress. Trends Ecol. Evol. 13, 387–390.

Hughes, G.L., Koga, R., Xue, P., Fukatsu, T., Rasgon, J.L., 2011. *Wolbachia* infections are virulent and inhibit the human malaria parasite *Plasmodium falciparum* in *Anopheles gambiae*. PLoS Pathog. 7. e1002043.

Huntley, B., Barnard, P., Altwegg, R., Chambers, L., Coetzee, B.W.T., Gibson, L., Hockey, P.A.R., Hole, D.G., Midgley, G.F., Underhill, L.G., Willis, S.G., 2010. Beyond bioclimatic envelopes: dynamic species' range and abundance modelling in the context of climatic change. Ecography 33, 621–626.

Ibanez, I., Diez, J.M., Miller, L.P., Olden, J.D., Sorte, C.J.B., Blumenthal, D.M., Bradley, B.A., D'Antonio, C.M., Dukes, J.S., Early, R.I., Grosholz, E.D., Lawler, J.J., 2014. Integrated assessment of biological invasions. Ecol. Appl. 24, 25–37.

Iwase, S., Tani, S., Saeki, Y., Tuda, M., Haran, J., Skuhrovec, J., Takagi, M., 2015. Dynamics of infection with *Wolbachia* in *Hypera postica* (Coleoptera: Curculionidae) during invasion and establishment. Biol. Invasions 17, 3639–3648.

Jaenike, J., Unckless, R., Cockburn, S.N., Boelio, L.M., Perlman, S.J., 2010. Adaptation *via* symbiosis: recent spread of a *Drosophila* defensive symbiont. Science 329, 212–215.

Jeschke, J.M., 2014. General hypotheses in invasion ecology. Divers. Distrib. 20, 1229–1234.

Jeschke, J.M., Gomez Aparicio, L., Haider, S., Heger, T., Lortie, C.J., Pysek, P., Strayer, D.L., 2012. Support for major hypotheses in invasion biology is uneven and declining. NeoBiota 14, 1–20.

Johnson, K.P., Clayton, D.H., 2004. Untangling coevolutionary history. Syst. Biol. 53, 92–94.

Johnson, N.C., Graham, J.H., Smith, F.A., 1997. Functioning of mycorrhizal associations along the mutualism-parasitism continuum. New Phytol. 135, 575–586.

Johnson, P.T.J., Hartson, R.B., Larson, D.J., Sutherland, D.R., 2008. Diversity and disease: community structure drives parasite transmission and host fitness. Ecol. Lett. 11, 1017–1026.

Johnson, N.C., Angelard, C., Sanders, I.R., Kiers, E.T., 2013. Predicting community and ecosystem outcomes of mycorrhizal responses to global change. Ecol. Lett. 16, 140–153.

Jones, C.M., Brown, M.J.F., 2014. Parasites and genetic diversity in an invasive bumblebee. J. Anim. Ecol. 83, 1428–1440.

Jousselin, E., Coeur d'Acier, A., Vanlerberghe-Masutti, F., Duron, O., 2013. Evolution and diversity of *Arsenophonus* endosymbionts in aphids. Mol. Ecol. 22, 260–270.

Julien, M.H., Griffiths, M.W., 1998. Biological Control of Weeds: World Catalogue of Agents and Their Target Weeds, fourth ed. CAB International, Wallingford, Oxon, UK. 223 pp.

Keane, R.M., Crawley, M.J., 2002. Exotic plant invasions and the enemy release hypothesis. Trends Ecol. Evol. 17, 164–170.

Kelehear, C., Cabrera-Guzman, E., Shine, R., 2012. Inadvertent consequences of community-based efforts to control invasive species. Conserv. Lett. 5, 360–365.

Kelehear, C., Brown, G.P., Shine, R., 2013. Invasive parasites in multiple invasive hosts: the arrival of a new host revives a stalled prior parasite invasion. Oikos 122, 1317–1324.

Kelly, D.W., Paterson, R.A., Townsend, C.R., Poulin, R., Tompkins, D.M., 2009. Parasite spillback: a neglected concept in invasion ecology? Ecology 90, 2047–2056.

Kerr, P.J., Liu, J., Cattadori, I., Ghedin, E., Read, A.F., Holmes, E.C., 2015. Myxoma virus and the Leporipoxviruses: an evolutionary paradigm. Viruses 7, 1020–1061.

Kestrup, A.M., Thomas, S.H., Ricciardi, A., Duffy, M.A., 2011. Differential infection of exotic and native freshwater amphipods by a parasitic water mold in the St. Lawrence River. Biol. Invasions 13, 769–779.

Klironomos, J.N., 2002. Feedback with soil biota contributes to plant rarity and invasiveness in communities. Nature 417, 67–70.

Lacey, L.A., Frutos, R., Kaya, H.K., Vail, P., 2001. Insect pathogens as biological control agents: do they have a future? Biol. Control 21, 230–248.

Lambrinos, J.G., 2004. How interactions between ecology and evolution influence contemporary invasion dynamics. Ecology 85, 2061–2070.

Leduc, A., Traoré, Y.N., Boyer, K., Magne, M., Grygiel, P., Juhasz, C.C., Boyer, C., Guerin, F., Wonni, I., Ouedraogo, L., Vernière, C., Ravigné, V., Pruvost, O., 2015. Bridgehead invasion of a monomorphic plant pathogenic bacterium: Xanthomonas citri pv. citri, an emerging Citrus pathogen in Mali and Burkina Faso. Environ. Microbiol. 17, 4429–4442.

Lee, K., Klasing, K., 2004. A role of immunology in invasion biology. Trends Ecol. Evol. 19, 523–529.

Lerat, E., Daubin, V., Ochman, H., Moran, N.A., 2005. Evolutionary origins of genomic repertoires in bacteria. PLoS Biol. 3. e130.

Lester, P.J., Bosch, P.J., Gruber, M.A.M., Kapp, E.A., Peng, L., Brenton-Rule, E.C., Buchanan, J., Stanislawek, W.L., Archer, M., Corley, J.C., Masciocchi, M., Van Oystaeyen, A., Wenseleers, T., 2015. No evidence of enemy release in pathogen and microbial communities of common wasps (Vespula vulgaris) in their native and introduced range. PLoS One 10. e0121358.

Lewicki, K.E., Huyvaert, K.P., Piaggio, A.J., Diller, L.V., Franklin, A.B., 2015. Effects of barred owl (Strix varia) range expansion on Haemoproteus parasite assemblage dynamics and transmission in barred and northern spotted owls (Strix occidentalis caurina). Biol. Invasions 17, 1713–1727.

Liebhold, A., Bascompte, J., 2003. The Allee effect, stochastic dynamics and the eradication of alien species. Ecol. Lett. 6, 133–140.

Linnenbrink, M., Wang, J., Hardouin, E.A., Kunzel, S., Metzler, D., Baines, J.F., 2013. The role of biogeography in shaping diversity of the intestinal microbiota in house mice. Mol. Ecol. 22, 1904–1916.

Liu, X., Marshall, J.L., Stary, P., Edwards, O., Puterka, G.J., Dolatti, L., El Bouhssini, M., Malinga, J., Lage, J., Smith, C.M., 2010. Global phylogenetics of Diuraphis noxia (Hemiptera: Aphididae), an invasive aphid species: evidence for multiple invasions into North America. J. Econ. Entomol. 103, 958–965.

Lodge, D.M., 1993. Biological invasions: lessons for ecology. Trends Ecol. Evol. 8, 133–137.

Lombaert, E., Guillemaud, T., Lundgren, J., Koch, R., Facon, B., Grez, A., Loomans, A., Malausa, T., Nedved, O., Rhule, E., Staverlokk, A., Steenberg, T., Estoup, A., 2014. Complementarity of statistical treatments to reconstruct worldwide routes of invasion: the case of the Asian ladybird Harmonia axyridis. Mol. Ecol. 23, 5979–5997.

Lopez-Darias, M., Ribas, A., Feliu, C., 2008. Helminth parasites in native and invasive mammal populations: comparative study on the Barbary ground squirrel *Atlantoxerus getulus* L. (Rodentia, Sciuridae) in Morocco and the Canary Islands. Acta Parasitol. 53, 296–301.

Lymbery, A.J., Morine, M., Kanani, H.G., Beatty, S.J., Morgan, D.L., 2014. Co-invaders: the effects of alien parasites on native hosts. Int. J. Parasitol. 3, 171–177.

MacLeod, C.J., Paterson, A.M., Tompkins, D.M., Duncan, R.P., 2010. Parasites lost—do invaders miss the boat or drown on arrival? Ecol. Lett. 13, 516–527.

Makarova, K.S., Aravind, L., Grishin, N.V., Rogozin, I.B., Koonin, E.V., 2002. A DNA repair system specific for thermophilic Archaea and bacteria predicted by genomic context analysis. Nucleic Acids Res. 30, 482–496.

Malmstrom, C.M., Hughes, C.C., Newton, L.A., Stoner, C.J., 2005. Virus infection in remnant native bunchgrasses from invaded California grasslands. New Phytol. 168, 217–230.

Mangla, S.I., Callaway, R.M., 2008. Exotic invasive plant accumulates native soil pathogens which inhibit native plants. J. Ecol. 96, 58–67.

Maron, J.L., Kliromonos, J., Waller, L., Callaway, R.M., 2014. Invasive plants escape from suppressive soil biota at regional scales. J. Ecol. 102, 19–27.

Maron, J.L., Luo, W., Callaway, R.M., Pal, R.W., 2015. Do exotic plants lose resistance to pathogenic soil biota from their native range? A test with *Solidago gigantea*. Oecologia 179, 447–454.

Martin, L.B., Coon, C.A.C., Liebl, A.L., Schrey, A.W., 2014. Surveillance for microbes and range expansion in house sparrows. Proc. R. Soc. Lond. B Biol. Sci. 281, 20132690.

Marzal, A., Ricklefs, R.E., Valkiūnas, G., Albayrak, T., Arriero, E., Bonneaud, C., Czirják, G.A., Ewen, J., Hellgren, O., Hořáková, D., Iezhova, T.A., Jensen, H., Križanauskienė, A., Lima, M.R., de Lope, F., Magnussen, E., Martin, L.B., Møller, A.P., Palinauskas, V., Pap, P.L., Pérez-Tris, J., Sehgal, R.N.M., Soler, M., Szöllősi, E., Westerdahl, H., Zetindjiev, P., Bensch, S., 2011. Diversity, loss, and gain of malaria parasites in a globally invasive bird. PLoS One 6. e21905.

Mastitsky, S.E., Karatayev, A.Y., Burlakova, L.E., Molloy, D.P., 2010. Parasites of exotic species in invaded areas: does lower diversity mean lower epizootic impact? Divers. Distrib. 16, 798–803.

Mazé-Guilmo, E., Blanchet, S., McCoy, K.D., Loot, G., 2016. Host dispersal as the driver of parasite genetic structure: a paradigm lost? Ecol. Lett. 19, 336–347.

McColl, K.A., Cooke, B.D., Sunarto, A., 2014. Viral biocontrol of invasive vertebrates: lessons from the past applied to cyprinid herpesvirus-3 and carp (*Cyprinus carpio*) control in Australia. Biol. Control 72, 109–117.

McMeniman, C.J., Lane, R.V., Cass, B.N., Fong, A.W., Sidhu, M., Wang, Y.F., O'Neill, S.L., 2009. Stable introduction of a life-shortening *Wolbachia* infection into the mosquito *Aedes aegypti*. Science 323, 141–144.

Mei, L., Zhu, M., Zhang, D.Z., Wang, Y.Z., Guo, J., Zhang, H.B., 2014. Geographical and temporal changes of foliar fungal endophytes associated with the invasive plant *Ageratina adenophora*. Microb. Ecol. 67, 402–409.

Meinhardt, K.A., Gehring, C.A., 2012. Disrupting mycorrhizal mutualisms: a potential mechanism by which exotic tamarisk outcompetes native cottonwoods. Ecol. Appl. 22, 532–549.

Messing, R.H., Wright, M.G., 2006. Biological control of invasive species: solution or pollution. Front. Ecol. Environ. 4, 132–140.

Meusnier, I., Olsen, J.L., Stam, W.T., Destombe, C., Valero, M., 2001. Phylogenetic analyses of *Caulerpa taxifolia* (Chlorophyta) and of its associated bacterial microflora provide clues to the origin to the Mediterranean introduction. Mol. Ecol. 10, 931–946.

Meyer, J.Y., Fourdrigniez, M., 2011. Conservation benefits of biological control: the recovery of a threatened plant subsequent to the introduction of a pathogen to contain an invasive tree species. Biol. Conserv. 144, 106–113.

Miller, L.K., Lingg, A.J., Bulla, L.A., 1983. Bacterial, viral and fungal insecticides. Science 219, 715–721.

Minard, G., Tran, F.H., Van, V.T., Goubert, C., Bellet, C., Lambert, G., Kim, K.L.H., Thuy, T.H.T., Mavingui, P., Moro, C.V., 2015. French invasive Asian tiger mosquito populations harbor reduced bacterial microbiota and genetic diversity compared to Vietnamese autochthonous relatives. Front. Microbiol. 6, 970.

Mitchell, C.E., Power, A., 2003. Release of invasive plants from fungal and viral pathogens. Nature 421, 625–627.

Mitchell, C.E., Agrawal, A.A., Bever, J.D., Gilbert, G.S., Hufbauer, R.A., Klironomos, J.N., Maron, J.L., Morris, W.F., Parker, I.M., Power, A.G., Seabloom, E.W., Torchin, M.E., Vazquez, D.P., 2006. Biotic interactions and plant invasions. Ecol. Lett. 9, 726–740.

Mitchell, C.E., Blumenthal, D., Jarosik, V., Puckett, E.E., Pysek, P., 2010. Controls on pathogen species richness in plants' introduced and native ranges: roles of residence time, range size and host traits. Ecol. Lett. 13, 1525–1535.

Mnif, I., Ghribi, D., 2015. Potential of bacterial derived biopesticides in pest management. Crop Prot. 77, 52–64.

Montllor, C.B., Maxmen, A., Purcell, A.H., 2002. Facultative bacterial endosymbionts benefit pea aphids *Acyrthosiphon pisum* under heat stress. Ecol. Entomol. 27, 189–195.

Moran, N.A., Russell, J.A., Koga, R., Fukatsu, T., 2005. Evolutionary relationships of three new species of Enterobacteriaceae living as symbionts of aphids and other insects. Appl. Environ. Microbiol. 71, 3302–3310.

Moran, N.A., McCutcheon, J.P., Nakabachi, A., 2008. Genomics and evolution of heritable bacterial symbionts. Annu. Rev. Genet. 42, 165–190.

Moreira, L.A., Iturbe-Ormaetxe, I., Jeffery, J.A., Lu, G., Pyke, A.T., Hedges, L.M., Rocha, B.C., Hall-Mendelin, S., Day, A., Riegler, M., Hugo, L.E., Johnson, K.N., Kay, B.H., McGraw, E.A., van den Hurk, A.F., Ryan, P.A., O'Neill, S.L., 2009. A *Wolbachia* symbiont in *Aedes aegypti* limits infection with dengue, Chikungunya, and *Plasmodium*. Cell 139, 1268–1278.

Murphy, G.E.P., Romanuk, T.N., 2014. A meta-analysis of declines in local species richness from human disturbances. Ecol. Evol. 4, 91–103.

Nakaew, N., Rangjaroen, C., Sungthong, R., 2015. Utilization of rhizospheric *Streptomyces* for biological control of *Rigidoporus* sp. causing white root disease in rubber tree. Eur. J. Plant Pathol. 142, 93–105.

Nandasena, K.G., O'Hara, G.W., Tiwari, R.P., Howieson, J.G., 2006. Rapid *in situ* evolution of nodulating strains for *Biserrula pelecinus* L. through lateral transfer of a symbiosis island from the original mesorhizobial inoculant. Appl. Environ. Microbiol. 72, 7365–7367.

Ndlovu, J., Richardson, D.M., Wilson, J.R.U., Le Roux, J.J., 2013. Co-invasion of South African ecosystems by an Australian legume and its rhizobial symbionts. J. Biogeogr. 40, 1240–1251.

Nguyen, D.T., Spooner-Hart, R.N., Riegler, M., 2016. Loss of *Wolbachia* but not *Cardinium* in the invasive range of the Australian thrips species, *Pezothrips kellyanus*. Biol. Invasions 18, 197–214.

Nieberding, C.M., Olivieri, I., 2007. Parasites: proxies for host genealogy and ecology. Trends Ecol. Evol. 22, 156–165.

Nunez, M.A., Horton, T.R., Simberloff, D., 2009. Lack of belowground mutualisms hinders Pinaceae invasions. Ecology 90, 2352–2359.

O'Brien, V.A., Moore, A.T., Young, G.R., Komar, N., Reisen, W.K., Brown, C.R., 2011. An enzootic vector-borne virus is amplified at epizootic levels by an invasive avian host. Proc. R. Soc. Lond. B Biol. Sci. 278, 239–246.

Oliveira, N.M., Hilker, F.M., 2010. Modelling disease introduction as biological control of invasive predators to preserve endangered prey. Bull. Math. Biol. 72, 444–468.

Oliver, K.M., Russell, J.A., Moran, N.A., Hunter, M.S., 2003. Facultative bacterial symbionts in aphids confer resistance to parasitic wasps. Proc. Natl. Acad. Sci. U.S.A. 100, 1803–1807.

Oliver, K.M., Moran, N.A., Hunter, M.S., 2005. Variation in resistance to parasitism in aphids is due to symbionts not host genotype. Proc. Natl. Acad. Sci. U.S.A. 102, 12795–12800.

Oliver, K.M., Campos, J., Moran, N.A., Hunter, M.S., 2008. Population dynamics of defensive symbionts in aphids. Proc. R. Soc. Lond. B Biol. Sci. 275, 293–299.

Oliver, K.M., Degnan, P.H., Burke, G.R., Moran, N.A., 2010. Facultative symbionts in aphids and the horizontal transfer of ecologically important traits. Annu. Rev. Entomol. 55, 247–266.

Oliver, K.M., Smith, A.H., Russell, J.A., 2014. Defensive symbiosis in the real world—advancing ecological studies of heritable, protective bacteria in aphids and beyond. Funct. Ecol. 28, 341–355.

Palma, L., Munoz, D., Berry, C., Murillo, J., Caballero, P., 2014. *Bacillus thuringiensis* toxins: an overview of their biocidal activity. Toxins (Basel) 6, 3296–3325.

Parker, M.A., 2001. Mutualism as a constraint on invasion success for legumes and rhizobia. Divers. Distrib. 7, 125–136.

Parker, M.A., Malek, W., Parker, I.M., 2006. Growth of an invasive legume is symbiont limited in newly occupied habitats. Divers. Distrib. 12, 563–571.

Parker, I.M., Saunders, M., Bontrager, M., Weitz, A.P., Hendricks, R., Magarey, R., Suiter, K., Gilbert, G.S., 2015. Phylogenetic structure and host abundance drive disease pressure in communities. Nature 520, 542–544.

Pearson, R.G., Dawson, T.P., 2003. Predicting the impacts of climate change on the distribution of species: are bioclimate envelope models useful? Glob. Ecol. Biogeogr. 12, 361–371.

Perkins, S.E., Altizer, S., Bjornstad, O., Burdon, J.J., Clay, K., Gómez-Aparicio, L., Jeschke, J.M., Johnson, P.T.J., Lafferty, K.D., Malmstrom, C.M., Martin, P., Power, A., Strayer, D.L., Thrall, P.H., Uriarte, M., 2008. Invasion biology and parasitic infections. In: Ostfeld, R.S., Keesing, F., Eviner, V.T. (Eds.), Infectious Disease Ecology: Effects of Ecosystems on Disease and of Disease on Ecosystems. Princeton University Press, Princeton, NJ, pp. 179–204 (Chapter 8).

Phillips, B.L., Kelehear, C., Pizzato, L., Brown, G.P., Barton, D., Shine, R., 2010. Parasites and pathogens lag behind their host during periods of host range advance. Ecology 91, 872–881.

Pimentel, D., McNair, S., Janecka, J., Wightman, J., Simmonds, C., O'Connell, C., Wong, E., Russel, L., Zern, J., Aquino, T., Tsomondo, T., 2001. Economic and environmental threats of alien plant, animal, microbe invasions. Agric. Ecosyst. Environ. 84, 1–20.

Polin, S., Simon, J.-C., Outreman, Y., 2014. An ecological cost associated with protective symbionts of aphids. Ecol. Evol. 4, 836–840.

Portales-Reyes, C., Van Doornik, T., Schultheis, E.H., Suwa, T., 2015. A novel impact of a novel weapon: allelochemicals in *Alliaria petiolata* disrupt the legume-rhizobia mutualism. Biol. Invasions 17, 2779–2791.

Porter, S.S., Simms, E.L., 2014. Selection for cheating across disparate environments in the legume-*rhizobium* mutualism. Ecol. Lett. 17, 1121–1129.

Porter, S.S., Rice, K.J., Stanton, M.L., 2011. Mutualism and adaptive divergence: co-invasion of a heterogeneous grassland by an exotic legume-*rhizobium* symbiosis. PLoS One 6. e27935.

Power, A.G., Mitchell, C.E., 2004. Pathogen spillover in disease epidemics. Am. Nat. 164, S79–S89.

Prenter, J., MacNeil, C., Dick, J.T.A., Dunn, A.M., 2004. Roles of parasites in animal invasions. Trends Ecol. Evol. 19, 385–390.

Pringle, A., Bever, J.D., Gardes, M., Parrent, J.L., Rillig, M.C., Klironomos, J.N., 2009. Mycorrhizal symbioses and plant invasions. Annu. Rev. Ecol. Evol. Syst. 40, 699–715.

Prior, K.M., Powell, T.H.Q., Joseph, A.L., Hellmann, J.J., 2015. Insights from community ecology into the role of enemy release in causing invasion success: the importance of native enemy effects. Biol. Invasions 17, 1283–1297.

Reinhart, K.O., Callaway, R.M., 2006. Soil biota and invasive plants. New Phytol. 170, 445–457.

Reuter, M., Pedersen, J.S., Keller, L., 2005. Loss of *Wolbachia* infection during colonisation in the invasive Argentine ant *Linepithema humile*. Heredity 94, 364–369.

Rex Consortium, 2013. Heterogeneity of selection and the evolution of resistance. Trends Ecol. Evol. 28, 110–118.

Rey, O., Estoup, A., Facon, B., Loiseau, A., Aebi, A., Duron, O., Vavre, F., Foucaud, J., 2013. Distribution of endosymbiotic reproductive manipulators reflects invasion process and not reproductive system polymorphism in the little fire ant *Wasmannia auropunctata*. PLoS One 8. e58467.

Ricciardi, A., Hoopes, M.F., Marchetti, M.P., Lockwood, J.L., 2013. Progress toward understanding the ecological impacts of nonnative species. Ecol. Monograph. 83, 263–282.

Richardson, D.M., Allsopp, N., D'Antonio, C.M., Milton, S.J., Rejmanek, M., 2007. Plant invasions—the role of mutualisms. Biol. Rev. 75, 65–93.

Rizzo, L., Fraschetti, S., Alifano, P., Pizzolante, G., Stabili, L., 2016. The alien species *Caulerpa cylindracea* and its associated bacteria in the Mediterranean Sea. Mar. Biol. 163, 1–12.

Rode, N.O., Lievens, E.J.P., Segard, A., Flaven, E., Jabbour-Zahab, R., Lenormand, T., 2013. Cryptic microsporidian parasites differentially affect invasive and native *Artemia* spp. Int. J. Parasitol. 43, 795–803.

Rodriguez-Echeverria, S., 2010. Rhizobial hitchhikers from Down Under: invasional meltdown in a plant-bacteria mutualism? J. Biogeogr. 37, 1611–1622.

Rodriguez-Echeverria, S., Fajardo, S., Ruiz-Diez, B., Fernandez-Pascual, M., 2012. Differential effectiveness of novel and old legume-rhizobia mutualisms: implications for invasion by exotic legumes. Oecologia 170, 253–261.

Röhrich, C.R., Ngwa, C.J., Wiesner, J., Schmidtberg, H., Degenkolb, T., Kollewe, C., Fischer, R., Pradel, G., Vilcinskas, A., 2012. Harmonine, a defence compound from the harlequin ladybird, inhibits mycobacterial growth and demonstrates multi-stage antimalarial activity. Biol. Lett. 8, 308–311.

Romeis, J., Meissle, M., Bigler, F., 2006. Transgenic crops expressing *Bacillus thuringiensis* toxins and biological control. Nat. Biotechnol. 24, 63–71.

Rosenberg, N.A., Nordborg, M., 2002. Genealogical trees, coalescent theory, and the analysis of genetic polymorphisms. Nat. Rev. Genet. 3, 380–390.

Roy, H.E., Lawson Handley, L.-J., 2012. Networking: a community approach to invaders and their parasites. Funct. Ecol. 26, 1238–1248.

Roy, B.A., Hudson, K., Visser, M., Johnson, B.R., 2014. Grassland fires may favor native over introduced plants by reducing pathogen loads. Ecology 95, 1897–1906.

Rúa, M.A., Pollina, E.C., Power, A.G., Mitchell, C.E., 2011. The role of viruses in biological invasions: friend or foe? Curr. Opin. Virol. 1, 68–72.

Rudgers, J.A., Koslow, J.M., Clay, K., 2004. Endophytic fungi alter relationships between diversity and ecosystem properties. Ecol. Lett. 7, 42–51.

Russell, J.A., Moran, N.A., 2006. Costs and benefits of symbiont infection in aphids: variation among symbionts and across temperatures. Proc. R. Soc. Lond. B Biol. Sci. 273, 603–610.

Russell, J.A., Latorre, A., Sabater-Munoz, B., Moya, A., Moran, N.A., 2003. Side-stepping secondary symbionts: widespread horizontal transfer across and beyond the Aphidoidea. Mol. Ecol. 12, 1061–1075.

Saikkonen, K., Lehtonen, P., Helander, M., Koricheva, J., Faeth, S.H., 2006. Model systems in ecology: dissecting the endophyte–grass literature. Trends Plant Sci. 11, 428–433.

Sanahuja, G., Banakar, R., Twyman, R.M., Capell, T., Christou, P., 2011. *Bacillus thuringiensis*: a century of research development and commercial applications. Plant Biotechnol. J. 9, 283–300.

Sander, J.D., Joung, J.K., 2014. CRISPR-Cas systems for editing, regulating and targeting genomes. Nat. Biotechnol. 32, 347–355.

Santini, A., Ghelardini, L., De Pace, C., Desprez-Loustau, M.L., Capretti, P., Chandelier, A., Cech, T., Chira, D., Diamandis, S., Gaitniekis, T., Hantula, J., Holdenrieder, O., Jankovsky, L., Jung, T., Jurc, D., Kirisits, T., Kunca, A., Lygis, V., Malecka, M., Marcais, B., Schmitz, S., Schumacher, J., Solheim, H., Solla, A., Szabò, I., Tsopelas, P., Vannini, A., Vettraino, A.M., Webber, J., Woodward, S., Stenlid, J., 2013. Biogeographical patterns and determinants of invasion by forest pathogens in Europe. New Phytol. 197, 238–250.

Saunders, G., Cooke, B., McColl, K., Shine, R., Peacock, T., 2010. Modern approaches for the biological control of vertebrate pests: an Australian perspective. Biol. Control 52, 288–295.

Schmid-Hempel, P., 2011. Evolutionary Parasitology. Oxford University Press, Oxford.

Schmidtberg, H., Röhrich, C.R., Vogel, H., Vilcinskas, A., 2013. A switch from constitutive chemical defence to inducible innate immune responses in the invasive ladybird *Harmonia axridis*. Biol. Lett. 8, 308–311.

Schoemaker, D.D., Ross, K.G., Keller, L., Vargo, E.L., Werren, J.H., 2000. *Wolbachia* infections in native and introduced populations of fire ants (*Solenopsis* spp.). Insect Mol. Biol. 9, 661–663.

Schwartz, M.W., Hoeksema, J.D., Gehring, C.A., Johnson, N.C., Klironomos, J.N., Abbott, L.K., Pringle, A., 2006. The promise and the potential consequences of the global transport of mycorrhizal fungal inoculum. Ecol. Lett. 9, 501–515.

Seddon, P.J., Griffiths, C.J., Soorae, P.S., Armstrong, D.P., 2014. Reversing defaunation: restoring species in a changing world. Science 345, 406–412.

Seifert, E.K., Bever, J.D., Maron, J.M., 2009. Evidence for evolution of reduced mycorrhizal dependence during plant invasion. Ecology 90, 1055–1062.

Shah, M.A., Reshi, Z.A., Khasa, D.P., 2009. Arbuscular mycorrhizas: drivers or passengers of alien plant invasion. Bot. Rev. 75, 397–417.

Sheath, D., Williams, C., Reading, A., Britton, J.R., 2015. Parasites of non-native freshwater fishes introduced into England and Wales suggest enemy release and parasite acquisition. Biol. Invasions 17, 2235–2246.

Simberloff, D., Von Holle, B., 1999. Positive interactions of nonindigenous species: invasional meltdown? Biol. Invasions 1, 21–32.

Simberloff, D., Martin, J.L., Genovesi, P., Maris, V., Wardle, D.A., Aronson, J., Courchamp, F., Galil, B., Garcia-Berthou, E., Pascal, M., Pysek, P., Sousa, R., Tabacchi, E., Vila, M., 2013. Impacts of biological invasions: what's what and the way forward. Trends Ecol. Evol. 28, 58–66.

Simon, J.-C., Boutin, S., Tsuchida, T., Koga, R., Le Gallic, J.-F., Frantz, A., Outreman, Y., Fukatsu, T., 2011. Facultative symbiont infections affect aphid reproduction. PLoS One 6. e21831.

Slothouber Galbreath, J.G., Smith, J.E., Becnel, J.J., Butlin, R.K., Dunn, A.M., 2010. Reduction in post-invasion genetic diversity in *Crangonyx pseudogracilis* (Amphipoda: Crustacea): a genetic bottleneck or the work of hitchhiking vertically transmitted microparasites? Biol. Invasions 12, 191–209.

Smith, K.F., Carpenter, S.M., 2006. Potential spread of introduced black rat (*Rattus rattus*) parasites to endemic deer mice (*Peromyscus maniculatus*) on the California Channel Islands. Divers. Distrib. 12, 742–748.

Söderhäll, K., Cerenius, L., 1999. The crayfish plague fungus: history and recent advances. Freshw. Crayfish 12, 11–35.

Stock, C.A., McLoughlin, T.J., Klein, J.A., Adang, M.J., 1990. Expression of a *Bacillus thuringiensis* crystal protein gene in *Pseudomonas cepacia* 526. Can. J. Microbiol. 36, 879–884.

Strauss, A., White, A., Boots, M., 2012. Invading with biological weapons: the importance of disease-mediated invasions. Funct. Ecol. 26, 1249–1261.

Stricker, K.B., Harmon, P.F., Goss, E.M., Clay, K., Flory, S.L., 2016. Emergence and accumulation of novel pathogens suppress an invasive species. Ecol. Lett. 19, 469–477.

Strong Jr., D.R., Levin, D.A., 1975. Species richness of the parasitic fungi of British Trees. Proc. Natl. Acad. Sci. U.S.A. 72, 2116–2119.

Stukenbrock, E.H., Banke, S., Javan-Nikkhah, M., McDonald, B.A., 2007. Origin and domestication of the fungal wheat pathogen *Mycosphaerella graminicola* via sympatric speciation. Mol. Biol. Evol. 24, 398–411.

Suding, K.N., Stanley Harpole, W., Fukami, T., Kulmatiski, A., MacDougall, A.S., Stein, C., Putten, W.H., 2013. Consequences of plant–soil feedbacks in invasion. J. Ecol. 101, 298–308.

Sullivan, J.T., Patrick, H.N., Lowther, W.L., Scott, D.B., Ronson, C.W., 1995. Nodulating strains of *Rhizobium loti* arise through chromosomal symbiotic gene transfer in the environment. Proc. Natl. Acad. Sci. U.S.A. 92, 8985–8989.

Sullivan, J.T., Trzebiatowski, J.R., Cruickshank, R.W., Gouzy, J., Brown, S.D., Elliot, R.M., Fleetwood, D.J., McCallum, N.G., Rossbach, U., Stuart, G.S., Weaver, J.E., Webby, R.J., de Bruijn, F.J., Ronson, C.W., 2002. Comparative sequence analysis of the symbiosis island of *Mesorhizobium loti* strain R7A. J. Bacteriol. 184, 3086–3095.

Tabashnik, B.E., Brévault, T., Carrière, Y., 2013. Insect resistance to *Bt* crops: lessons from the first billion acres. Nat. Biotechnol. 31, 510–521.

Thomas, M., Reid, A., 2007. Are exotic natural enemies an effective way of control invasive plants? Trends Ecol. Evol. 22, 447–453.

Thompson, J.N., 1998. Rapid evolution as an ecological process. Trends Ecol. Evol. 13, 329–332.

Torchin, M.E., Lafferty, K.D., Dobson, A.P., McKenzie, V.J., Kuris, A.M., 2003. Introduced species and their missing parasites. Nature 421, 628–630.

Traveset, A., Richardson, D.M., 2014. Mutualistic interactions and biological invasions. Annu. Rev. Ecol. Evol. Syst. 45, 89–113.

Tsuchida, T., Koga, R., Fukatsu, T., 2004. Host plant specialization governed by facultative symbiont. Science 303, 1989.

Tsutsui, N.D., Kauppinen, S.N., Oyafuso, A.F., Grosberg, R.K., 2003. The distribution and evolutionary history of *Wolbachia* infection in native and introduced populations of the invasive argentine ant (*Linepithema humile*). Mol. Ecol. 12, 3057–3068.

van Elsas, J.D., Chiurazzi, M., Mallon, C.A., Elhottova, D., Kristufek, V., Salles, J.F., 2012. Microbial diversity determines the invasion of soil by a bacterial pathogen. Proc. Natl. Acad. Sci. U.S.A. 109, 1159–1164.

van Grunsven, R.H.A., van der Putten, W.H., Bezemer, T.M., Tamis, W.L.M., Berendse, F., Veenendaal, A.M., 2007. Reduced plant–soil feedback of plant species expanding their range as compared to natives. J. Ecol. 95, 1050–1057.

van Kleunen, M., Fischer, M., 2009. Release from foliar and floral fungal pathogen species does not explain the geographic spread of naturalized North American plants in Europe. J. Ecol. 97, 385–392.

Vandegrift, R., Blaser, W., Campos-Cerda, F., Heneghan, A.F., Carroll, G.C., Roy, B.A., 2015. Mixed fitness effects of grass endophytes modulate impact of enemy release and rapid evolution in an invasive grass. Biol. Invasions 17, 1239–1251.

Vavre, F., Charlat, S., 2012. Making (good) use of Wolbachia: what the models say. Curr. Opin. Microbiol. 15, 263–268.

Vavre, F., Fleury, F., Lepetit, D., Fouillet, P., Bouletreau, M., 1999. Phylogenetic evidence for horizontal transmission of Wolbachia in host-parasitoid associations. Mol. Biol. Evol. 16, 1711–1723.

Vercken, E., Fontaine, M.C., Gladieux, P., Hood, M.E., Jonot, O., Giraud, T., 2010. Glacial refugia in pathogens: European genetic structure of anther smut pathogens on Silene latifolia and Silene dioica. PLoS Pathog. 6. e1001229.

Vilcinskas, A., 2015. Pathogens as biological weapons of invasive species. PLoS Pathog. 11. e1004714.

Vilcinskas, A., Mukherjee, K., Vogel, H., 2013a. Expansion of the antimicrobial peptide repertoire in the invasive ladybird Harmonia axyridis. Proc. R. Soc. Lond. B Biol. Sci. 80, 2012–2113.

Vilcinskas, A., Stoecker, K., Schmidtberg, H., Rohrich, C.R., Vogel, H., 2013b. Invasive harlequin ladybird carries biological weapons against native competitors. Science 340, 862–863.

Vilcinskas, A., Schmidtberg, H., Estoup, A., Tayeh, A., Facon, B., Vogel, H., 2015. Evolutionary ecology of microsporidia associated with the invasive ladybird Harmonia axyridis. Insect Sci. 22, 313–324.

Vinale, F., Sivasithamparamb, K., Ghisalbertic, E.L., Marraa, R., Wooa, S.L., Lorito, M., 2008. Trichoderma–plant–pathogen interactions. Soil Biol. Biochem. 40, 1–10.

Vogelsang, K.M., Bever, J.D., 2009. Mycorrhizal densities decline in association with nonnative plants and contribute to plant invasion. Ecology 90, 399–407.

Walker, S.F., Bosch, J., James, T.Y., Litvintseva, A.P., Valls, J.A.O., Pina, S., Garcia, G., Rosa, G.A., Cunningham, A.A., Hole, S., Griffiths, R., Fisher, M.C., 2008. Invasive pathogens threaten species recovery programs. Curr. Biol. 18, R853–R854.

Wattier, R.A., Haine, E.R., Beguet, J., Martin, G., Bollache, L., Musko, I.B., Platvoet, D., Rigaud, T., 2007. No genetic bottleneck or associated microparasite loss in invasive populations of a freshwater amphipod. Oikos 116, 1941–1953.

Webber, B.L., Raghu, S., Edwards, O.R., 2015. Is CRISPR-based gene drive a biocontrol silver bullet or global conservation threat? Proc. Natl. Acad. Sci. U.S.A. 112, 10565–10567.

Weeks, A.R., Turelli, M., Harcombe, W.R., Reynolds, K.T., Hoffmann, A.A., 2007. From parasite to mutualist: rapid evolution of Wolbachia in natural populations of Drosophila. PLoS Biol. 5, 997–1005.

Wei, G., Chen, W., Zhu, W., Chen, C., Young, J.P.W., Bontemps, C., 2009. Invasive Robinia pseudoacacia in China is nodulated by Mesorhizobium and Sinorhizobium species that share similar nodulation genes with native American symbionts. FEMS Microbiol. Ecol. 68, 320–328.

Wernegreen, J.J., 2012. Endosymbiosis. Curr. Biol. 22, R555–R561.

White, T.A., Perkins, S.E., 2012. The ecoimmunology of invasive species. Funct. Ecol. 26, 1313–1323.

Whiteman, N.K., Parker, P.G., 2005. Using parasites to infer host population history: a new rationale for parasite conservation. Anim. Conserv. 8, 175–181.

Wilfert, L., Jiggins, F.M., 2014. Flies on the move: an inherited virus mirrors Drosophila melanogaster's elusive ecology and demography. Mol. Ecol. 23, 2093–2104.

Wilson, J.R.U., Dormontt, E.E., Prentis, P.J., Lowe, A.J., Richardson, D.M., 2009. Something in the way you move: dispersal pathways affect invasion success. Trends Ecol. Evol. 24, 136–144.

Wirth, T., Meyer, A., Achtman, M., 2005. Deciphering host migrations and origins by means of their microbe. Mol. Ecol. 14, 3289–3306.

Wolfe, B.E., Rodgers, V.L., Stinson, K.A., Pringle, A., 2008. The invasive plant *Alliaria petiolata* (garlic mustard) inhibits ectomycorrhizal fungi in its introduced range. J. Ecol. 96, 777–783.

Yang, C.-C., Yu, Y.-C., Valles, S.M., Oi, D.H., Chen, Y.-C., Schoemaker, D., Wu, W.-J., Shih, C.-J., 2010. Loss of microbial (pathogen) infections associated with recent invasions of the red imported fire ant *Solenopsis invicta*. Biol. Invasions 12, 3307–3318.

Zabalou, S., Riegler, M., Theodorakopoulou, M., Stauffer, C., Savakis, C., Bourtzis, K., 2004. *Wolbachia*-induced cytoplasmic incompatibility as a means for insect pest population control. Proc. Natl. Acad. Sci. U.S.A. 101, 15042–15045.

Zavaleta, E.S., Hobbs, R.J., Mooney, H.A., 2001. Viewing invasive species removal in a whole-ecosystem context. Trends Ecol. Evol. 16, 454–459.

Zchori-Fein, E., Brown, J.K., 2002. Diversity of prokaryotes associated with *Bemisia tabaci* (Gennadius) (Hemiptera: Aleyrodidae). Ann. Entomol. Soc. Am. 95, 711–718.

Zélé, F., Nicot, A., Duron, O., Rivero, A., 2012. Infection with *Wolbachia* protects mosquitoes against *Plasmodium*-induced mortality in a natural system. J. Evol. Biol. 25, 1243–1252.

Zhang, Q., Yang, R., Tang, J., Yang, H., Hu, S., Chen, X., 2010. Positive feedback between mycorrhizal fungi and plants influences plant invasion success and resistance to invasion. PLoS One 5. e12380.

Zhang, B., Edwards, O., Kang, L., Fuller, S., 2014. A multi-genome analysis approach enables tracking of the invasion of a single Russian wheat aphid (*Diuraphis noxia*) clone throughout the New World. Mol. Ecol. 23, 1940–1951.

Zindel, R., Gottlieb, Y., Aebi, A., 2011. Arthropod symbioses: a neglected parameter in pest- and disease-control programmes. J. Appl. Ecol. 48, 864–872.

Zug, R., Hammerstein, P., 2012. Still a host of hosts for *Wolbachia*: analysis of recent data suggests that 40% of terrestrial arthropod species are infected. PLoS One 7. e38544.

Zwart, M.P., van der Werf, W., van Oers, M.M., Hemerik, L., van Lent, J.M.V., de Visser, J.A.G.M., Vlak, J.M., Cory, J.S., 2009. Mixed infections and the competitive fitness of faster-acting genetically modified viruses. Evol. Appl. 2, 209–221.

CHAPTER FOUR

Massively Introduced Managed Species and Their Consequences for Plant–Pollinator Interactions

B. Geslin*,1, B. Gauzens†, M. Baude‡, I. Dajoz§, C. Fontaine¶, M. Henry‖, L. Ropars*,§, O. Rollin#,**, E. Thébault§, N.J. Vereecken††

*Institut Méditerranéen de Biodiversité et d'Ecologie marine et continentale (IMBE-UMR-CNRS-IRD 7263), Equipe Ecologie de la Conservation et Interactions Biotiques, Aix Marseille Univ, Univ Avignon, CNRS, IRD, IMBE, Marseille, France
†German Centre for Integrative Biodiversity Research (iDiv) Halle-Jena-Leipzig, Leipzig, Germany
‡Laboratoire Biologie des Ligneux et des Grandes Cultures (EA 1207), Equipe entomologie et biologie intégrée, Université d'Orléans, Orléans, France
§Institut d'écologie et des sciences de l'environnement de Paris (iEES-Paris UMR CNRS 7618), Equipe Ecologie et évolution des réseaux d'interactions, Université Paris-Diderot–CNRS—UPMC, Paris, France
¶Centre d'Ecologie et des Sciences de la Conservation (CESCO UMR CNRS 7204), Equipe Socio-écosystèmes, CNRS-Muséum national d'histoire naturelle, Paris, France
‖INRA, UR406 Abeilles et Environnement, Avignon, France
#ITSAP-Institut de l'Abeille, Avignon, France
**UMT PrADE, Avignon, France
††Agroecology & Pollination Group, Landscape Ecology & Plant Production Systems (LEPPS/EIB), Boulevard du Triomphe CP 264/2, Université Libre de Bruxelles (ULB), Brussels, Belgium
1Corresponding author: e-mail address: benoit.geslin@imbe.fr

Contents

Abstract

Since the rise of agriculture, human populations have domesticated plant and animal species to fulfil their needs. With modern agriculture, a limited number of these species has been massively produced over large areas at high local densities. Like invasive

Advances in Ecological Research, Volume 57
ISSN 0065-2504
http://dx.doi.org/10.1016/bs.aecr.2016.10.007

147

species, these Massively Introduced Managed Species (MIMS) integrate local commu-nities and can trigger cascading effects on the structure and functioning of ecosystems. Here, we focus on plant and insect MIMS in the context of plant–pollinator systems. Several crop species such as mass flowering crops (e.g. *Brassica napus*) and domesti-cated pollinating insects (e.g. *Apis mellifera*, *Bombus terrestris*) have been increasingly introduced worldwide and their impact on natural communities is addressed by an increasing number of scientific studies.

First, we review the impacts of major insect and plant MIMS on natural comm-unities by identifying how they affect other species through competition (direct and apparent competition) or facilitation (attraction, spillover). Second, we show how MIMS can alter the structure of plant–pollinator networks. We specifically analysed the posi-tion of *A. mellifera* from 63 published plant–pollinator webs to illustrate that MIMS can occupy a central position in the networks, leading to functional consequences. Finally, we present the features of MIMS in sensitive environments ranging from oceanic islands to protected areas, as a basis to discuss the impacts of MIMS in urban context and agrosystems. Through the case study of MIMS in plant–pollinator interactions, we thus provide here a first perspective of the role of MIMS in the functioning of ecosystems.

1. INTRODUCTION

Since the rise of agriculture, humans have selected and introduced plant and animal species in their environment to cover their needs. A set of species has thus been favoured inside and outside their native geo-graphic ranges. With the agricultural intensification of the 20th century, an unprecedented amplification in the breeding of these species has been observed (Hoekstra and Wiedmann, 2014; MEA, 2005). Here we define as Massively Introduced Managed Species (hereafter MIMS) all plant and animal species introduced voluntarily and abundantly in a given location for agricultural and/or domestic purposes. We specifically focus on MIMS involved in plant–pollinator interactions because of their critical impor-tance for agricultural production but also for native plant reproduction in natural or urban habitats (see Boxes 1–6). Indeed, the cultivated area of pollinator-dependent crops (i.e. Mass Flowering Crops—hereafter MFC) has strongly expanded in relation with the increase in their demand (Aizen and Harder, 2009), notably for human food supply (Eilers et al., 2011), but also for biofuels (Stanley and Stout, 2013). In parallel, the demand for biotic pollination has increased to ensure sufficient MFC yield (Aizen et al., 2008) and this has been amplified with the current worldwide decline of wild pollinators (Goulson et al., 2015; Potts et al., 2010;

BOX 1 Oceanic Islands

Most small (several tens to hundreds of km²) oceanic islands and archipelagos are areas of exceptional concentration of endemic species (Brooks et al., 2006; Myers et al., 2000). Species communities on islands are usually fragile and sensitive to introductions and invasions (Kaiser-Bunbury et al., 2010, see also Massol et al., 2017). The reported cases of native species extinctions after introduction of alien species on islands are usually due to predation or pathogens rather than exploitative competition (Sax and Gaines, 2008; Sugiura, 2016). Still, exploitative competition with introduced species can reduce the fitness and lead to local extinctions of the native island species, including bees (Sugiura, 2016).

Several cases of exploitative competition for floral resources have been evidenced or suspected in oceanic island systems of various sizes. In the small oceanic Bonin archipelago (Western Pacific, four island groups totalling 106 km² only), native bees suffered competition with the European honeybee introduced in the 1880s for beekeeping (Kato et al., 1999). Some of them became rare or locally extinct on islands on which managed and feral honeybees became dominant on native flowers, even in well-conserved forest tracts. This was most probably due to an increased competition for nectar and pollen. Given the drastic change of pollination networks in this case study, conservation biologists recommend the removal of managed honeybees and eradication of feral colonies to restore native bee populations.

Similar cases were inferred in other oceanic island systems of various sizes, including the Canary islands for honeybees (Dupont et al., 2004) or Hawaii (Miller et al., 2015)—see also Hansen et al. (2002) for negative effects of introduced honeybees on endemic nectarivorous birds in Mauritius, and Kaiser-Bunbury et al. (2010) for additional insular examples involving introduced bumblebees.

The effects of honeybee introductions on small oceanic islands are, however, difficult to assess due to the lack of suitable control sites without honeybees (Dupont et al., 2004; Hansen et al., 2002). In some insular contexts in which native bees are locally rare or extinct, the introduction of honeybees may help overcome the pollination deficiency of native plants (Hanna et al., 2012), but can also promote the establishment and development of invasive plant populations (Abe et al., 2011).

Vanbergen, 2013). To compensate for the losses of wild pollinators, or lack thereof in intensively managed farmland areas, modern agricultural practices usually rely on massively introduced managed pollinators (mainly *Apis mellifera* and *Bombus* spp.) that sometimes become the unique pollinating species of the targeted crops (Cunningham et al., 2016).

BOX 2 Natural and Protected Habitats

As outlined by Torné-Noguera et al. (2015), there is currently no specific legisla-
tion for the preventive exclusion or limitation of beekeeping activity in most
protected areas worldwide. Beekeeping usually conveys the positive image of
a traditional breeding activity that contributes to plant community sustainability
through pollination. Therefore, some European countries support agri-
environmental schemes (AES) promoting the seasonal introduction of apiaries
in "interesting biodiversity areas" through subsidies to beekeepers. Yet, this naïve
reasoning may be inappropriate in most situations and should be reevaluated
with respect to the context of interest. Protected areas are classified into different
categories (e.g. strict nature reserves, Wilderness areas, National parks, Habitat or
species management areas; Dudley, 2008), depending on their objectives, history, anthropogenic influence, governance and management, or on the outstanding species diversity or aggregation they hold. The beekeeping
innocuousness is unlikely to be supported in the protected areas of highest
concern. For instance, 21 apiaries (475 colonies) are held in the 32 km^2 El Garraf
Natural Park, Spain, leading to significant reductions in the pollen and nectar
availability, and eventually in the biomass of the native bee community
(Torné-Noguera et al., 2015). Authors estimate that native bee communities
are likely to be affected at densities beyond 3.5 honeybee colonies per km^2. They,
however, outline the difficulty to establish consistent recommendations for maximal admissible honeybee colony density given the highly heterogeneous nature
of floral resource carrying capacity in space and time. Similarly, Shavit et al. (2009)
found significant decreases of native bee flower visitation rates, at local scales,
when introducing ten honeybee colonies in two Israelian Natural Parks of a
few km^2. They highlighted the possible aggravation of competition outcomes
during drought years and seasonal food scarcity periods, and recommended
keeping the nature reserves in Israel out of bounds for honeybee colonies.

As any new species that integrates a natural community, the introduction
of MIMS can potentially lead to modifications of the interactions among
other cooccurring species, the structure of networks and ultimately the functioning of ecosystems (Tylianakis, 2008; Tylianakis et al., 2010, see also
Fig. 1). While there are several studies focusing on invasive species and their
impacts on plant and pollinator communities (e.g. Aizen et al., 2008; Stout
and Morales, 2009; Traveset and Richardson, 2011, 2014, see the meta-
analysis of Mollot et al., 2017), and on interaction networks (e.g. Aizen
et al., 2008; Morales and Aizen, 2006; Stouffer et al., 2014; Traveset and
Richardson, 2014), few studies have investigated the potential impacts of

BOX 3 Agrosystems

Agrosystems are highly disturbed landscapes, characterized by a heterogeneous mosaic of habitats, strong spatial and temporal instability (Garder, 1996) and substantial reduction of the quality and quantity of seminatural areas (Benton et al., 2003; Duelli and Obrist, 2003; Tscharntke et al., 2005).

Agrosystems provide high concentrations of both plant and animal MIMS. Many MFC need insect pollination (Klein et al., 2007; Williams, 1994). Even though non-bee insect pollinators play a significant role in crop production, they are less efficient that bees per flower visit (Rader et al., 2009, 2015). Managed honeybees are the most economically valuable pollinators for monofloral crops and some seed, fruit and nut crops worldwide (Klein et al., 2007; Morse and Calderone, 2000), but other managed bee species are widely used owing to their more specific pollination service, such as bumblebees for tomatoes (Graystock et al., 2016; Van Engelsdorp and Meixner, 2010; Velthuis and Van Doorn, 2006). However, wild native species are highly efficient pollinators, especially the most common and abundant species (Kleijn et al., 2015; Rader et al., 2009) and complementarity between managed and wild native bee species can increase quality of MFC pollination (Garibaldi et al., 2013, 2016).

In intensive agricultural habitats, feeding resources for pollinators are mainly provided by wild flowering plants in seminatural remnants (e.g. grasslands, field margins, hedgerows; Ricketts et al., 2008; Rollin et al., 2013; Steffan-Dewenter and Tscharntke, 2001) and MFC for a short duration (Herrmann et al., 2007; Westphal et al., 2009). While native wild bees are mostly observed in seminatural habitats (Rollin et al., 2013, 2015), managed honeybees have strong preferences for MFC and tend to use seminatural areas less intensively when alternative floral resources increase in the landscape (Henry et al., 2012; Rollin et al., 2013). However, when MFC are not available, honeybees switch to natural areas with a significant increase of their abundance in these habitats (Rollin and Decourtye, 2015). During this period of food scarcity, wild bee communities show a significant decrease of their local scale diversity as compared to MFC flowering periods, while their local abundance and landscape-scale species diversity are the highest (Rollin and Decourtye, 2015; Rollin et al., 2015). These results provide indirect evidence that honeybees outcompete native bees in agrosystems during periods of food scarcity, with a spatial reorganization of wild bee species at the local scale in seminatural remnants.

In intensive agricultural landscapes, farming practices more friendly to bees are needed to alleviate potential competition for floral resources between wild and managed bee species. For example, intercropping practices with plant mixtures of interest for honeybees and beekeepers are being developed, even if their implementation still generates various agronomic constraints (Biniaś et al., 2015; Labreuche and Tosser, 2014).

BOX 4 Emergent MIMS—Sown Flower Strips

The intensification of agriculture has caused dramatic declines in farmland bio-diversity (Carvalheiro et al., 2013; Senapathi et al., 2015). Since the 1990s, agricultural policies have been developed in Europe to mitigate this loss through agri-environmental schemes (AES). One AES is "sown wildflower strips", the aim of which is to create new ecological infrastructures by sowing attractive wild flowers on arable land (a few % of the cultivated area). These ecological infrastructures fall within our definition of MIMS since they represent a massive introduction of managed species in the landscape.

Wildflower strips usually include mixtures of annual and biennial flowering species known to offer pollen and nectar rewards. Their aim is to promote pollination services, biological pest control, plant diversity and to support farmland bird populations by providing seeds as food resources and invertebrates as preys (Dicks et al., 2014).

Evaluations of the effects of wildflower strips in situ have yielded mixed results with only marginal to moderately positive effects of these AES on biodiversity (e.g. Albrecht et al., 2007; Dicks et al., 2014; Kleijn et al., 2006). Most studies acknowledged that common insects are the main beneficiaries of wildflower strips, which were not designed to have positive impacts on threatened or specialized species (Potts et al., 2006). Haaland et al. (2011) showed that sown wildflower strips support higher insect abundances and diversity than cropped habitats or other field margin management types such as sown grass margins and natural regeneration. These effects are perceptible at the plot scale but also in the wider landscape (Jönsson et al., 2015).

Although these results are encouraging, they also highlight the challenges ahead for the restoration of plant–pollinator interactions. Agro-ecological management practices must take into account common, but also rare species that require more conservation efforts. This "community approach" to pollinator conservation needs a stronger focus on the botanical composition of sown wildflower strips which should be tailored to the identity of pollinators and their ecological requirements (Blaauw and Isaacs, 2014; Blackmore and Goulson, 2014; Garratt et al., 2014; Tschumi et al., 2016) without systematically overlooking the need to conserve *all* species of pollinators (Kleijn et al., 2015).

MIMS (see Gill et al., 2016). The term MIMS refers both to alien species (*Bombus terrestris* later) and to geographically native ones (*Brassica napus* see later). Thus, whether native or alien, MIMS are plants or animals voluntarily introduced in high quantities for the sake of human needs. This might partly explain the lack of consideration of their potential effects on ecosystems. For

BOX 5 Buzz in the City

Cities are scheduled to increase by more than 250% in the next 15 years, with urbanization being one of the main drivers of habitat fragmentation and associated biodiversity losses (Concepcion et al., 2015; Geslin et al., 2016a; Nieto et al., 2014). Many towns are taking into account the importance of biodiversity and try to set up conservation measures. However, regarding pollinators, the public policies and awareness campaigns are mainly focused on the introductions of numerous colonies of *A. mellifera*. Urban inhabitants associate pollination and *A. mellifera* colonies to the quality of their living environment, with beekeeping becoming increasingly popular in cities (Geslin et al., 2013). Honeybees have thus become a symbol of biodiversity for the general public, which reinforces the growth of urban apiculture. For example, in Paris (an area of $\sim$105 km^2), more than 700 hives have been introduced in the last years (http://www.paris.fr/actualites/paris-se-mobilise-pour-les-abeilles-3488).

An important diversity of wild bees might persist in cities and competition and/or spillover might occur between these species and *Apis mellifera*. Indeed, urban habitats support a relatively high biodiversity of wild bees (Poznan, 104 spp., Banaszak-Cibicka and Żmihorski, 2012; 291 spp. in the large urban area of Lyon Fortel et al., 2014; Paris, 44 spp., Geslin et al., 2016b; Brussels >100 spp., N. Vereecken et al., unpublished data). Therefore, caution should be taken when introducing high densities of *A. mellifera* in cities. Torné-Noguera et al. (2015) showed a negative impact on wild bee abundance when *A. mellifera* reached a density of 3.5 colonies/km^2 in a natural park in Spain. The city of Paris already harbours twice this density of beehives ($\sim$7 colonies/km^2) and the city of London even more ($\sim$10 colonies/km^2; Alton and Ratnieks, 2016).

We argue here that MIMS in cities might be a counter-productive conservation measure for urban biodiversity, creating new pressures for wild species instead of preserving them. Other conservation practices might be developed in cities towards plant and pollinators such as sowing flowering communities and ecological gardening (Blackmore and Goulson, 2014), reduced use of pesticides (Muratet and Fontaine, 2015) or installing nesting habitats for pollinating insects (Fortel et al., 2016). We do not seek to oppose *Apis* and *non-Apis* conservation strategies (Aebi et al., 2012), but a priority is to determine the optimal densities of domesticated and wild pollinators to preserve both biodiversity and beekeeping activities. Additional work on this topic is strongly and urgently needed.

BOX 6 Emergent MIMS—*Megachile*

The most intensively managed wild bee is certainly the Leafcutting bee *Megachile rotundata* (Pitts-Singer and Cane, 2011). Mostly used in North America (Canada, USA) for the pollination of alfalfa, whose production increases by 50% as a result of *M. rotundata* pollination, this solitary bee species has gregarious nesting habits facilitating its commercialization. *M. rotundata* fits in the definition of a MIMS, with an easy and cheap production allowing mass commercialization (the density of individuals may reach 150,000 bees/ha during alfalfa flowering, Pitts-Singer and Bosch, 2010; Pitts-Singer and Cane, 2011). *M. rotundata* has nesting requirements similar to those of many wild species in the United States, which could foster a competition for nesting sites (Barthell et al., 1998, see also Pitts-Singer and Cane, 2011). However, in the study of Barthell et al. (1998), *M. rotundata* only occupied a small percentage (3–4%) of monitored artificial nests. Donovan (1980) argued that the competition for nesting resources between *M. rotundata* and native Hylaeinae might be very unlikely as nesting holes could not be a limiting factor.

To date, about 20 species of social and solitary bees are commercially reared for pollination services (Stout and Morales, 2009). To a lesser degree *Osmia* species are used to pollinate fruit crops such as pears and apples, including *Osmia cornuta* (Maccagnani et al., 2007), *Osmia cornifrons* (Matsumoto et al., 2009) and *Osmia lignaria* (Sheffield, 2014). The magnitude of introductions in the orchards varies from a few hundred individuals (Sheffield, 2014) to a few thousands (four boxes of 600 individuals in two pear orchards in Maccagnani et al., 2007).

Finally, many alternatives to classic commercialized pollinators have been developed in the last 20 years and the development of new commercial pollinator species is increasingly promoted by local managers. This is the case for stingless bees such as *Trigona* and *Melipona* species, which are increasingly used for greenhouse pollination (reviewed in Slaa et al., 2006).

instance, MFC have positive economic impacts and managed bees are positively perceived thanks to their role in the production of honey and in the pollination of plants (Goulson, 2003). Yet, both insect and plant MIMS can interact with other wild species, rearrange pollination networks at the landscape scale (Spiesman and Gratton, 2016), and either facilitate or impair interactions with coflowering wild plant communities (Holzschuh et al., 2011; Rollin et al., 2015).

Here we review the literature to analyse how MIMS can integrate plant–pollinator systems. First, we introduce the main plant and animal MIMS

Without hives

Solitary bees   Bumblebees   Diptera   Honeybees   Other groups

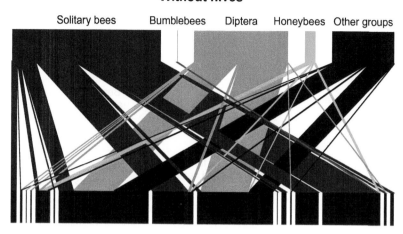

With hives

Solitary bees   Bumblebees   Diptera   Honeybees   Other groups

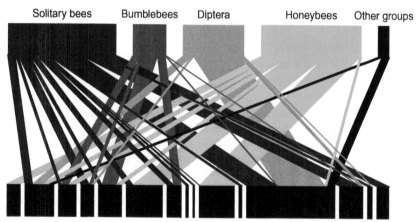

Fig. 1 Modifications of a plant–pollinator network after the introduction of five *Apis mellifera* colonies in an experimental site in Paris. *Top squares* represent flower–visitor morphogroups and *bottom squares* represent flowering plant species. The link widths represent the number of observed interactions between visitors and flowering plants. *Data retrieved from L. Ropars et al. (unpublished data).*

involved in pollination processes and we review the mechanisms by which they interact with other species. We analyse their role as either direct and indirect competitors, or facilitators in plant–pollinators communities (Fig. 2). In a second part, we analyse a published dataset of plant–pollinator networks (Fontaine and Thébault, 2015; Thébault and Fontaine, 2010) to

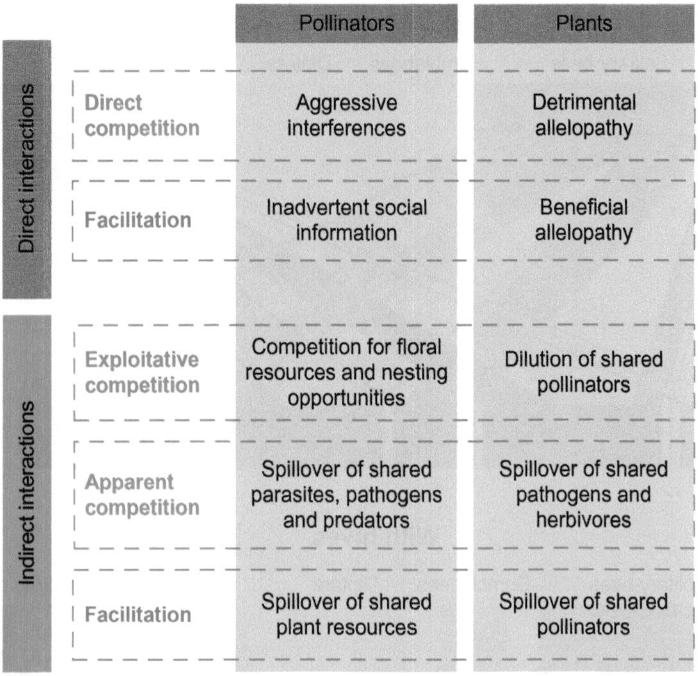

	Pollinators	Plants
Direct interactions		
Direct competition	Aggressive interferences	Detrimental allelopathy
Facilitation	Inadvertent social information	Beneficial allelopathy
Indirect interactions		
Exploitative competition	Competition for floral resources and nesting opportunities	Dilution of shared pollinators
Apparent competition	Spillover of shared parasites, pathogens and predators	Spillover of shared pathogens and herbivores
Facilitation	Spillover of shared plant resources	Spillover of shared pollinators

Fig. 2 Synthesis of possible ecological interactions between MIMS and wild pollinators or plants (see text for definitions). The introduction of managed plant or pollinator species may interfere with native interaction networks following a variety of ecological processes. Some of them are referred to as *direct* interactions because they occur with the simultaneous physical presence of both introduced and native organisms (e.g. aggressive interferences). Others are termed *indirect* interactions because they are indirectly mediated by a third party such as resources in exploitative competition or pathogens and predators in spillover events. MIMS may exert either detrimental (competitive) or beneficial (facilitative) effects on native interaction networks. Facilitation typically occurs when the focus native network already expressed any deficiency at the time of managed species introduction.

understand the position of MIMS in mutualistic networks and their impacts on network structure in terms of nestedness, connectance and modularity; and to infer their potential consequences on community functioning. We further present in detail several case studies on the introduction of MIMS in sensitive environments [islands (Box 1), natural and protected areas (Box 2), seminatural areas in agrosystems (Box 3)]. Finally, we also discuss the avenues of future research on new management practices and environmental policies such as sown flower strips (Box 4) and urban beekeeping (Box 5) as well as emergent MIMS (Box 6), and how those practices may lead to the spread of MIMS.

2. FIRST PART: IMPACTS OF MIMS IN PLANT AND POLLINATOR COMMUNITIES

2.1 The Case of Pollinators

2.1.1 Pollinating MIMS

2.1.1.1 Introduction

The use of managed pollinators to pollinate crops has become more and more common with the increased dependency of agriculture crop production on animal pollination (Aizen et al., 2008; Lautenbach et al., 2012). Further, numerous managed honeybee colonies are brought into agricultural landscapes for honey production (Graystock et al., 2016; Morse, 1991). This translates into important fluxes of pollinators into agricultural and adjacent seminatural habitats during the flowering season of crops (Box 3). Globally, these managed pollinator species represent around 20 species among all social and solitary bee species (Stout and Morales, 2009). Three of these species, *A. mellifera*, *B. terrestris* and *Megachile rotundata*, have particularly extended their geographic range either due to voluntary or accidental introduction (Goulson, 2003; Box 6).

2.1.1.2 The Honeybee, *A. mellifera*

A. mellifera is used for honey production and to pollinate MFC such as almonds (Cunningham et al., 2016), mango (Geslin et al., 2016c), apples (Ramírez and Davenport, 2013), pears (Stern et al., 2004) and many others (Garibaldi et al., 2013, 2016).

Contrary to other *Apis* species that are restricted to Asia,[a] the native geographical range of *A. mellifera* spans from Scandinavia to Central Asia and the African continent (Ruttner, 1988; Seeley, 1985; Sheppard and Meixner, 2003). However, its current geographical distribution encompasses almost all habitats on the planet, at the exception of deserts and areas of permafrost. Such a wide distribution is largely due to deliberate human introduction (Crane, 1975, 1999) by settlers on every newly colonized continent, combined with the fact that honeybees are highly polylectic (pollen generalist) with a very wide ecological (here, dietary) niche. Evidence from behavioural and palynological studies reveal that although workers may regularly specialize on the collection of pollen from a single locally abundant host plant, at

---

[a] Note that *A. florea* has recently expanded its geographic range in Jordan (Haddad et al., 2009) and is now spreading in Sudan (Lord and Nagi, 1987) and Ethiopia (Pauly and Hora, 2013).

the colony level, they collect pollen from a very wide taxonomical range of flowering plants (e.g. Requier et al., 2015).

2.1.1.3 The Large Earth Bumblebee, *B. terrestris*

B. terrestris is currently the most widespread managed bumblebee species for pollination services. Since the 19th century, colonies of *B. terrestris* have been introduced in many parts of the world for commercial purpose (Graystock et al., 2016; Lye et al., 2011). It has originally a West Palearctic distribution, but populations are now present from the tip of the Argentinean Patagonia (Geslin and Morales, 2015) to the North of the Arctic Circle (Martinet et al., 2015). This bumblebee species also owes its success to its foraging skills. As summarized in Dafni and Shmida (1996), *B. terrestris* is a generalist pollinator, capable of foraging at low temperatures (Stelzer and Chittka, 2010), visiting deep corollas, presenting a variety of foraging behaviours (buzz pollination; nectar robbing), and is relatively easy to breed and to produce commercially (Velthuis and Van Doorn, 2006). *B. terrestris* does not produce honey, so it is used solely for pollination service in greenhouses, orchards (e.g. apples, pears, raspberry and avocado) or other highly valuable crops (e.g. peppers) to replace or supplement honeybee populations (Dafni et al., 2010; Goulson, 2003; Goulson and Hughes, 2015).

Today, the mass commercialization of *B. terrestris* colonies reaches more than two millions of traded colonies per year, representing an economic value of ~€55 million (Graystock et al., 2016; Velthuis and Van Doorn, 2006). Following this massive global trade of *B. terrestris* colonies, many countries reported cases of invasions (mostly due to queens escaped from greenhouses) such as in New Zealand, Israel, Japan, South Korea, China, Chile and Argentina (see Acosta et al., 2016 for a review). Today, *B. terrestris* is considered as one of the most globally invasive bee species, and, given the forecasts provided by species distribution modelling, the invasion is likely to amplify (Acosta et al., 2016).

In addition to *B. terrestris*, four other bumblebee species are reared worldwide and commercially distributed in high quantities for pollination service: *B. lucorum*, *B. occidentalis*, *B. ignitus* and *B. impatiens* (Velthuis and Van Doorn, 2006). The production is, however, substantially lower than for *B. terrestris*, with 55 000 colonies produced per year for *B. impatiens*, the second most commonly reared species. Despite warnings regarding its use outdoor in nonnative areas, *B. impatiens* is used for pollination services as an alternative to *A. mellifera* (Artz and Nault, 2011) either in open fields or in greenhouse like in Mexico (Velthuis and Van Doorn, 2006). More recently, other bumblebee species have been used for pollination service, with a special interest in favouring local species including *B. patagiatus* in China (Zhang et al., 2015) or *B. hypocrita* in Japan (Williams et al., 2012).

2.1.2 Direct Interactions Between MIMS and Wild Pollinators
2.1.2.1 Interference Competition

There is little evidence of interference competition through aggressive behaviour between pollinating MIMS and wild pollinators. In Mexico, Cairns et al. (2005) observed competitive behaviour between *A. mellifera* and stingless bees, including aggressive contacts between species. There are descriptions of pollen theft by *A. mellifera* workers directly from bodies of native bees (*Bombus* and *Megachile* species), but these are considered as rare events (Jean, 2005; see also on *Bombus*, Brian, 1957; Inouye, 1978). Studies on commercial hives in New Zealand have shown that workers of *A. mellifera* could reduce wasp densities through aggressive interaction in forests where both bees and wasps foraged on honeydew exudates of the scale insect *Ultracoelostoma brittini* (Hempitera: Margarodidae) (Markwell et al., 1993).

Conversely, Frankie et al. (2005) did not observe any aggressive behaviour between *Apis* and non-*Apis* during a 5-year survey of bee visitation within gardens in a city of California (USA). Roubik (1978) reported on exploitative competition between honeybees and wild stingless bees, but no aggressive behaviour between bees for resources of either natural or artificial flowers. Studies on other MIMS such as the invasive *B. terrestris* on the island of Tasmania did not provide evidence of aggressive behaviour towards the local wild bee fauna (Hingston and McQuillan, 1999).

Overall, these studies show that aggressive interactions between MIMS pollinators on the one hand, and other unmanaged wild bees or wasps on the other hand are relatively seldom reported in the scientific literature. This suggests that interference competition with MIMS pollinators is unlikely to have significant consequences on unmanaged pollinating species.

2.1.2.2 Facilitation Through Inadvertent Social Information

Interactions between pollinator MIMS and wild pollinators can also involve social information (Danchin et al., 2004). Insect pollinators have been shown to use social information to optimize their foraging, either to locate rewarding patches (Baude et al., 2011) or to avoid predation (Dawson and

Chittka, 2014). Although mainly studied at intra-specific level, information flow among pollinators also occurs between species (Dawson and Chittka, 2014; Goodale and Nieh, 2012). For example, Goodale and Nieh (2012) showed that *B. impatiens* was able to interpret *A. mellifera* alarm signals (presence of *A. mellifera* haemolymph) as an indication of a predation event. It remains to be investigated if massively introduced pollinators pheromones may modify wild pollinator behaviours (interactions with flower or predators avoidance).

2.1.3 Indirect Interactions Between MIMS and Wild Pollinators

As MIMS and wild pollinators are part of the same interaction network, the effects of one on the other can be indirect, i.e. mediated by shared partners, either plants leading to exploitative competition or parasites and pathogens leading to apparent competition.

2.1.3.1 Exploitative Competition: Competition for Floral Resources

As expressed by Stout and Morales (2009), exploitative competition for floral resources occurs when the consumption of limiting floral resources overlap between species, resulting in reduced population size, fecundity or survival for at least one of the interacting species.

We expect strong asymmetric competition for floral resources between the honeybee and wild bees for several reasons. First, there is a potential for strong numeric imbalance between wild and domestic bees. The number of workers within a single honeybee colony (20,000–60,000) is more or less equivalent to, or even surpasses, the local density of all wild bees (a viable wild bee population reaches several tens to hundreds of individuals). Second, honeybee colonies are able to harvest large quantities of floral resources, either nectar or pollen, from various plant species. The annual amount of pollen collected by a single honeybee colony ranges from 20 to 50 kg (see Vereecken et al., 2015). Given that a honeybee visits up to 80 flowers per trip (representing 8–20 mg of pollen), we can estimate that 80–200 million of flowers will be visited for pollen by a single colony annually. Thus, Cane and Tepedino (2016) recently estimated that, during 3 months, a strong honeybee colony gathers as much pollen as could produce 100,000 progeny of *M. rotundata*. Third, honeybee colonies remain active throughout the year (except during cold months), while the vast majority of solitary wild bee species are only active for few weeks or months. Finally, contrary to wild bees, which are often described with small foraging flight distances (100–500 m, Greenleaf et al., 2007; Zurbuchen et al., 2010), honeybees can cover large areas for resources (mean distance: 1.5 km; range from few metres to 10 km; Steffan-Dewenter and Kuhn, 2003).

Goulson and Sparrow (2009) showed that the introduction of honeybees decreased the average body size of native bumblebees, a phenomenon observed when a sudden shortage of local food resources occurs, e.g. towards the end of the season. Other studies found that the maintenance of honeybee colonies had a significant impact on the fitness of bumblebees as measured by their colony weight, and the number and the weight of their reproducing caste (Elbgami et al., 2014; Thomson, 2004). Also detrimental effect of competition for resources between honeybee and bumblebee could be exacerbated by unfavourable climatic conditions (Thomson, 2016) or reduce resources availability in homogeneous landscapes (Herbertsson et al., 2016).

The introduction of honeybee colonies can also decrease the visitation frequency of wild bees on plants, as reported by studies on Mediterranean plants (Shavit et al., 2009), in the Japanese archipelago of Okasawara (Kato et al., 1999), in Australia (Paini and Roberts, 2005) and in Mexico (Badano and Vergara, 2011). Wild bees competing for floral resources with honeybees might also shift towards alternative plants that are sometimes taxonomically and chemically distant from their preferred plants (Roubi and Villanueva-Guttiérez, 2009; Schaffer et al., 1983; Thorp, 1996; Walther-Hellwig et al., 2006). While this shift in visited plants might not hinder the local maintenance of polylectic species, which can exploit alternative food sources, it might represent a major obstacle for oligolectic species. Indeed, a shift towards plants belonging to different genera or families is strictly impossible for the latter, since they are physiologically and behaviourally adapted to a narrow spectrum of plants, which makes them overall more vulnerable (Biesmeijer et al., 2006) and prone to extinction (Nieto et al., 2014) in a context of land-use change affecting their host plant(s) (Biesmeijer et al., 2006; Carvalheiro et al., 2013).

However, competition for floral resources with honeybees will likely depend on the density of honeybee colonies. Steffan-Dewenter and Tscharntke (2000) reported a lack of competitive interactions between introduced honeybees and wild bees on a grassland with a density of 3.1 colonies/km^2. To date, this density is one of the very few threshold values (with Torné-Noguera et al., 2015; 3.5 colonies/km^2; see Box 2) recommended to conservation managers in natural protected areas. Yet, this study might have significantly underestimated competition because it focused on a single season, whereas the deleterious effects of competition through the analysis of the reproductive success of species interacting together may be detected several years after honeybee introduction. Moreover, this study only focused on polylectic species that nest in newly installed artificial shelters (bee hotels) and the effect on the community of soil-nesting bees was not investigated.

Many studies have reported high levels of resource overlap between *B. terrestris* and the local insect fauna such as in Argentina (*B. dahlbomii*; Morales et al., 2013), Japan (*B. ardens*, *B. hypocrita* and *B. diversus*; Dohzono et al., 2008; Inari et al., 2005; Matsumura et al., 2004; Nishikawa and Shimamura, 2016) and Tasmania (Hingston and McQuillan, 1998, 1999; Hingston et al., 2002). However, the presence of resource overlap does not necessarily indicate competition and evidence of such competition is difficult to get in natural conditions (reduced population, fecundity or survival of one or both species; Goulson, 2003).

A field removal experiment of *B. terrestris* queens conducted in Japan showed a subsequent increase of the queen's populations of the two *Bombus* species that overlapped with *B. terrestris* in their diet breadth (Nagamitsu et al., 2009). In Tasmania, Hingston and McQuillan (1999) found a reduced visitation rate of local *Megachile* (*Chalicodoma*) species and argued that a displacement of local species might be due to resource depletion by *B. terrestris*.

On the contrary, Goulson et al. (2002) found that, although *A. mellifera* presence was negatively correlated to wild bee presence, there was no correlation between the presence of *B. terrestris* and the abundance and richness of native pollinators. Similarly, Nishikawa and Shimamura (2016) and Nagamitsu et al. (2007) did not detect negative competitive effects between *B. terrestris* and local bumblebee species despite apparent resource overlaps.

The challenge to draw clear conclusions lies mostly in the difficulty to acquire information on the population dynamics of wild bees. Studies are often focused on correlative measures regarding spatial presence/absence of wild bees and MIMS (Stout and Morales, 2009) without direct measurement of bee fitness. Taken together, few studies truly described competition for floral resources between MIMS and wild bee species, especially oligolectic ones, and because such competition might occur in highly invaded locations, this question calls for future research in the near future.

2.1.3.2 Exploitative Competition: Competition for Nesting Opportunities

Although most studies on the competitive interactions between managed and wild bees have focused on floral resources, a few studies have also addressed the issue of the availability and partitioning of nesting sites between species.

Inoue et al. (2008) have suggested that competition for nesting sites is the main mechanism for the displacement of native bumblebees by B. terrestris. Likewise, Matsumura et al. (2004) have highlighted the similarity between B. terrestris and local bumblebee species nests. Morales et al. (2013) suggested a potential competition for nesting resources between B. terrestris and other bumblebee species. In the study by Inoue et al. (2008), most bumblebee nests (native and terrestris confounded) were constructed in abandoned rodent nests. Authors suggested that B. terrestris might outcompete and exclude other species by occupying available nests earlier in the year. However, the competition for floral or nesting resources is unlikely to explain the collapse of some bumblebee populations such as the sudden regression of the native B. dahlbomii after the introduction of B. terrestris in South America (Morales et al., 2013).

2.1.3.3 Apparent Competition: Spillover of Shared Parasites and Pathogens

MIMS and wild pollinators can also interact negatively through pathogen transmission from infected to sympatric healthy populations (pathogen spillover; Daszak et al., 2000). The recent review of Graystock et al. (2016) reveals that pathogen spillover between pollinating MIMS and wild pollinators might be a strong cause of wild bee decline worldwide (see also Fürst et al., 2014; Meeus et al., 2011).

To date, studies on pollinators' pathogens and viruses have focused on honeybees. Honeybees are infected by a diversity of pathogens, such as the Varroa mite, Nosema apis (Microsporidia), the deformed wing RNA virus (DWV), and Nosema ceranae. N. ceranae was until recently specifically associated to the Asian honeybee A. cerana, but it is now considered as one of the drivers behind the massive decline of honeybee in Spain and in other parts of Europe (Higes et al., 2006, 2010). Some studies have also shown that honeybee pathogens and viruses can infect other pollinator species. For example, the DWV has been reported in wild specimens of B. terrestris and B. pascuorum from Germany and the United Kingdom (Evison et al., 2012; Genersch et al., 2006), as well as in B. huntii in the United States (Li et al., 2011). Recent studies carried out in the United Kingdom have shown that DWV and N. ceranae infected simultaneously

honeybees and six native *Bombus* species. These infections appeared inti-
mately associated, suggesting pathogen spillover from the honeybee colonies to the wild bees.

Bumblebees, particularly *B. terrestris*, are infected by several other
pathogens such as *Apicystis bombi* (protozoan), *Crithidia bombi* (protozoan),
Nosema bombi (Microsporidia), *Locustacarus buchneri* (Acari) (Cameron
et al., 2016; Colla et al., 2006; Graystock et al., 2014, 2015, 2016;
Otterstater and Thompson, 2008; Stout and Morales, 2009). There is
now clear evidence of a high prevalence of parasites in commercially reared
bumblebees (Colla et al., 2006; Graystock et al., 2013a, 2015, 2016). Tests
on *B. terrestris* colonies retrieved from commercial supply showed that
25% of colonies were infected with both *Crithidia* spp. and *N. bombi*
(Murray et al., 2013) and 77% of colonies were infested by at least one
parasite (Graystock et al., 2013a). Parasite spillover between managed and
wild bumblebees has been highlighted in Japan, Europe, United States
and South America (Graystock et al., 2016). Indeed, wild bumblebees foraging close to greenhouses supplied with commercial bumblebees showed
higher prevalence of parasite infection than bumblebees collected further
away (Colla et al., 2006; Murray et al., 2013; Otterstater and Thompson,
2008). In Argentina, Arbetman et al. (2012) provided evidence for a link
between invasion of *B. terrestris* and infection by *A. bombi* in *B. ruderatus*
(alien) and *B. dahlbomii* (native). Parasite spillover consequences for wild pollinators could be huge and remain largely underestimated (Fürst et al., 2014),
because infections can strongly reduce fitness, survival and even foraging
behaviour in wild populations (Graystock et al., 2013b, 2016, Garibaldi
et al., 2016; Meeus et al., 2011). In Patagonia, the spectacular regression
of the endemic *B. dahlbomii* following *B. terrestris* invasion has been suspected
to be caused by pathogen spillover (Morales et al., 2013; Schmid-Hempel
et al., 2014). In the United States, Cameron et al. (2011) observed a correlation between *N. bombi* infections and declining populations of eight native
bumblebee species. The impact of individual drivers has received considerable interest from contemporary researchers, but the idea of synergistic
effects between individual drivers—a more realistic approach to the drivers
of decline—is increasingly discussed and investigated (Tylianakis et al.,
2008). This represents a particularly promising avenue for future research
not only on honeybees but also on non-*Apis* species, since several of
the prominent pathogens or antagonists have a worldwide distribution;
interactions between multiple drivers therefore occur de facto under field
conditions (see Chen and Siede, 2007).

2.1.3.4 Apparent Competition: Spillover of Shared Predators

Apparent competition between MIMS and wild pollinators might also occur through shared predators. The Asian Hornet *Vespa velutina nigrithorax* is an invasive species in Europe that mainly preys upon *A. mellifera* individuals (Villemant et al., 2006, 2011a). Analyses of prey residuals revealed that *V. velutina* can also prey upon wild Hymenoptera (Villemant et al., 2011b), and particularly on wild bees from the Halictidae family (Perrard et al., 2009). Thus, we might expect apparent competition between *A. mellifera* and wild bees mediated by *V. velutina* predation, but to date, no study has focused on the impact of *V. velutina* on wild pollinator populations.

2.1.3.5 Facilitation: Spillover of Shared Plant Resources

To our knowledge, there is no published evidence of facilitative effects between pollinating MIMS and wild pollinators. Such facilitative effects have been reported among coflowering plants to attract shared pollinators (e.g. Rathcke, 1983, see later). This could theoretically occur among pollinators, if a plant species is pollinated sequentially by a pollinating MIMS and a wild pollinator, leading to increased plant reproduction. Facilitative effects might also arise from asymmetry in plant–pollinator interactions: if a MIMS pollinator efficiently pollinates plant species upon which wild pollinators are dependent (Bascompte et al., 2006; Vázquez and Aizen, 2004). However, both *A. mellifera* and *B. terrestris* have long periods of foraging activity and these facilitative effects might be buffered by within season competition for resources with wild pollinators.

2.2 The Case of Plants

Here we review the impacts of plant MIMS on nearby natural plant communities, by examining their direct or indirect interactions through shared species (pollinators and herbivores/pathogens), and how these effects translate into the reproductive success and stability of wild plant communities. We focus on cultivated species of intensive farmland habitats, especially MFC such as oilseed rape (*B. napus*), sunflower (*Helianthus annuus*), field bean (*Vicia faba*), phacelia (*Phacelia tanacetifolia*), wild mustard (*Sinapis arvensis*), potato (*Solanum tuberosum*) and clover (*Trifolium* spp.). However, most of the following examples are extracted from studies based on oilseed rape fields, for which land cover has increased by 49.9% between 2000 and 2010 in Europe (European Commission, 2011), largely due to an increased demand for biofuel.

Given the increase in pollinator densities commonly ascribed to MFC, from the local to the landscape scale (Holzschuh et al., 2013; Jauker et al., 2012; Knight et al., 2009; Westphal et al., 2003, 2009; Box 3), it seems likely that competition or facilitation between coflowering plants via shared pollinators would be influenced by MFC. For example, in oilseed rape (MFC), a single field provides 350,000–700,000 plants per hectare, each producing more than 100 flowers (Hoyle et al., 2007) during a flowering period of about 4 weeks. This huge, temporary peak of resources available to pollinators can entirely rearrange pollination networks at the landscape scale (Spiesman and Gratton, 2016), and either facilitate or impair pollination networks with coflowering wild plant communities (Rollin et al., 2015).Thus, it may indirectly influence a coflowering plant reproduction via the modification of shared pollinators foraging activity (Carvalheiro et al., 2014).

2.2.1 Direct Interaction Between MIMS and Natural Plant Communities: Allelopathy

Direct interactions between crops and native plant communities can occur through competition for space or light and are thus not specific of MFC. On the contrary, allelopathy, which has been defined as "direct or indirect harmful (or beneficial) effects of one plant on another through the production of chemical compounds that escape into the environment" (Rice, 1984) has been reported for several MFC. Many allelochemicals from plant tissues, usually secondary plant products, can cause germination or growth inhibition of the target wild species. Most research on allelopathy has focused on the effect of interactions among weeds (Newman and Rovira, 1975), among crops (Hegde and Miller, 1990), but also among weeds and crops (Rice, 1984; Turk and Tawaha, 2003). Researchers have evaluated the allelopathic potential of crops for weed control in order to discover novel natural plant compounds with herbicidal properties (Massantini et al., 1977; Maun, 1977; Rawat et al., 2012). There are several examples of allelopathy from MFC to weeds. *Brassica* species have been frequently cited as allelopathic crops (Bell and Muller, 1973). For example, Turk and Tawaha (2003) experimentally demonstrated that a plant extract from *Brassica nigra* inhibited the germination, seedling length and weight of *Avena fatua*. There are also examples of allelopathic effects of other crops such as white Lupin (*Lupinus albus*), sweetcorn (*Zea mays*) and sunflower (*H. annus*) on the surrounding vegetation. Rawat et al. (2012) demonstrated the allelopatic effect of sunflower on a weed species (*Trianthema portulacastrum*) and Leather (1983) reported that dried sunflower leaf and stem tissues inhibited broadleaf weed seedling growth but had little effect on grass weeds. Even if it is difficult to disentangle allelopathy from other mutual plant interference effects such as competition for light, whether or not plant MIMS can influence the composition of natural plant communities within field boundaries through allelopathy remains to be investigated. In addition to the effects on seed germination or the development of vegetative parts in the targeted plants, plant competition also affects floral traits (corolla size, nectar content), i.e. the primary advertisement cues in plant–pollinator interactions (Baude et al., 2011; Flacher et al., 2015).

2.2.2 Indirect Interactions Between MIMS and Wild Plants
2.2.2.1 Facilitation: Spillover of Shared Pollinators
Facilitation of pollination takes place when the presence of a plant species increases the rate of pollinator visitation and reproductive success of a second species sharing pollinators. While facilitation through spillover of pollinators from natural habitats to agricultural ones has been well documented, there are fewer studies addressing the effects of pollinators from managed habitats on natural ones (Blitzer et al., 2012). Spillover from managed to natural habitats occurs if pollinators benefit from the high productivity of managed habitats and then move to natural habitats, which are assumed to be less productive in floral rewards for pollinators (pollen and nectar). In addition to Westphal et al. (2003), who reported that increased bumblebee densities in MFC resulted in higher visitation rates in *P. tanacetifolia*, other studies using phytometers[b] found that MFC facilitated a pollinator spillover. Cussans et al. (2010) found that fruit set of *Lotus corniculatus* increased when planted close to oilseed rape and Hanley et al. (2011) observed higher effectiveness of bumblebees visiting flower margins adjacent to MFC during the flowering period. However, the positive effect of MFC on native communities appears to be limited to a restricted number of insect species (Le Féon et al., 2010; Rollin et al., 2013, 2015) and to be limited in time (Hanley et al., 2011; Westphal et al., 2009). Contrary to studies on alien plant species, the underlying mechanisms of facilitation are rarely investigated. For example, in *Osmia bicornis* (= *O. rufa*), a wild mason bee species active at the time of the MFC flowering season, larger amounts of oilseed rape led to increased

---

[b] A phytometer being a plant or group of plants grown usually under controlled conditions and used as a measure of the responses to various environmental factors. Following Albrecht et al. (2007), phytometers are obligatory outcrossing plant species relying on insect for pollination and commonly used to estimate pollination success.

nest building and number of produced offspring, but no evidence was available concerning the spillover of these beneficial aspects on the pollination of natural habitats (Jauker et al., 2012). Overall, this spillover may occur in conditions where agricultural habitats are subject to biodiversity-friendly management practices such as organic farming. Orford et al. (2016) have shown that experimentally enhancing plant species diversity of conventional agricultural grasslands was associated to increases in functional diversity, species richness and abundance of pollinators. This, in turn, led to increased pollination efficiency of a wild phytometer species, the red campion (*Silene dioica*). Also, Hardman et al. (2016) focused on the impact of wildlife-friendly farming practices on the pollination service delivered to a non-MIMS phytometer species (*Eschscholtzia californica*) and showed that organic farming practices supported higher densities of flowers and a better fruit set through increased diversity of floral resources in crop habitats.

2.2.2.2 Exploitation Competition: Dilution of Shared Pollinators
Competition for pollination occurs when the presence of a plant species induces pollen limitation, leading to a reduced reproductive success, in a second species sharing pollinators.

Negative effects of coflowering plant species on flower visitation and reproductive success of a focal species are known for plants within a local community (e.g. Bell et al., 2005), but less evidence is available for coflowering plant species interacting with plant MIMS.

Holzschuh et al. (2011) analysed whether abundance of oilseed rape leads to transient dilution of pollinators in the landscape and to increased competition for pollination between this crop MIMS and wild plant species. They showed that bumblebee abundance in grasslands decreased when the proportion of oilseed rape within a 1 km radius increased from 5% to 15%, and this led to a 20% decrease in seed set of *Primula veris*, a coflowering wild plant species. However, proximate mechanisms for reduced reproductive success of native plant communities confronted to coflowering MIMS are rarely investigated. Following the examples of alien species literature, reduced reproductive success of native plant communities may be related to (i) reduced pollinator visitation to native plants in the presence of preferred cultivated plants, so that conspecific pollen deposition is diminished in native plants (Campbell and Motten, 1985), or (ii) switches of pollinators between cultivated and native flowers, which can lead to increased heterospecific pollen deposition and/or decreased conspecific pollen deposition (Morales and Traveset, 2009). This has been reported by Marrero et al.

(2016) who recorded an increase in heterospecific pollen deposition on stigmas of native species in landscapes subjected to agricultural management. Also, Diekötter et al. (2010) investigated changes in frequency of plant–pollinator interactions resulting from pollinator community shifts towards the species that would benefit from plant MIMS recurrent resource pulses. In a phytometer study with *Trifolium pratense*, these authors observed a change in bumblebee visitation behaviour and bumblebee communities composition associated with increased oilseed rape cultivation. Increasing amounts of oilseed rape in the landscape led to a decrease in long-tongued bumblebees visiting *T. pratense* plants that have deep corollas. The simultaneous increase of nectar robbing by short-tongued bees suggests that resource depletion is a likely explanation for the abundance decline of long-tongued bumblebees. However, this decline did not translate in an effect on seed set of *T. pratense*.

2.2.2.3 Apparent Competition: Spillover of Shared Herbivores or Pathogens

Apparent competition mediated by shared herbivores or pathogens is susceptible to occur between wild plants and plant MIMS (reviewed in Rand et al., 2006). Apparent facilitation between some MIMS and wild plants might also arise if they share predators/parasites of their herbivores or pathogens (Rand et al., 2006). Indirect effects mediated by antagonists (e.g. herbivores, pathogens) have been mainly investigated between wild and cultivated plants. Indeed, numerous studies have considered the spillover of herbivores, pathogens or natural enemies from natural to agricultural habitats and its consequences on crop yields (Blitzer et al., 2012; Power and Mitchell, 2004). Few studies have considered spillovers from agricultural areas to natural habitats, and the related indirect effects of plant MIMS on wild plants.

We expect strong apparent competition mediated by plant MIMS on wild plants due to differences in productivity between agricultural and natural habitats. Indeed, the high abundance of MIMS (see earlier) could maintain high abundance of herbivores, natural enemies and pathogens, and thus strongly affect native communities sharing these antagonist species. Several studies have emphasized the potential consequences of existing shared pathogens and viruses between crops and wild plants on native plant diversity and pathogen evolution (Burdon and Thrall, 2008; Jones and Coutts, 2015). A few studies have shown herbivore spillover and consequent apparent competition from crops to natural vegetation (Blitzer et al., 2012). McKone et al. (2001) found that adult corn-rootworm beetles (*Diabrotica barberi*), the larvae of which feed on corn, had higher densities in tall-grass prairie located close to fields and reduced the seed set of native species in these endangered tall-grass ecosystems. More recently, Squires et al. (2009) also showed the spillover of diamondback moth (*Plutella xylostella*) from agricultural Brassicaceae to wild native Brassicaceae species (*Braya longii* and *Braya fernaldii*), leading to a 60% decrease in seed set among damaged plants (about 50% of *B. longii* and *B. fernaldii* individuals). Chamberlain et al. (2013) also looked at the consequence of proximity to crop sunflower for abundance of antagonists of crop wild relatives. However, they found higher abundances of herbivores and seed predators far from the crop, which might be due to use of pesticides in crops leading to fewer enemies nearby crops or due to preference of antagonists for the crop (more attractive than the wild species). Indirect interactions between MIMS and wild plants might also vary temporally. Spillover could be related to temporal shifts in resources with active emigration of herbivores or predators once crops senesce or are harvested. Adler et al. (2014) studied the dispersion of two generalist crop herbivores (sweet potato whitefly, *Bemisia tabaci* and western flower thrips, *Frankliniella occidentalis*) to nearby desert habitat during the cropping season. They found whiteflies on 6 plant species and thrips on 19 out of 36 desert plant species and they further showed that the spillover of whitefly depended on the cropping stage (i.e. planting, growing, presanitation, sanitation stages). Last, spillover from crop MIMS may also affect trophic interactions on wild plant species of surrounding habitats, as shown by Gladbach et al. (2011) who investigated how oilseed rape affects the wild species *S. arvensis* through its pollen beetles (*Meligethes aeneus*) and their parasitoids (*Tersilochus heterocerus*). They found a spillover effect only for the parasitoids, but not for the pollen beetle, with parasitism rates benefiting from increasing presence of oilseed rape in the landscape.

2.3 MIMS as Competitors or Facilitators—Consequences for Communities

2.3.1 Pollinators

Pollinator MIMS impact on plant–pollinator interactions through both other pollinating species and the plant community with which they interact (Fig. 2). In Boxes 1–3, we developed the potential consequences of pollinating MIMS introductions in vulnerable ecosystems as well as in agrosystems. Traveset and Richardson (2006, 2011, 2014) and Dohzono and Yokoyama (2010) reviewed the consequences of the introduction of pollinating MIMS on plant communities and highlighted three main mechanisms which may disrupt mutualistic interactions: (i) a decrease in the quantity or quality of conspecific pollen received by wild plants due to insect diet preferences or cheating behaviour (nectar robbing), (ii) a reinforcement of invasive plant pollination through invasion meltdown[c] and (iii) a lower reproductive success for plant species suffering of pollination deficit, especially in areas with scarce pollinators (Sanguinetti and Singer, 2014). Taken together, these mechanisms could have major collateral effects on native pollinators through modifications of plant resource availability.

To date, few studies have addressed the issue of the introduction of honeybee hives on the dynamics of native plant communities. It has been documented that introduced colonies enhance the visitation frequency of honeybees on invasive plant species, enhancing their reproduction (e.g. the effect on the purple loosestrife, *Lythrum salicaria*, in America: Barthell et al., 2001; Mal et al., 1992). By contrast, other studies have demonstrated that the introduction of honeybee hives can negatively affect the reproductive success of plants through increased pollen theft (Hargreaves et al., 2009), nectar robbing (Kenta et al., 2007), physical damage to the flowers (Dohzono et al., 2008) or through the disruptive influence of honeybee workers on the patterns of pollen transfer among compatible plants that have otherwise evolved highly specialized interactions with native pollinators (Gross and Mackay, 1998; Vaughton, 1996; Watts et al., 2012).

Other studies have shown that these effects are not restricted to honeybees. *B. terrestris* can also have collateral effects on the local floral communities and, in turn, on the wild bee species visiting with them. For example, Sanguinetti and Singer (2014) evidenced better reproductive success of a native orchid from Argentina due to more frequent visits by *B. terrestris* than by the native *B. dahlbomii*. However, deleterious effects for the local fauna following invasions of *B. terrestris* are frequently reported. In a cage experiment, Kenta et al. (2007) showed a decrease in reproductive success for five native plants following *B. terrestris* visits compared with native bumblebee visits. Also, *B. terrestris* showed a preference for exotic species in Argentina (Montalva et al., 2011) and promoted the invasion of a weed (*Lupinus arboreus*) in Tasmania through pollination (Stout et al., 2002). *B. terrestris* is able to rob flowers by perforating floral tubes which can damage flowers and reduce their reproductive success by repelling their native pollinators (Dohzono et al., 2008). Moreover, Sáez et al. (2014) showed that

---

[c] Following Traveset and Richardson (2014) invasion meltdown are "community-level phenomenon whereby alien species enhance one another's establishment, spread and impacts".

B. terrestris can also alter the reproductive success and production of some MFC (*Rubus idaeus*) by damaging flowers and stigmas through overvisitation (see also Aizen et al., 2014).

2.3.2 Plants

The range of reported impacts of plant MIMS on natural plant communities vary from facilitation to competition, occurring directly or indirectly through shared pollinators and herbivores/pathogens (Fig. 2). The magnitude of the effect of plant MIMS on natural plant community and the outcome of the interaction are likely to be mediated by several parameters. First, phylogenetic constraints influencing the proximity in plant traits like flower morphology (for pollinators) or secondary metabolites (against herbivores) between the cultivated and wild plants probably matters, as closely related plants tend to share pollinators and herbivores (Carvalheiro et al., 2014; Fontaine and Thébault, 2015). Yet, most studies on the impacts of MFC on natural communities are based on oilseed rape (see Box 3 and Table S1 (http://dx.doi.org/10.1016/bs.aecr.2016.10.007)); and this limits our ability to generalize the results with other crops. Second, the spatial scale considered when studying the interactions between plant MIMS and natural plant communities is important. Holzschuh et al. (2013) suggest that small-scale effects of oilseed rape are much stronger than landscape-scale effects, at least for solitary bees, which perceive their environment at smaller scales than bumblebees (Westphal et al., 2006). Third, in view of the contrasting responses of pollinators to MFC during or after flowering, studies should also consider the temporal scales. Short-term effects (during vs after MFC flowering) as well as long-term effects via crop rotations that result in annual changes in the distribution of MFC fields might modify the interactions between plant MIMS and natural plant communities. For example, we may observe transient impact of MFC on pollinator populations and wild plant pollination during the flowering that does not necessarily remain after the flowering season (Hanley et al., 2011; Jauker et al., 2012). Furthermore, by influencing pollinator population dynamics, plant MIMS might induce competition for pollinators during mass flowering that potentially translate into facilitation at larger time scales. Fourth, the pollinating insect species involved in indirect interactions between MIMS and wild plants might differ in their phenology and resource requirements. Agricultural MIMS provide food for pollinators at a given moment in the season, but not later in the season, when reproduction of many pollinating species occurs (Requier et al., 2015; Westphal et al., 2009). Early season pollinating species are thus more likely to be favoured by nectar and pollen resource pulses from MFC. Increasing surfaces occupied by agricultural MIMS can also lead to a shortage of nesting sites for wild pollinators. For example, oilseed rape may increase competition among cavity-nesting bee species when nesting sites are the most limiting resource (Steffan-Dewenter and Schiele, 2008). Many solitary species depend on preexisting above-ground cavities or specific soil micro-habitats often associated with seminatural habitats (Cane et al., 2007). These habitats have become increasingly scarce in modern agroecosystems (Potts et al., 2010), leading to pollination deficit in agricultural systems. Overall, agricultural MIMS may favour nonsocial, early reproducing pollinator species (Jauker et al., 2012), and ultimately early reproducing wild plant species, but be globally detrimental to diversity of wild plant communities. Thus, it is questionable whether the exclusive provision of food resources by transient MFC sustainably promotes pollinator reproduction in these systems. An increase in the amount of food resources in the landscape through plant MIMS is expected to benefit wild bees only if the amount of nesting habitats is simultaneously increasing. Last, both the enhancement of facilitation processes and the buffering of competitive interactions appear to depend on biodiversity-friendly management practices. The species diversity within a plant community can enhance pollination due to positive relations between plant and pollinator diversities (Ebeling et al., 2012). It also renders plant–pollinator interaction networks more resilient to changes in floral resources (Tiedeken and Stout, 2015) that are prevalent in MIMS-dominated communities. Indeed, Kovács-Hostyánszki et al. (2013) underlined that resources provided by MFC are most beneficial for wild populations of bees and plants if seminatural habitats are available, providing continuous nesting and food resources during the season. In nonagricultural habitats, buffering the impacts of plant MIMS, be they ornamental or invasive, is also linked to the preservation of floral diversity (Blackmore and Goulson, 2014; Kaluza et al., 2016).

3. SECOND PART: MIMS IN PLANT–POLLINATOR NETWORKS

3.1 Impacts of MIMS on the Structure of Plant–Pollinator Networks

3.1.1 The Case of Pollinators

Few studies have investigated the impact of pollinator MIMS on plant–pollinator network structure. During the past 5 years, two studies have, however, started to assess the consequences of *A. mellifera* on pollination networks. Santos et al. (2012) studied six plants—flower visitors networks in a region of Brazil where beekeeping is intensive and *A. mellifera* is feral and considered as an invasive species. All six networks had *A. mellifera* present and showed that removal of *A. mellifera* and associated links in these networks lead to decreasing nestedness and increasing modularity. However, network connectance was not affected by *A. mellifera* removal and authors did not find any correlation between network structure and the proportion of interactions made by *A. mellifera* in the networks. Giannini et al. (2015) studied 21 plant–bee weighted interaction networks from different ecosystem types, also in Brazil. They found that *A. mellifera* generalism and mean interaction strength was correlated to nestedness and plant niche overlap in their networks.

Being a hub responsible of a large proportion of connections, it can be expected that removal of *A. mellifera* would affect functional properties such as network robustness to extinctions (sensu Burgos et al., 2007). Surprisingly, Santos et al. (2012) found that robustness was unaffected by the removal of *A. mellifera*. However, authors reached this conclusion by considering simulated node deletion in networks in which *A. mellifera* was always present instead of comparing actually observed networks with and without *A. mellifera*. To assess the consequences of pollinator MIMS on plant–pollinator networks, other potential effects should be considered such as pollinator niche rewiring (Kaiser-Bunbury et al., 2010). Pollinators are able to switch to alternative plant species in response to changes in local species floral abundances (Tiedeken and Stout, 2015) or to changes in pollinator densities (Fontaine et al., 2008). To our knowledge, no study has yet compared plant–pollinator networks in the presence and absence of pollinator MIMS, and this clearly deserves future attention.

3.1.2 The Case of Plants

Very few studies have considered the consequences of MFC on pollinator communities and on the reproduction of some focal natural plant species from a network perspective. A few studies have considered the consequences of distance from natural habitats on within-crop plant–pollinator networks (Carvalheiro et al., 2010, 2012). These studies demonstrate that the diversity of pollinators on crops declines with increasing distance to natural habitats, both in mango plantations (Carvalheiro et al., 2010) and in sunflower fields (Carvalheiro et al., 2012, see also Ricketts et al., 2008). When distance to natural habitats increases, the decrease in pollinator diversity also leads to a reduction of network complexity up to the single couple MFC—managed honeybees (see Box 3). In Carvalheiro et al. (2012), this reduction of pollinator diversity and network complexity is correlated with a reduction of crop production, mainly because managed honeybees move less between flowers at lower pollinator diversity. However, these two studies also show that interaction networks of crop fields are more complex and diverse during MIMS flowering period, probably due to the huge floral cover offered, although this effect is mostly present on the edges. Stanley and Stout (2014) suggested that field edges provide alternative resources to pollinators even during crop mass flowering. They found that oilseed rape and coflowering plant species shared pollinators, which is in agreement with the results of Carvalheiro et al. (2010, 2012).

To our knowledge, no study has precisely investigated the impact of MFC on the structure of plant–pollinator networks, focusing on network descriptors, and at a larger scale. Plant MIMS can be diverse, with different morphologies or phenologies (e.g. oilseed rape, sunflowers, peas). We might thus expect that their position in plant–pollination networks will be variable, depending on their floral traits and attractiveness. All MFC are present in high densities at some time of the year. This high abundance argues in favour of some characteristics of networks. We have evidence that species density or biomass is correlated with several structural properties like generalism (Sauve et al., 2016; Spiesman and Gratton, 2016) through pollinator rewiring capacities or higher encounter probabilities (Fort et al., 2016). A recent study suggests that oilseed rape is central in plant–pollinator networks in crop fields when it is flowering (Stanley, 2013). However, overall network structure is unaffected by mass flowering of oilseed rape in this case (i.e. network descriptors such as nestedness and connectance are similar during and after flowering), which suggests that network structure might be robust to such pulses of resources (Stanley, 2013).

3.2 Case Studies of the Position of MIMS in Pollination Webs

3.2.1 A. mellifera *in Plant–Pollinator Networks*

We analysed the position of *A. mellifera* in plant–pollinator networks from a database consisting of 63 community-wide qualitative pollination webs from 38 publications (Fontaine and Thébault, 2015; Thébault and Fontaine, 2010). We compared the position of *A. mellifera* in the network to that of other pollinators in terms of generalism or contribution to networks nestedness (how niche of specialized species are included in the ones of generalists) and modularity (how networks are divided in groups with few interactions between themselves). We chose to consider *A. mellifera* because it is the most widespread managed pollinator species and it occurs in many networks of the database. No information was available on whether *A. mellifera* populations were managed or not in the network studied. However, *A. mellifera* position in the networks can still give some important insights on the potential impact of this important MIMS on the structure of pollination networks. We calculated the degree (i.e. number of links), the contribution to nestedness and the topological role within modules of all pollinator species in the networks. The nestedness contribution of a species quantifies the degree to which observed network nestedness is modified when only the interactions of the focal species are randomized (Saavedra et al., 2011). Modularity defines the degree of compartmentalization of networks and species can be characterized according to their within-module degree (z) and their among-module connectivity (c, Guimera and Amaral, 2005; Olesen et al., 2007). Species with high z values are important for connecting species inside the same module and species with high c values connect different modules.

   A. mellifera was present in about 75% of the network datasets (47 webs out of 63). In those 47 networks, the proportion of interactions due to *A. mellifera* in the web (i.e. number of links of *A. mellifera* over the total number of links in the network) ranged between 0.23% and 23% (mean = 4.88 and sd = 4.48). *A. mellifera* was on average more generalist than other pollinators in the network (i.e. *A. mellifera* degree is in most cases in the 90th percentile, meaning that only 10% of the pollinator species in the network have the same or a higher degree; see Fig. 3B). *A. mellifera* also contributed more to nestedness than the average pollinator (Fig. 3C), and it had in most cases higher within-module degree (Fig. 3D) and higher among-module connectivity (Fig. 3E) than most pollinators in the network. These results thus support the idea that *A. mellifera* tends to be a highly generalist pollinator species that contributes to network nestedness and acts both as a connector within its own module as well as between modules (Giannini et al., 2015; Santos et al., 2012).

3.2.2 MFC in Plant–Pollinator Networks

We analysed the position of plant MIMS in two available datasets describing interactions between plants and flower visitors in farmlands (Fig. 4). Both

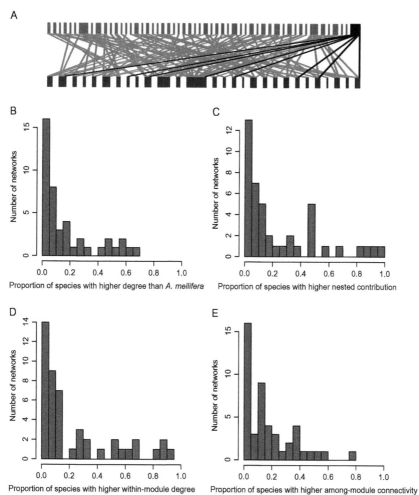

Fig. 3 Example of plant–pollinator network with *Apis mellifera* taken from Ramirez and Brito (1992) (A) position of *A. mellifera* in plant–pollinator networks for degree (B), nestedness contribution (C) and modularity (D–E). In (A), *rectangles* at the bottom of the web correspond to plant (pollinator) species and the size of the *rectangles* is proportional to species degree. *A. mellifera* and its interactions are in *black*, while other pollinators are represented in *grey*. In (B and C), the position of *Apis mellifera* in a given network is represented as the proportion of pollinators in the network with similar or higher degree (B) and nestedness contribution (C) than itself. In (D and E), the position of *Apis mellifera* in a given network is represented as the proportion of pollinators in the network with similar or higher within-module degree z (D) and among-module connectivity c (C) than itself.

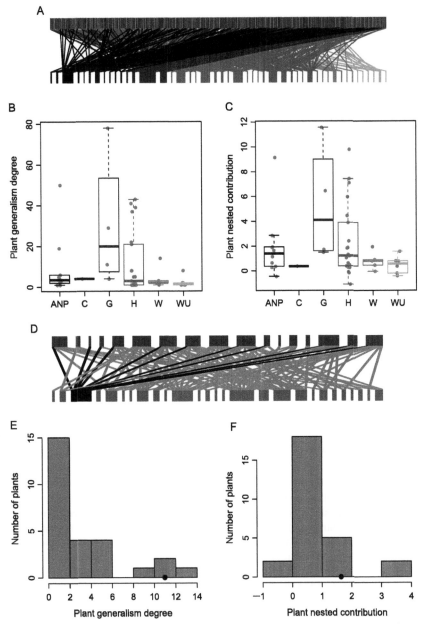

Fig. 4 Position of plant MIMs in two examples of plant–pollinator networks. (A) Plant–pollinator network at Norwood Farm, Somerset, UK based on data from Pocock et al. (2012). *Rectangles* at the *bottom* (*top*) of the web correspond to plant (pollinator) species. Habitats to which plant species belong are represented by *different rectangle* and *link colours*, with the same colour legend as in boxplots in panels (B and C). (B and C)

datasets had the advantage of describing not only the visits on crops and associated weeds (as done in Carvalheiro et al., 2010, 2012) but also the flower visits in field margins (Stanley, 2013) or in all managed and nonmanaged farmland habitats (Pocock et al., 2012). We assessed the position of managed and unmanaged plants in the two networks regarding degree and contribution to nestedness (Fig. 4). In Pocock et al. (2012), the only flowering crop was *Medicago sativa*, which was cultivated for silage. Our analyses reveal that this plant had a low degree in the network, as well as a low contribution to nestedness, contrary to what we expected (Fig. 4B and C). To the contrary, plant species in grass fields, which were managed as pasture, and in hedges or field margins, tended to occupy more central positions in the network, with high degrees and high contributions to nestedness (the plant species with highest degree and contribution to nestedness was the grassland species *Ranunculus repens*). In the dataset collected by Stanley (2013), the flowering crop was *B. napus* and it was one of the most generalist plant species in the network as well as one of the greatest contributors to nestedness (Fig. 4D and E). The analysis of these two datasets thus shows contrasting results regarding the position of MFC in pollination networks. Although such analyses would obviously need to be repeated on other datasets, this result suggests that plant MIMs, despite their common feature of being abundant, might have highly variable positions in plant–pollination networks depending on their floral morphology and attractiveness relative to other plants in nearby seminatural habitats. Thus, comparison of managed and nonmanaged species with similar traits would allow to better understanding the impact of management practices on species positions in networks. A recent study showed that arable lands are poor in nectar resources relatively to grasslands (Baude et al., 2016). Our results from the dataset of Pocock et al. (2012) are in agreement with this finding since plant species in grasslands seem attractive to a larger diversity of flower visitors than the crop.

Distribution of plant degrees (B) and nested contributions (C) in Norwood Farm dataset as a function of habitat types as defined in Pocock et al. (2012) (APN, arable not planted fields; C, crop; G, grass fields; H, hedges and field margins; W, woody species; WU, woodland understorey). (D) Plant–pollinator network in oilseed rape fields and bordering margins in South-East Ireland farms, based on data from Stanley (2013). Same legend as in (A), except that *black rectangle* and *links* are related to oilseed rape, while *light grey rectangles* and *links* correspond to noncultivated plants. (E and F) Distribution of plant generalism degrees (E) and nested contributions (F) in the network of oilseed rape fields, the *black dot* giving the corresponding values for oilseed rape.

3.3 MIMS in Plant–Pollinator Networks: Consequences for Community Dynamics

The few existing studies, as well as our data analysis, suggest that both plant and pollinator MIMS may effectively interfere with native plant–pollinator networks (Fig. 1). In some cases, MIMS such as *A. mellifera* and oilseed rape occupy a central position in the networks, being characterized by high degree and a high contribution to nestedness. This position is comparable to that often found for invasive pollinator and plant species in mutualistic networks (Albrecht et al., 2014; Stouffer et al., 2014; Traveset and Richardson, 2014; Traveset et al., 2013), and have been highlighted as an important driver of eco-evolutionary feedbacks favouring trait complementarity between the two levels of a mutualistic network (for example, plants and pollinators) and increasing trait convergence within levels (Guimaraes et al., 2011). For *A. mellifera*, this similarity with invasive species is not surprising since it is considered as invasive in some parts of the world (Goulson, 2003). The observed high generalism of MIMS in networks can also be related with their potentially high abundance in ecosystems relative to wild species. Indeed, abundant species generally are more likely to be generalists in observed mutualistic networks, due to higher encounter probability (Fort et al., 2016). MIMS might thus, in some cases, increase connectance and nestedness of plant–pollinator networks as well as decrease network modularity, as has been suggested for invasive species (Albrecht et al., 2014; Stouffer et al., 2014; Traveset et al., 2013). Increased network connectance and nestedness might favour species persistence and community stability (Rohr et al., 2014; Thébault and Fontaine, 2010), but it could also decrease species diversity if competition between pollinators for resources and between plants for pollinator access is important (Benadi et al., 2012; Valdovinos et al., 2013). However, it remains unclear whether MIMS will actually modify overall network properties. While invasive species have central positions in mutualistic networks, the connectance and nestedness of invaded and noninvaded networks often do not differ, invasion leading to rearrangement of links within the network without affecting its overall structure (Traveset and Richardson, 2014; Vilà et al., 2009). Tiedeken and Stout (2015) assessed the impact of flower pulse from alien *Rhododendron ponticum* and found few differences between pollination networks during and after the flowering period. Only the mean generalism of the network increased after the flowering period, as most pollinators increased their diet breadth to compensate for the loss of *R. ponticum* availability. Thus, we might expect that MFC will have similar consequences on plant–pollinator networks, by mainly modifying link arrangements within networks during flowering rather than affecting overall network properties. The results of Stanley (2013) suggest that networks seem to be more prone to modification after the introduction of specific crops (even introduction of nonflowering crops) rather than to changes in resource abundances following mass flowering (Stanley, 2013).

Future studies will need to assess how plant–pollinator networks change depending on MIMS characteristics, and they will also need to investigate more closely the functional consequences of link rearrangements following MIMS introduction.

4. CONCLUSION

By reviewing the literature on the impacts of MIMS in pollination systems, we showed that such species can affect native plant–pollinator communities with consequences for ecosystem functioning. To sum up, both competition for floral resources or for pollinators and pathogen spillover, appear as the main mechanisms by which MIMS can have a negative impact on native species (Fig. 2). Further, pollinating MIMS seem also more prone to visit invasive or exotic plant species which might favour these plants at the cost of natural species (invasion meltdown). The analysis of how *A. mellifera*, and MFC fit into pollination networks showed that MIMS tend to be integrated in interaction networks in a way similar to invasive species, although this definitely needs further investigation (see Pantel et al., 2017 for questions relating to invasions in ecological networks).

We stress that the consequences of massive introduction of managed species should not be overlooked, even if such species are domesticated, well known and could have positive economic impact. This is particularly important for ecosystems with high biodiversity such as oceanic islands (Box 1) or protected habitats (Box 2) that should probably be preserved from MIMS, but it is also the case for anthropogenic habitat such as cities where the density of *A. mellifera* colonies is currently skyrocketing (Box 5) or agricultural landscapes where the balance between native and managed pollinators is critically needed (Box 3). Our review focused on *A. mellifera*, *B. terrestris* and MFC since these are currently the most widespread MIMS. However, recent practices tend to diversify the set of managed pollinator species with some incentive to use local species (Box 6).Other measures such as the creation of flower-rich patches within cities and agricultural habitats (Box 4) might support the functioning of the pollination systems. One should, however, be careful that such new managed species do not become victims of their own success and become the new MIMS. Indeed, while we did not discuss the case of ornamental plants here, they also could be considered as MIMS and their consequences on plant–pollinator networks needs to be addressed in the future. As an example, Garbuzov et al. (2015) have shown that the majority (77%) of ornamental flower communities grown in urban parks in a UK city were poorly or totally unattractive to insect floral visitors, and some ornamental plants can also behave as invasive species if they spread outside of the patches they were initially sown.

As a conclusion, we argue here that practices aimed at introducing new species, be they managed pollinators or sown flowering species, should be careful in their choice of species and their methodology. We are also convinced that attention must be focused on wild species regarding conservation measures. The preservation of natural habitats, the development of agroecology (through intercropping, for example) and the generalization of friendly practices towards wild pollinators should be encouraged. Moreover, with increasing evidences of the importance of wild bees for crop pollination (Garibaldi et al., 2016), this could also be beneficial for agrosystems and prevent excessive introduction of managed pollinating species.

ACKNOWLEDGEMENTS

We would like to thank Michel Aubert and Eric Dufrêne for their contribution to an earlier literature overview (Vereecken et al., 2015) upon which the present work is partly based. We also thank Floriane Flacher for her advice on earlier versions of the draft.

REFERENCES

Abe, T., Wada, K., Kato, Y., Makino, S., Okochi, I., 2011. Alien pollinator promotes invasive mutualism in an insular pollination system. Biol. Invasions 13, 957–967.

Acosta, A.L., Giannini, T.C., Imperatriz-Fonseca, V.L., Saraiva, A.M., 2016. Worldwide alien invasion: a methodological approach to forecast the potential spread of a highly invasive pollinator. PLoS One 11, e0148295. http://dx.doi.org/10.1371/journal.pone.0148295.

Adler, V.H., Lubin, Y., Coll, M., 2014. Spillover of crop herbivores into adjacent desert habitats. Agric. Ecosyst. Environ. 193, 117–124.

Aebi, A., Vaissière, B.E., Vanengelsdorp, D., Delaplane, K.S., Roubik, D.W., Neumann, P., 2012. Back to the future: *Apis* versus non-*Apis* pollination. Trends Ecol. Evol. 27, 142–143.

Aizen, M.A., Harder, L.D., 2009. The global stock of domesticated honeybees is growing slower than agricultural demand for pollination. Curr. Biol. 19, 915–918. http://dx.doi.org/10.1016/j.cub.2009.03.071.

Aizen, M.A., Garibaldi, L.A., Cunningham, S.A., Klein, A.M., 2008. Long-term global trends in crop yield and production reveal no current pollination shortage but increasing pollinator dependency. Curr. Biol. 18, 1572–1575. http://dx.doi.org/10.1016/j.cub.2008.08.066.

Aizen, M.A., Morales, C.L., Vázquez, D.P., Garibaldi, L.A., Sáez, A., Harder, L.D., 2014. When mutualism goes bad: density-dependent impacts of introduced bees on plant reproduction. New Phytol. 204, 322–328. http://dx.doi.org/10.1111/nph.12924.

Albrecht, M., Duelli, P., Müller, C., Kleijn, D., Schmid, B., 2007. The Swiss agri-environment scheme enhances pollinator diversity and plant reproductive success in nearby intensively managed farmland. J. Appl. Ecol. 44, 813–822. http://dx.doi.org/10.1111/j.1365-2664.2007.01306.x.

Albrecht, J., Gertrud Berens, D., Jaroszewicz, B., Selva, N., Brandl, R., Farwig, N., 2014. Correlated loss of ecosystem services in coupled mutualistic networks. Nat. Commun. 5, 3810. http://dx.doi.org/10.1038/ncomms4810.

Alton, K., Ratnieks, F.L.W., 2016. To Bee or not to Bee. Biologist 60, 12–15.

Arbetman, M.P., Meeus, I., Morales, C.L., Aizen, M.a., Smagghe, G., 2012. Alien parasite hitchhikes to Patagonia on invasive bumblebee. Biol. Invasions 15, 489–494. http://dx.doi.org/10.1007/s10530-012-0311-0.

Artz, D.R., Nault, B.A., 2011. Performance of *Apis mellifera*, *Bombus impatiens*, and *Peponapis pruinosa* (Hymenoptera: Apidae) as Pollinators of Pumpkin. J. Econ. Entomol. 104, 1153–1161. http://dx.doi.org/10.1603/EC10431.

Badano, E.I., Vergara, C.H., 2011. Potential negative effects of exotic honeybees on the diversity of native pollinators and yield of highland coffee plantations. Agric. For. Entomol. 13, 365–372. http://dx.doi.org/10.1111/j.1461-9563.2011.00527.x.

Banaszak-Cibicka, W., Żmihorski, M., 2012. Wild bees along an urban gradient: winners and losers. J. Insect. Conserv. 16, 331–343. http://dx.doi.org/10.1007/s10841-011-9419-2.

Barthell, J.F., Frankie, G.W., Thorp, R.W., 1998. Invader effects in a community of cavity nesting megachilid bees (Hymenoptera : Megachilidae). Environ. Entomol. 27, 240–247. http://dx.doi.org/10.1093/ee/27.2.240.

Barthell, J.F., Randall, J.M., Thorp, R.W., Wenner, A.M., 2001. Promotion of seed set in yellow star-thistle by honeybees: evidence of an invasive mutualism. Ecol. Appl. 11, 1870–1883.

Bascompte, J., Jordano, P., Olesen, J.M., 2006. Asymmetric coevolutionary networks facilitate biodiversity maintenance. Science 312, 431–433.

Baude, M., Leloup, J., Suchail, S., Allard, B., Benest, D., Mériguet, J., Nunan, N., Dajoz, I., Raynaud, X., 2011. Litter inputs and plant interactions affect nectar sugar content. J. Ecol. 99, 828–837. http://dx.doi.org/10.1111/j.1365-2745.2011.01793.x.

Baude, M., Kunin, W.E., Boatman, N.D., Conyers, S., Davies, N., Gillespie, M.A.K., Morton, R.D., Smart, S.M., Memmott, J., 2016. Historical nectar assessment reveals the fall and rise of floral resources in Britain. Nature 530, 85–88. http://dx.doi.org/10.1038/nature16532.

Bell, D.T., Muller, C.H., 1973. Dominance of California annual grasslands by *Brassica nigra*. Am. Midl. Nat. 90, 227–299.

Bell, J.M., Karron, J.D., Mitchell, R.J., 2005. Interspecific competition for pollination lowers seed production and outcrossing in *Mimulus ringens*. Ecology 83, 762–771.

Benadi, G., Blüthgen, N., Hovestadt, T., Poethke, H.-J., 2012. Population dynamics of plant and pollinator communities: stability reconsidered. Am. Nat. 179, 157–168. http://dx.doi.org/10.1086/663685.

Benton, T.G., Vickery, J.A., Wilson, J.D., 2003. Farmland biodiversity: is habitat heterogeneity the key? Trends Ecol. Evol. 18, 182–188.

Biesmeijer, J.C., Roberts, S.P.M., Reemer, M., Ohlemüller, R., Edwards, M., Peeters, T., Schaffers, A.P., Potts, S.G., Kleukers, R., Thomas, C.D., Settele, J., Kunin, W.E., 2006. Parallel declines in pollinators and insect-pollinated plants in Britain and the Netherlands. Science 313, 351–354. http://dx.doi.org/10.1126/science.1127863.

Biniaś, B., Gospodarek, J., Rusin, M., 2015. The effect of intercropping of broad bean (*Vicia faba* L.) with sweet alyssum (*Lobularia maritima* L.) and white mustard (*Synapis alba* L.) on the energy and the ability of seed germination. J. Res. Appl. Agric. Eng. 60, 11–15.

Blaauw, B.R., Isaacs, R., 2014. Flower plantings increase wild bee abundance and the pollination services provided to a pollination-dependent crop. J. Appl. Ecol. 180, 120–126. http://dx.doi.org/10.1111/1365-2664.12257.

Blackmore, L.M., Goulson, D., 2014. Evaluating the effectiveness of wildflower seed mixes for boosting floral diversity and bumblebee and hoverfly abundance in urban areas. Insect Conserv. Divers. 7, 480–484.

Blitzer, E.J., Dormann, C.F., Holzschuh, A., Klein, A.M., Rand, T.A., Tscharntke, T., 2012. Spillover of functionally important organisms between managed and natural habitats. Agric. Ecosyst. Environ. 146, 34–43. http://dx.doi.org/10.1016/j.agee.2011.09.005.

Brian, A., 1957. Differences in the flowers visited by four species of bumble-bees and their causes. J. Anim. Ecol. 26, 71–98.

Brooks, T.M., Mittermeier, R.A., da Fonseca, G.A.B., Gerlach, J., Hoffmann, M., Lamoreux, J.F., Mittermeier, C.G., Pilgrim, J.D., Rodrigues, A.S.L., 2006. Global biodiversity conservation priorities. Science 313, 58–61. http://dx.doi.org/10.1126/science.1127609.

Burdon, J.J., Thrall, P.H., 2008. Pathogen evolution across the agro-ecological interface: implications for disease management. Evol. Appl. 1, 57–65. http://dx.doi.org/10.1111/j.1752-4571.2007.00005.x.

Burgos, E., Ceva, H., Perazzo, R.P.J., Devoto, M., Medan, D., Zimmermann, M., Delbue, A.M., 2007. Why nestedness in mutualistic networks? J. Theor. Biol. 249, 307–313.

Cairns, C.E., Villanueva-Gutiérrez, R., Koptur, S., Bray, D.B., 2005. Bee populations, forest disturbance, and Africanization in Mexico. Biotropica 37, 686–692. http://dx.doi.org/10.1111/j.1744-7429.2005.00087.x.

Cameron, S.A., Lozier, J.D., Strange, J.P., Koch, J.B., Cordes, N., Solter, L.F., Griswold, T.L., 2011. Patterns of widespread decline in North American bumblebees. Proc. Natl. Acad. Sci. U.S.A. 108, 662–667. http://dx.doi.org/10.1073/pnas.1014743108.

Cameron, S.A., Lim, H.C., Lozier, J.D., Duennes, M.A., Thorp, R., 2016. Test of the invasive pathogen hypothesis of bumble bee decline in North America. Proc. Natl. Acad. Sci. U.S.A. 113, 4386–4391. http://dx.doi.org/10.1073/pnas.1525266113.

Campbell, D.R., Motten, A.F., 1985. The mechanism of competition of pollination between two forest herbs. Ecology 66, 554–563.

Cane, J.H., Tepedino, V.J., 2016. Gauging the effect of honey bee pollen collection on native bee communities. Conserv. Lett. 53, 1–30. http://dx.doi.org/10.1111/conl.12263.

Cane, J.H., Griswold, T., Parker, F.D., 2007. Substrates and materials used for nesting by North American Osmia bees (Hymenoptera: Apiformes: Megachilidae). Ann. Entomol. Soc. Am. 100, 350–358.

Carvalheiro, L.G., Seymour, C.L., Veldtman, R., Nicolson, S.W., 2010. Pollination services decline with distance from natural habitat even in biodiversity-rich areas. J. Appl. Ecol. 47, 810–820. http://dx.doi.org/10.1111/j.1365-2664.2010.01829.x.

Carvalheiro, L.G., Seymour, C.L., Nicolson, S.W., Veldtman, R., 2012. Creating patches of native flowers facilitates crop pollination in large agricultural fields: mango as a case study. J. Appl. Ecol. 49, 1373–1383. http://dx.doi.org/10.1111/j.1365-2664.2012.02217.x.

Carvalheiro, L.G., Kunin, W.E., Keil, P., Aguirre-Gutiérrez, J., Ellis, W.N., Fox, R., Groom, Q., Hennekens, S., Van Landuyt, W., Maes, D., Van de Meutter, F., Michez, D., Rasmont, P., Ode, B., Potts, S.G., Reemer, M., Roberts, S.P.M., Schaminée, J., WallisDeVries, M.F., Biesmeijer, J.C., 2013. Species richness declines and biotic homogenisation have slowed down for NW-European pollinators and plants. Ecol. Lett. 16, 870–878. http://dx.doi.org/10.1111/ele.12121.

Carvalheiro, L.G., Biesmeijer, J.C., Benadi, G., Fründ, J., Stang, M., Bartomeus, I., Kaiser-Bunbury, C.N., Baude, M., Gomes, S.I.F., Merckx, V., Baldock, K.C.R., Bennett, A.T.D., Boada, R., Bommarco, R., Cartar, R., Chacoff, N., Dänhardt, J., Dicks, L.V., Dormann, C.F., Ekroos, J., Henson, K.S.E., Holzschuh, A., Junker, R.R., Lopezaraiza-Mikel, M., Memmott, J., Montero-Castaño, A., Nelson, I.L., Petanidou, T., Power, E.F., Rundlöf, M., Smith, H.G., Stout, J.C., Temitope, K., Tscharntke, T., Tscheulin, T., Vilà, M., Kunin, W.E., 2014. The potential for indirect effects between co-flowering plants via shared pollinators depends on resource abundance, accessibility and relatedness. Ecol. Lett. 17, 1389–1399. http://dx.doi.org/10.1111/ele.12342.

Chamberlain, S.A., Whitney, K.D., Rudgers, J.A., 2013. Proximity to agriculture alters abundance and community composition of wild sunflower mutualists and antagonists. Ecosphere 4, 96. http://dx.doi.org/10.1890/ES13-00026.1.

Chen, Y.P., Siede, R., 2007. Honeybee viruses. Adv. Virus Res. 70, 33–80.

Colla, S.R., Otterstatter, M.C., Gegear, R.J., Thomson, J.D., 2006. Plight of the bumblebee: pathogen spillover from commercial to wild populations. Biol. Conserv. 129, 461–467. http://dx.doi.org/10.1016/j.biocon.2005.11.013.

Concepcion, E.D., Moretti, M., Altermatt, F., Nobis, M.P., Obrist, M.K., 2015. Impacts of urbanisation on biodiversity: the role of species mobility, degree of specialisation and spatial scale. Oikos 124, 1571–1582. http://dx.doi.org/10.1111/oik.02166.

Crane, E., 1975. Honey: A Comprehensive Survey. Heinemann in Co-Operation with International Bee Research Association, London.

Crane, E., 1999. Recent research on the world history of beekeeping. Bee World 80, 174–186.

Cunningham, S.A., Fournier, A., Neave, M.J., Le Feuvre, D., 2016. Improving spatial arrangement of honeybee colonies to avoid pollination shortfall and depressed fruit set. J. Appl. Ecol. 53, 350–359. http://dx.doi.org/10.1111/1365-2664.12573.

Cussans, J., Goulson, D., Sanderson, R., Goffe, L., Darvill, B., Osborne, J.L., 2010. Two bee-pollinated plant species show higher seed production when grown in gardens compared to arable farmland. PLoS One 5, e11753. http://dx.doi.org/10.1371/journal.pone.0011753.

Dafni, A., Shmida, A., 1996. The possible ecological implications of the invasion of Bonlbus terrestris (L.) (Apidae) at Mt Carmel, Israel. In: Matheson, A., Buchmann, M., O' Toole, C., Westrich, P., Williams, I.H. (Eds.), The Conservation of Bees. The Linnean Society of London and the International Bee Research Association, London, pp. 84–199.

Dafni, A., Kevan, P., Gross, C.L., Goka, K., 2010. Bombus terrestris, pollinator, invasive and pest: an assessment of problems associated with its widespread introductions for commercial purposes. Appl. Entomol. Zool. 45, 101–113. http://dx.doi.org/10.1303/aez.2010.101.

Danchin, É., Giraldeau, L.A., Valone, T.J., Wagner, R.H., 2004. Public information: from nosy neighbors to cultural evolution. Science 305 (5683), 487–491.

Daszak, P., Cunningham, A.A., Hyatt, A.D., 2000. Emerging infectious diseases of wildlife-threats to biodiversity and human health. Science 287, 443–449.

Dawson, E.H., Chittka, L., 2014. Bumblebees (*Bombus terrestris*) use social information as an indicator of safety in dangerous environments. Proc. R. Soc. Lond. B 281, 20133174.

Dicks, L.V., Ashpole, J.E., Dänhardt, J., James, K., Jönsson, A., Randall, N., Showler, D.A., Smith, R.K., Turpie, S., Williams, D., Sutherland, W.J., 2014. Farmland Conservation: Evidence for the Effects of Interventions in Northern and Western Europe. Pelagic Publishing Ltd., Exeter, p. 504.

Diekötter, T., Kadoya, T., Peter, F., Wolters, V., Jauker, F., 2010. Oilseed rape crops distort plant-pollinator interactions. J. Appl. Ecol. 47, 209–214. http://dx.doi.org/10.1111/j.1365-2664.2009.01759.x.

Dohzono, I., Yokoyama, J., 2010. Impacts of alien bees on native plant-pollinator relationships: a review with special emphasis on plant reproduction. Appl. Entomol. Zool. 45, 37–47. http://dx.doi.org/10.1303/aez.2010.37.

Dohzono, I., Kunitake, Y.K., Yokoyama, J., Goka, K., 2008. Alien bumble bee affects native plant reproduction through interactions with native bumble bees. Ecology 89, 3082–3092. http://dx.doi.org/10.1890/07-1491.1.

Donovan, B.J., 1980. Interactions between native and introduced bees in New Zealand. N. Z. J. Ecol. 3, 104–116.

Dudley, N., 2008. Guidelines for applying protected area management categories. IUCN, Gland, Switzerland, x + 86pp. Accessible at: https://cmsdata.iucn.org/downloads/guidelines_for_applying_protected_area_management_categories.pdf.

Duelli, P., Obrist, M.K., 2003. Biodiversity indicators: the choice of values and measures. Agric. Ecosyst. Environ. 2063, 1–12.

Dupont, Y.L., Hansen, D.M., Valido, A., Olesen, J.M., 2004. Impact of introduced honey-bees on native pollination interactions of the endemic *Echium wildpretii* (Boraginaceae) on Tenerife, Canary Islands. Biol. Conserv. 118, 301–311. http://dx.doi.org/10.1016/j.biocon.2003.09.010.

Ebeling, A., Klein, A.-M., Weisser, W.W., Tscharntke, T., 2012. Multitrophic effects of experimental changes in plant diversity on cavity-nesting bees, wasps, and their parasitoids. Oecologia 169, 453–465. http://dx.doi.org/10.1007/s00442-011-2205-8.

Eilers, E.J., Kremen, C., Smith Greenleaf, S., Garber, A.K., Klein, A.-M., 2011. Contribution of pollinator-mediated crops to nutrients in the human food supply. PLoS One 6, e21363. http://dx.doi.org/10.1371/journal.pone.0021363.

Elbgami, T., Kunin, W.E., Hughes, W.O.H., Biesmeijer, J.C., 2014. The effect of proximity to a honeybee apiary on bumblebee colony fitness, development, and performance. Apidologie 45, 504–513. http://dx.doi.org/10.1007/s13592-013-0265-y.

European Commission, 2011. Our Life Insurance, Our Natural Capital: An EU Biodiversity Strategy to 2020. Brussels, Belgium.

Evison, S.E.F., Roberts, K.E., Laurenson, L., Pietravalle, S., Hui, J., Biesmeijer, J.C., Smith, J.E., Budge, G., Hughes, W.O.H., 2012. Pervasiveness of parasites in pollinators. PLoS One 7, e30641. http://dx.doi.org/10.1371/journal.pone.0030641.

Flacher, F., Raynaud, X., Hansart, A., Motard, E., Dajoz, I., 2015. Competition with wind-pollinated plant species alters floral traits of insect-pollinated plant species. Sci. Rep. 5, 13345. http://dx.doi.org/10.1038/srep13345.

Fontaine, C., Thébault, E., 2015. Comparing the conservatism of ecological interactions in plant-pollinator and plant-herbivore networks. Popul. Ecol. 57, 29–36. http://dx.doi.org/10.1007/s10144-014-0473-y.

Fontaine, C., Collin, C.L., Dajoz, I., 2008. Generalist foraging of pollinators: diet expansion at high density. J. Ecol. 96, 1002–1010. http://dx.doi.org/10.1111/j.1365-2745.2008.01405.x.

Fort, H., Vázquez, D.P., Lan, B.L., 2016. Abundance and generalisation in mutualistic networks: solving the chicken-and-egg dilemma. Ecol. Lett. 19, 4–11. http://dx.doi.org/10.1111/ele.12535.

Fortel, L., Henry, M., Guilbaud, L., Guirao, A.L., Kuhlmann, M., Mouret, H., Rollin, O., Vaissière, B.E., 2014. Decreasing abundance, increasing diversity and changing structure of the wild bee community (Hymenoptera: Anthophila) along an urbanization gradient. PLoS One 9, e104679. http://dx.doi.org/10.1371/journal.pone.0104679.

Fortel, L., Henry, M., Guilbaud, L., Mouret, H., Vaissière, B.E., 2016. Use of human-made nesting structures by wild bees in an urban environment. J. Insect Conserv. 20, 239–253. http://dx.doi.org/10.1007/s10841-016-9857-y.

Frankie, G., Thorp, R., Schindler, M., Hernandez, J.L., Ertter, B., Rizzardi, M.A., 2005. Ecological patterns of bees and their host ornamental flowers in two northern California cities. J. Kansas Entomol. Soc. 78, 227–246.

Fürst, M.A., McMahon, D.P., Osborne, J.L., Paxton, R.J., Brown, M.J.F., 2014. Disease associations between honeybees and bumblebees as a threat to wild pollinators. Nature 506, 364–366. http://dx.doi.org/10.1038/nature12977.

Garbuzov, M., Schürch, R., Ratnieks, F.L.W., 2015. Eating locally: dance decoding demonstrates that urban honeybees in Brighton, UK, forage mainly in the surrounding urban area. Urban Ecosys. 18, 411–418. http://dx.doi.org/10.1007/s11252-014-0403-y.

Garder, B., 1996. European Agriculture: Policies, Production, and Trade, Gardner, Brian. Routledge Publications, New York, USA. 244 pp.

Garibaldi, L.A., Steffan-Dewenter, I., Winfree, R., Aizen, M.A., Bommarco, R., Cunningham, S.A., Kremen, C., Carvalheiro, L.G., Harder, L.D., Afik, O., Bartomeus, I., Benjamin, F., Boreux, V., Cariveau, D., Chacoff, N.P., Dudenhöffer, J.H., Freitas, B.M., Ghazoul, J., Greenleaf, S., Hipólito, J., Holzschuh, A., Howlett, B., Isaacs, R., Javorek, S.K., Kennedy, C.M., Krewenka, K.M., Krishnan, S., Mandelik, Y., Mayfield, M.M., Motzke, I., Munyuli, T., Nault, B.A., Otieno, M., Petersen, J., Pisanty, G., Potts, S.G., Rader, R., Ricketts, T.H., Rundlöf, M., Seymour, C.L., Schüepp, C., Szentgyörgyi, H., Taki, H., Tscharntke, T., Vergara, C.H., Viana, B.F., Wanger, T.C., Westphal, C., Williams, N., Klein, A.M., 2013. Wild pollinators enhance fruit set of crops regardless of honeybee abundance. Science 339, 1608–1611. http://dx.doi.org/10.1126/science.1230200.

Garibaldi, L.A., Carvalheiro, L.G., Vaissiere, B.E., Gemmill-Herren, B., Hipolito, J., Freitas, B.M., Ngo, H.T., Azzu, N., Saez, A., Astrom, J., An, J., Blochtein, B., Buchori, D., Garcia, F.J.C., Oliveira da Silva, F., Devkota, K., Ribeiro, M.D.F., Freitas, L., Gaglianone, M.C., Goss, M., Irshad, M., Kasina, M., Filho, A.J.S.P., Kiill, L.H.P., Kwapong, P., Pires, C., Pires, V., Rawal, R.S., Rizali, A., Saraiva, A.M., Veldtman, R., Viana, B.F., Witter, S., Zhang, H., 2016. Mutually beneficial pollinator diversity and crop yield outcomes in small and large farms. Science 351, 388–391. http://dx.doi.org/10.1126/science.aac7287.

Garratt, M.P.D., Breeze, T.D., Jenner, N., Polce, C., Biesmeijer, J.C., Potts, S.G., 2014. Avoiding a bad apple: insect pollination enhances fruit quality and economic value. Agric. Ecosyst. Environ. 184, 34–40. http://dx.doi.org/10.1016/j.agee.2013.10.032.

Genersch, E., Yue, C., Fries, I., de Miranda, J.R., 2006. Detection of Deformed wing virus, a honeybee viral pathogen, in bumble bees (*Bombus terrestris* and *Bombus pascuorum*) with wing deformities. J. Invertebr. Pathol. 91, 61–63.

Geslin, B., Morales, C.L., 2015. New records reveal rapid geographic expansion of *Bombus terrestris* Linnaeus, 1758 (Hymenoptera: Apidae), an invasive species in Argentina. Check List 11, 3–5. http://dx.doi.org/10.15560/11.3.1620.

Geslin, B., Gauzens, B., Thébault, E., Dajoz, I., 2013. Plant pollinator networks along a gradient of urbanisation. PLoS One 8, e63421. http://dx.doi.org/10.1371/journal.pone.0063421.

Geslin, B., Le Féon, V., Folschweiller, M., Flacher, F., Carmignac, D., Motard, E., Perret, S., Dajoz, I., 2016a. The proportion of impervious surfaces at the landscape scale structures wild bee assemblages in a densely populated region. Ecol. Evol. 1–17. http://dx.doi.org/10.1002/ece3.2374.

Geslin, B., Le Féon, V., Kuhlmann, M., Vaissière, B.E., Dajoz, I., 2016b. The bee fauna of large parks in downtown Paris, France. Ann. Soc. Entomol. Fr. 51, 487–493. http://dx.doi.org/10.1080/00379271.2016.1146632.

Geslin, B., Oddie, M., Folschweiller, M., Legras, G., Seymour, C.L., van Veen, F.J.F., Thébault, E., 2016c. Spatiotemporal changes in flying insect abundance and their functional diversity as a function of distance to natural habitats in a mass flowering crop. Agric. Ecosyst. Environ. 229, 21–29. http://dx.doi.org/10.1016/j.agee.2016.05.010.

Giannini, T.C., Garibaldi, L.A., Acosta, A.L., Silva, J.S., Maia, K.P., Saraiva, A.M., Guimarães, P.R., Kleinert, A.M.P., 2015. Native and non-native supergeneralist bee species have different effects on plant-bee networks. PLoS One 10, e0137198. http://dx.doi.org/10.1371/journal.pone.0137198.

Gill, R.J., Baldock, K.C.R., Brown, M.J.F., Cresswell, J.E., Dicks, L.V., Fountain, M.T., Garratt, M.P.D., Gough, L.A., Heard, M.S., Holland, J.M., Ollerton, J., Stone, G.N., Tang, C.Q., Vanbergen, A.J., Vogler, A.P., Woodward, G., Arce, A.N., Boatman, N.D., Brand-Hardy, R., Breeze, T.D., Green, M., Hartfield, C.M., O'Connor, R.S., Osborne, J.L., Phillips, J., Sutton, P.B., Potts, S.G., 2016. Protecting an ecosystem service: approaches to understanding and mitigating threats to wild insect pollinators. Adv. Ecol. Res. 53, 135–206. http://dx.doi.org/10.1016/bs.aecr.2015.10.007.

Gladbach, D.J., Holzschuh, A., Scherber, C., Thies, C., Dormann, C.F., Tscharntke, T., 2011. Crop–non-crop spillover: arable fields affect trophic interactions on wild plants in surrounding habitats. Oecologia 166, 433–441.

Goodale, E., Nieh, J.C., 2012. Public use of olfactory information associated with predation in two species of social bees. Anim. Behav. 84, 919–924. http://dx.doi.org/10.1016/j.anbehav.2012.07.016.

Goulson, D., 2003. Effects of introduced bees on native ecosystems. Annu. Rev. Ecol. Evol. Syst. 34, 1–26. http://dx.doi.org/10.1146/annurev.ecolsys.34.011802.132355.

Goulson, D., Hughes, W.O.H., 2015. Mitigating the anthropogenic spread of bee parasites to protect wild pollinators. Biol. Conserv. 191, 10–19. http://dx.doi.org/10.1016/j.biocon.2015.06.023.

Goulson, D., Sparrow, K.R., 2009. Evidence for competition between honeybees and bumblebees; effects on bumblebee worker size. J. Insect Conserv. 13, 177–181. http://dx.doi.org/10.1007/s10841-008-9140-y.

Goulson, D., Stout, J.C., Kells, A.R., 2002. Do exotic bumblebees and honeybees compete with native flower-visiting insects in Tasmania? J. Insect Conserv. 6, 179–189. http://dx.doi.org/10.1023/A:1023239221447.

Goulson, D., Nicholls, E., Botias, C., Rotheray, E.L., 2015. Bee declines driven by combined stress from parasites, pesticides, and lack of flowers. Science 347, 1255957. http://dx.doi.org/10.1126/science.1255957.

Graystock, P., Yates, K., Evison, S.E.F., Darvill, B., Goulson, D., Hughes, W.O.H., 2013a. The Trojan hives: pollinator pathogens, imported and distributed in bumblebee colonies. J. Appl. Ecol. 50, 1207–1215. http://dx.doi.org/10.1111/1365-2664.12134.

Graystock, P., Yates, K., Darvill, B., Goulson, D., Hughes, W.O.H., 2013b. Emerging dangers: deadly effects of an emergent parasite in a new pollinator host. J. Invertebr. Pathol. 114, 114–119. http://dx.doi.org/10.1016/j.jip.2013.06.005.

Graystock, P., Goulson, D., Hughes, W.O.H., 2014. The relationship between managed bees and the prevalence of parasites in bumblebees. PeerJ 2, e522. http://dx.doi.org/10.7717/peerj.522.

Graystock, P., Goulson, D., Hughes, W.O.H., 2015. Parasites in bloom: flowers aid dispersal and transmission of pollinator parasites within and between bee species. Proc. R. Soc. B 282, 20151371. http://dx.doi.org/10.1098/rspb.2015.1371.

Graystock, P., Blane, E.J., McFrederick, Q.S., Goulson, D., Hughes, W.O.H., 2016. Do managed bees drive parasite spread and emergence in wild bees? Int. J. Parasitol. Parasit. Wildl. 5, 64–75. http://dx.doi.org/10.1016/j.ijppaw.2015.10.001.

Greenleaf, S.S., Williams, N.M., Winfree, R., Kremen, C., 2007. Bee foraging ranges and their relationship to body size. Oecologia 153, 589–596. http://dx.doi.org/10.1007/s00442-007-0752-9.

Gross, C.L., Mackay, D., 1998. Honeybees reduce fitness in the pioneer shrub *Melastoma affine* (Melastomataceae). Biol. Conserv. 86, 169–178.

Guimaraes Jr., P.R., Jordano, P., Thompson, J.N., 2011. Evolution and coevolution in mutualistic networks. Ecol. Lett. 14 (9), 877–885.

Guimera, R., Amaral, L.A., 2005. Functional cartography of complex metabolic networks. Nature 433, 895–900.

Haaland, C., Naisbit, R.E., Bersier, L.-F., 2011. Sown wildflower strips for insect conservation: a review. Insect Conserv. Divers. 4, 60–80.

Haddad, N., Fuchs, S., Hepburn, H.R., Radloff, S.E., 2009. *Apis florea* in Jordan: source of the founder population. Apidologie 40, 508–512.

Hanley, M.E., Franco, M., Dean, C.E., Franklin, E.L., Harris, H.R., Haynes, A.G., Rapson, S.R., Rowse, G., Thomas, K.C., Waterhouse, B.R., Knight, M.E., 2011. Increased bumblebee abundance along the margins of a mass flowering crop: evidence for pollinator spill-over. Oikos 120, 1618–1624. http://dx.doi.org/10.1111/j.1600-0706.2011.19233.x.

Hanna, C., Foote, D., Kremen, C., 2012. Short- and long-term control of *Vespula pensylvanica* in Hawaii by fipronil baiting. Pest Manag. Sci. 68, 1026–1033.

Hansen, D.M., Olesen, J.M., Jones, C.G., 2002. Trees, birds and bees in Mauritius: exploitative competition between introduced honeybees and endemic nectarivorous birds? J. Biogeogr. 29, 721–734.

Hardman, C.J., Norris, K., Nevard, T.D., Hughes, B., Potts, S.G., 2016. Delivery of floral resources and pollination services on farmland under three different wildlife-friendly schemes. Agric. Ecol. Environ. 220, 142–151.

Hargreaves, A.L., Harder, L.D., Johnson, S.D., 2009. Consumptive emasculation: the ecological and evolutionary consequences of pollen theft. Biol. Rev. Camb. Philos. Soc. 84, 259–276. http://dx.doi.org/10.1111/j.1469-185X.2008.00074.x.

Hegde, R.S., Miller, D.A., 1990. Allelopathy and autotoxicity in alfalfa: characterization and effects of preceding crops and residue incorporation. Crop. Sci. 30, 1255–1259.

Henry, M., Béguin, M., Requier, F., Rollin, O., Odoux, J.-F., Aupinel, P., Aptel, J., Tchamitchian, S., Decourtye, A., 2012. A common pesticide decreases foraging success and survival in honeybees. Science 336, 348–350. http://dx.doi.org/10.1126/science.1215039.

Herbertsson, L., Lindström, S.A.M., Rundlöf, M., Bommarco, R., Smith, H.G., 2016. Competition between managed honeybees and wild bumblebees depends on landscape context. Basic Appl. Ecol. 17, 609–616. http://dx.doi.org/10.1016/j.baae.2016.05.001.

Herrmann, F., Westphal, C., Moritz, R.F.A., Steffan-Dewenter, I., 2007. Genetic diver- sity and mass resources promote colony size and forager densities of a social bee (*Bombus pascuorum*) in agricultural landscapes. Mol. Ecol. 16, 1167–1178.

Higes, M., Martin, R., Meana, A., 2006. *Nosema ceranae*, a new microsporidian parasite in honeybees in Euope. J. Invertebr. Pathol. 92, 93–95. http://dx.doi.org/10.1016/j.jip.2006.02.005.

Higes, M., Martín-Hernández, R., Martínez-Salvador, A., Garrido-Bailón, E., González-Porto, A.V., Meana, A., Bernal, J.L., del Noza, M.J., Bernal, J., 2010. A preliminary study of the epidemiological factors related to honeybee colony loss in Spain. Environ. Microbiol. Rep. 2, 243–250.

Hingston, A.B., McQuillan, P.B., 1998. Does the recently introduced bumblebee *Bombus terrestris* (Apidae) threaten Australian ecosystems? Aust. J. Ecol. 23, 539–549. http://dx.doi.org/10.1111/j.1442-9993.1998.tb00764.x.

Hingston, A.B., McQuillan, P.B., 1999. Displacement of Tasmanian native megachilid bees by the recently introduced bumblebee *Bombus terrestris* (Linnaeus, 1758) (Hymenoptera: Apidae). Aust. J. Zool. 47, 59–65.

Hingston, A.B., Marsden-Smedley, J., Driscoll, D.A., Corbett, S., Fenton, J., Anderson, R., Plowman, C., Mowling, F., Jenkin, M., Matsui, K., Bonham, K.J., Ilowski, M., Mcquillan, P.B., Yaxley, B., Reid, T., Storey, D., Poole, L., Mallick, S.A., Fitzgerald, N., Kirkpatrick, J.B., Febey, J., Harwood, A.G., Michaels, K.F., Russell, M.J., Black, P.G., Emmerson, L., Visoiu, M., Morgan, J., Breen, S., Gates, S., Bantich, M.N., Desmarchelier, J.M., 2002. Extent of invasion of Tasmanian native vegetation by the exotic bumblebee *Bombus terrestris* (Apoidea: Apidae). Austral Ecol. 27, 162–172. http://dx.doi.org/10.1046/j.1442-9993.2002.01179.x.

Hoekstra, A.Y., Wiedmann, T.O., 2014. Humanity's unsustainable environmental footprint. Science 344, 1114–1117. http://dx.doi.org/10.1126/science.1248365.

Holzschuh, A., Dormann, C.F., Tscharntke, T., Steffan-Dewenter, I., 2011. Expansion of mass-flowering crops leads to transient pollinator dilution and reduced wild plant pollination. Proc. Biol. Sci. 278, 3444–3451. http://dx.doi.org/10.1098/rspb.2011.0268.

Holzschuh, A., Dormann, C.F., Tscharntke, T., Steffan-Dewenter, I., 2013. Mass-flowering crops enhance wild bee abundance. Oecologia 172, 477–484. http://dx.doi.org/10.1007/s00442-012-2515-5.

Hoyle, M., Hayter, K., Cresswell, J.E., 2007. Effect of pollinator abundance on self-fertilization and gene flow: application to GM Canola. Ecol. Appl. 17, 2123–2135.

Inari, N., Nagamitsu, T., Kenta, T., Goka, K., Hiura, T., 2005. Spatial and temporal pattern of introduced *Bombus terrestris* abundance in Hokkaido, Japan, and its potential impact on native bumblebees. Popul. Ecol. 47, 77–82. http://dx.doi.org/10.1007/s10144-004-0205-9.

Inoue, M.N., Yokoyama, J., Washitani, I., 2008. Displacement of Japanese native bumblebees by the recently introduced *Bombus terrestris* (L.) (Hymenoptera: Apidae). J. Insect Conserv. 12, 135–146. http://dx.doi.org/10.1007/s10841-007-9071-z.

Inouye, D.W., 1978. Resource partitioning in bumblebees—experimental studies of foraging behavior. Ecology 59, 672–678. http://dx.doi.org/10.2307/1938769.

Jauker, F., Peter, F., Wolters, V., Diekötter, T., 2012. Early reproductive benefits of mass-flowering crops to the solitary bee *Osmia rufa* outbalance post-flowering disadvantages. Basic Appl. Ecol. 13, 268–276. http://dx.doi.org/10.1016/j.baae.2012.03.010.

Jean, R.P., 2005. Quantifying a rare event: pollen theft by honeybees from bumble bees and other bees (Apoidea: Apidae, Megachilidae) foraging at flowers. J. Kansas Entomol. Soc. 78, 172–175.

Jones, R.A.C., Coutts, B.A., 2015. Spread of introduced viruses to new plants in natural ecosystems and the threat this poses to plant biodiversity. Mol. Plant Pathol. 16, 541–545. http://dx.doi.org/10.1111/mpp.12268.

Jönsson, A., Ekroos, J., Dänhardt, J., Andersson, G., Olsson, O., Smith, H.G., 2015. Sown flower strips in southern Sweden increase abundances of wild bees and hoverflies in the wider landscape. Biol. Conserv. 184, 51–58.

Kaiser-Bunbury, C.N., Muff, S., Memmott, J., Müller, C.B., Caflisch, A., 2010. The robustness of pollination networks to the loss of species and interactions: a quantitative approach incorporating pollinator behaviour. Ecol. Lett. 13, 442–452. http://dx.doi. org/10.1111/j.1461-0248.2009.01437.x.

Kaluza, B.F., Wallace, H., Heard, T.A., Klein, A.-M., Leonhardt, S.D., 2016. Urban gardens promote bee foraging over natural habitats and plantations. Ecol. Evol. 6, 1304–1316. http://dx.doi.org/10.1002/ece3.1941.

Kato, M., Shibata, A., Yasui, T., Nagamasu, H., 1999. Impact of introduced honeybees, *Apis mellifera*, upon native bee communities in the Bonin (Ogasawara) Islands. Popul. Ecol. 41, 217–228. http://dx.doi.org/10.1007/s101440050025.

Kenta, T., Inari, N., Nagamitsu, T., Goka, K., Hiura, T., 2007. Commercialized European bumblebee can cause pollination disturbance: an experiment on seven native plant species in Japan. Biol. Conserv. 134, 298–309. http://dx.doi.org/10.1016/j.biocon. 2006.07.023.

Kleijn, D., Baquero, R.A., Clough, Y., Díaz, M., De Esteban, J., Fernández, F., Gabriel, D., Herzog, F., Holzschuh, A., Jöhl, R., Knop, E., Kruess, A., Marshall, E.J.P., Steffan-Dewenter, I., Tscharntke, T., Verhulst, J., West, T.M., Yela, J.L., 2006. Mixed biodiversity benefits of agri-environment schemes in five European countries. Ecol. Lett. 9, 243–254. http://dx.doi.org/10.1111/j.1461-0248.2005.00869.x. discussion 254–7.

Kleijn, D., Winfree, R., Bartomeus, I., Carvalheiro, L.G., Henry, M., Isaacs, R., Klein, A.-M., Kremen, C., M'Gonigle, L.K., Rader, R., Ricketts, T.H., Williams, N.M., Lee Adamson, N., Ascher, J.S., Báldi, A., Batáry, P., Benjamin, F., Biesmeijer, J.C., Blitzer, E.J., Bommarco, R., Brand, M.R., Bretagnolle, V., Button, L., Cariveau, D.P., Chifflet, R., Colville, J.F., Danforth, B.N., Elle, E., Garratt, M.P.D., Herzog, F., Holzschuh, A., Howlett, B.G., Jauker, F., Jha, S., Knop, E., Krewenka, K.M., Le Féon, V., Mandelik, Y., May, E.A., Park, M.G., Pisanty, G., Reemer, M., Riedinger, V., Rollin, O., Rundlöf, M., Sardiñas, H.S., Scheper, J., Sciligo, A.R., Smith, H.G., Steffan-Dewenter, I., Thorp, R., Tscharntke, T., Verhulst, J., Viana, B.F., Vaissière, B.E., Veldtman, R., Westphal, C., Potts, S.G., 2015. Delivery of crop pollination services is an insufficient argument for wild pollinator conservation. Nat. Commun. 6, 7414. http://dx.doi.org/10.1038/ncomms8414.

Klein, A.-M., Vaissière, B.E., Cane, J.H., Steffan-Dewenter, I., Cunningham, S.A., Kremen, C., Tscharntke, T., 2007. Importance of pollinators in changing landscapes for world crops. Proc. Biol. Sci. 274, 303–313. http://dx.doi.org/10.1098/rspb.2006.3721.

Knight, M.E., Osborne, J.L., Sanderson, R.A., Hale, R.J., Martin, A.P., Goulson, D., 2009. Bumblebee nest density and the scale of available forage. Insect Conserv. Divers. 2, 116–124.

Kovács-Hostyánszki, A., Haenke, S., Batáry, P., Jauker, B., Báldi, A., Tscharntke, T., Holzschuh, A., 2013. Contrasting effects of mass-flowering crops on bee pollination of hedge plants at different spatial and temporal scales. Ecol. Appl. 23, 1938–1946.

Labreuche, J., Tosser, V., 2014. Étude de la phénologie et du butinage de cultures intermédiaires: résultats du suivi de cinq expérimentations Colloque de restitution du projet InterAPI.

Lautenbach, S., Seppelt, R., Liebscher, J., Dormann, C.F., 2012. Spatial and temporal trends of global pollination benefit. PLoS One 7, e35954. http://dx.doi.org/10.1371/journal. pone.0035954.

Le Féon, V., Schermann-Legionnet, A., Delettre, Y., Aviron, S., Billeter, R., Bugter, R., Hendrickx, F., Burel, F., 2010. Intensification of agriculture, landscape composition and wild bee communities: a large scale study in four European countries. Agric. Ecosyst. Environ. 137, 143–150. http://dx.doi.org/10.1016/j.agee.2010.01.015.

Leather, G.R., 1983. Sunflowers (*Helianthus annuus*) are allelopathic to weeds. Weed Sci. 31, 37–42. http://dx.doi.org/10.2307/4043564.

Li, J., Peng, W., Wu, J., Strange, J.P., Boncristiani, H., Chen, Y., 2011. Cross-species infection of deformed wing virus poses a new threat to pollinator conservation. J. Econ. Entomol. 104, 732–739.

Lord, W.G., Nagi, S.K., 1987. *Apis florea* discovered in Africa. Bee World 68, 39–40.

Lye, G.C., Lepais, O., Goulson, D., 2011. Reconstructing demographic events from population genetic data: the introduction of bumblebees to New Zealand. Mol. Ecol. 20, 2888–2900. http://dx.doi.org/10.1111/j.1365-294X.2011.05139.x.

Maccagnani, B., Burgio, G., Stanisavljević, L.Ž., Maini, S., 2007. *Osmia cornuta* management in pear orchards. Bull. Insectol. 60, 77–82.

Mal, T.K., Lovett-Doust, J., Lovett-Doust, L., Mul-ligan, G.A., 1992. The biology of Canadian weeds. Number 100. *Lythrum salicaria*. Can. J. Plant Sci. 72, 1305–1330.

Markwell, T.J., Kelly, D., Duncan, K.W., 1993. Competition between honeybees (*Apis mellifera*) and wasps (*Vespula sp.*) in honeydew beech (*Nothofagus solandri var. solandri*) forest. N.Z. J. Ecol. 17, 85–93.

Marrero, H.J., Medan, D., Zarlavsky, G.E., Torretta, J.P., 2016. Agricultural land management negatively affects pollination service in Pampean agro-ecosystems. Agric. Ecosyst. Environ. 218, 28–32. http://dx.doi.org/10.1016/j.agee.2015.10.024.

Martinet, B., Rasmont, P., Cederberg, B., Evrard, D., Ødegaard, F., Paukkunen, J., Lecocq, T., 2015. Forward to the north: two Euro-Mediterranean bumblebee species now cross the Arctic Circle. Ann. Soc. Entomol. Fr., 1–7. http://dx.doi.org/10.1080/00379271.2015.1118357.

Massantini, F., Caporali, F., Zellin, G., 1977. Evidence for allelopathic control of weeds in lines of soybean. In: Proc. Eur. Weed Res. Soc. (EWRS) Symposium on the Different Methods of Weed Control and Their Integration, vol. 1, pp. 23–28.

Massol, F., Dubart, M., Calcagno, V., Cazelles, K., Jacquet, C., Kéfi, S., Gravel, D., 2017. Island biogeography of food webs. Adv. Ecol. Res. 56, 183–262.

Matsumoto, S., Abe, A., Maejima, T., 2009. Foraging behavior of *Osmia cornifrons* in an apple orchard. Sci. Hortic. (Amsterdam) 121, 73–79. http://dx.doi.org/10.1016/j.scienta.2009.01.003.

Matsumura, C., Yokoyama, J., Washitani, I., 2004. Invasion status and potential ecological impacts of an invasive alien bumblebee, Bombus terrestris L. (Hymenoptera: Apidae) naturalized in Southern Hokkaido, Japan. Glob. Environ. Res. 8, 51–66.

Maun, M.A., 1977. Suppressing effect of soybeans on barnyard grass. Can. J. Plant Sci. 57, 485–490.

McKone, M., McLauchlan, K.K., Lebrun, E.G., McCall, A.C., 2001. An edge effect caused by adult corn-rootworm beetles on sunflowers in tallgrass prairie remnants. Conserv. Biol. 15, 1315–1324.

Meeus, I., Brown, M.J.F., De Graaf, D.C., Smagghe, G., 2011. Effects of invasive parasites on bumble bee declines. Conserv. Biol. 25, 662–671. http://dx.doi.org/10.1111/j.1523-1739.2011.01707.x.

Millenium Ecosystem Assessment, 2005. Ecosystems and Human Well-Being: Synthesis. Island Press, Washington, DC.

Miller, A.E., Brosi, B.J., Magnacca, K., Daily, G.C., Pejchar, L., 2015. Pollen carried by native and nonnative bees in the large-scale reforestation of pastureland in Hawai'i: implications for pollination. Pac. Sci. 69, 67–79. http://dx.doi.org/10.2984/69.1.5.

Mollot, G., Pantel, J.H., Romanuk, T.N., 2017. The effects of invasive species on the decline in species richness: a global meta-analysis. Adv. Ecol. Res. 56, 61–83.

Montalva, J., Dudley, L., Arroyo, M.K., Retamales, H., Abrahamovich, A.H., 2011. Geographic distribution and associated flora of native and introduced bumble bees (*Bombus spp.*) in Chile. J. Apic. Res. 50, 11–21. http://dx.doi.org/10.3896/IBRA.1.50.1.02.

Morales, C.L., Aizen, M.A., 2006. Invasive mutualisms and the structure of plant-pollinator interactions in the temperate forests of north-west Patagonia, Argentina. J. Ecol. 94, 171–180. http://dx.doi.org/10.1111/j.1365-2745.2005.01069.x.

Morales, C.L., Traveset, L., 2009. A meta-analysis of impacts of alien vs. native plants on pollinator visitation and reproductive success of co-flowering native plants. Ecol. Lett. 12, 716–728.

Morales, C.L., Arbetman, M.P., Cameron, S.A., Aizen, M.A., 2013. Rapid ecological replacement of a native bumble bee by invasive species. Front. Ecol. Environ. 11, 529–534. http://dx.doi.org/10.1890/120321.

Morse, R.A., 1991. Honeybees forever. Trends Ecol. Evol. 6, 337–338. http://dx.doi.org/10.1016/0169-5347(91)90043-W. Personal Ed.

Morse, R.A., Calderone, N.W., 2000. The value of honeybees as pollinators of U.S. crops in 2000. Bee Cult. 132, 1–19.

Muratet, A., Fontaine, B., 2015. Contrasting impacts of pesticides on butterflies and bumblebees in private gardens in France. Biol. Conserv. 182, 148–154. http://dx.doi.org/10.1016/j.biocon.2014.11.045.

Murray, T.E., Coffey, M.F., Kehoe, E., Horgan, F.G., 2013. Pathogen prevalence in commercially reared bumble bees and evidence of spillover in conspecific populations. Biol. Conserv. 159, 269–276. http://dx.doi.org/10.1016/j.biocon.2012.10.021.

Myers, N., Mittermeier, R.A., Mittermeier, C.G., da Fonseca, G.A.B., Kent, J., 2000. Biodiversity hotspots for conservation priorities. Nature 403, 853–858. http://dx.doi.org/10.1038/35002501.

Nagamitsu, T., Kenta, T., Inari, N., Horita, H., Goka, K., Hiura, T., 2007. Foraging interactions between native and exotic bumblebees: enclosure experiments using native flowering plants. J. Insect Conserv. 11, 123–130. http://dx.doi.org/10.1007/s10841-006-9025-x.

Nagamitsu, T., Yamagishi, H., Kenta, T., Inari, N., Kato, E., 2009. Competitive effects of the exotic *Bombus terrestris* on native bumble bees revealed by a field removal experiment. Popul. Ecol. 52, 123–136. http://dx.doi.org/10.1007/s10144-009-0151-7.

Newman, E.I., Rovira, A.D., 1975. Allelopathy among some British grassland species. J. Ecol. 63, 727–737.

Nieto, A., Roberts, S.P.M., Kemp, J., Rasmont, P., Kuhlmann, M., Criado, M.G., Biesmeijer, J.C., Bogusch, P., Dathe, H.H., Rúa, P. De, 2014. European Red List of Bees. http://dx.doi.org/10.2779/77003.

Nishikawa, Y., Shimamura, T., 2016. Effects of alien invasion by *Bombus terrestris* L. (Apidae) on the visitation patterns of native bumblebees in coastal plants in northern Japan. J. Insect Conserv. 20, 1–14. http://dx.doi.org/10.1007/s10841-015-9841-y.

Olesen, J.M., Bascompte, J., Dupont, Y.L., Jordano, P., 2007. The modularity of pollination networks. Proc. Natl. Acad. Sci. U.S.A. 104, 19891–19896. http://dx.doi.org/10.1073/pnas.0706375104.

Orford, K.A., Murray, P.J., Vaughan, I.P., Memmott, J., 2016. Modest enhancements to conventional grassland diversity improve the provision of pollination services. J. Appl. Ecol. 53, 906–915. http://dx.doi.org/10.1111/1365-2664.12608.

Otterstater, M.C., Thompson, J.D., 2008. Does pathogen spill-over from commercially reared bumble bee threaten wild populations? PLoS One 3 (7), e2771.

Paini, D., Roberts, J., 2005. Commercial honeybees reduce the fecundity of an Australian native bee. Biol. Conserv. 123, 103–112. http://dx.doi.org/10.1016/j.biocon.2004.11.001.

Pantel, J.H., Bohan, D.A., Calcagno, V., David, P., Duyck, P.-F., Kamenova, S., Loeuille, N., Mollot, G., Romanuk, T.N., Thébault, E., Tixier, P., Massol, F., 2017. 14 questions for invasion in ecological networks. Adv. Ecol. Res. 56, 293–340.

Pauly, A., Hora, Z.A., 2013. Apini and Meliponini from Ethiopia (Hymenoptera: Apoidea: Apidae: Apinae). Belg. J. Entomol. 16, 1–36.

Perrard, A., Haxaire, J., Rortais, A., Villemant, C., 2009. Observations on the colony activity of the Asian hornet Vespa velutina Lepeletier 1836 (Hymenoptera: Vespidae: Vespinae) in France. Ann. Soc. Entomol. Fr. 45, 119–127. http://dx.doi.org/10.1080/00379271.2009.10697595.

Pitts-Singer, T.L., Bosch, J., 2010. Nest establishment, pollination efficiency, and reproductive success of Megachile rotundata (Hymenoptera: Megachilidae) in relation to resource availability in field enclosures. Environ. Entomol. 39, 149–158. http://dx.doi.org/10.1603/EN09077.

Pitts-Singer, T.L., Cane, J.H., 2011. The alfalfa leafcutting bee, Megachile rotundata: the world's most intensively managed solitary bee. Annu. Rev. Entomol. 56, 221–237. http://dx.doi.org/10.1146/annurev-ento-120709-144836.

Pocock, M.J.O., Evans, D.M., Memmott, J., 2012. The robustness and restoration of a network of ecological networks. Science 335, 973–977. http://dx.doi.org/10.1126/science.1214915.

Potts, S.G., Petanidou, T., Roberts, S., O'Toole, C., Hulbert, A., Willmer, P., 2006. Plant-pollinator biodiversity and pollination services in a complex Mediterranean landscape. Biol. Conserv. 129, 519–529. http://dx.doi.org/10.1016/j.biocon.2005.11.019.

Potts, S.G., Biesmeijer, J.C., Kremen, C., Neumann, P., Schweiger, O., Kunin, W.E., 2010. Global pollinator declines: trends, impacts and drivers. Trends Ecol. Evol. 25, 345–353. http://dx.doi.org/10.1016/j.tree.2010.01.007.

Power, A.G., Mitchell, C.E., 2004. Pathogen spillover in disease epidemics. Am. Nat. 164, S79–S89.

Rader, R., Howlett, B.G., Cunningham, S.A., Westcott, D.A., Newstrom-Lloyd, L.E., Walker, M.K., Teulon, D.A.J., Edwards, W., 2009. Alternative pollinator taxa are equally efficient but not as effective as the honeybee in a mass flowering crop. J. Appl. Ecol. 46, 1080–1087. http://dx.doi.org/10.1111/j.1365-2664.2009.01700.x.

Rader, R., Bartomeus, I., Garibaldi, L.A., Garratt, M.P.D., Howlett, B.G., Winfree, R., Cunningham, S.A., Mayfield, M.M., Arthur, A.D., Andersson, G.K.S., Bommarco, R., Brittain, C., Carvalheiro, L.G., Chacoff, N.P., Entling, M.H., Foully, B., Freitas, B.M., Gemmill-Herren, B., Ghazoul, J., Griffin, S.R., Gross, C.L., Herbertsson, L., Herzog, F., Hipólito, J., Jaggar, S., Jauker, F., Klein, A.-M., Kleijn, D., Krishnan, S., Lemos, C.Q., Lindström, S.A.M., Mandelik, Y., Monteiro, V.M., Nelson, W., Nilsson, L., Pattemore, D.E., de O. Pereira, N., Pisanty, G., Potts, S.G., Reemer, M., Rundlöf, M., Sheffield, C.S., Scheper, J., Schüepp, C., Smith, H.G., Stanley, D.A., Stout, J.C., Szentgyörgyi, H., Taki, H., Vergara, C.H., Viana, B.F., Woyciechowski, M., 2015. Non-bee insects are important contributors to global crop pollination. Proc. Natl. Acad. Sci. U.S.A. 113, 146–151. http://dx.doi.org/10.1073/pnas.1517092112.

Ramirez, N., Brito, Y., 1992. Pollination biology in a palm swamp community in the Venezuelan central plains. Bot. J. Linn. Soc. 110, 277–302.

Ramírez, F., Davenport, T.L., 2013. Apple pollination: a review. Sci. Hortic. 162, 188–203. http://dx.doi.org/10.1016/j.scienta.2013.08.007.

Rand, T.A., Tylianakis, J.M., Tscharntke, T., 2006. Spillover edge effects: the dispersal of agriculturally subsidized insect natural enemies into adjacent natural habitats. Ecol. Lett. 9, 603–614.

Rathcke, B., 1983. Competition and facilitation among plants for pollination. In: Real, L. (Ed.), Pollination Biol. Academic Press, Orlando, FL, pp. 305–329.

Rawat, L.S., Narwal, S.S., Kadiyan, H.S., Maikhuri, R.K., Negi, V.S., 2012. Allelopathic effects of sunflower on seed germination and seedling growth of Trianthema portulacastrum. Allelopath. J. 30, 11–21.

Requier, F., Odoux, J.F., Tamic, T., Moreau, N., Henry, M., Decourtye, A., Bretagnolle, V., 2015. Honeybee diet in intensive farmland habitats reveals an unexpectedly high flower richness and a major role of weeds. Ecol. Appl. 25, 881–890. http://dx.doi.org/10.1890/14-1011.1.

Rice, E.L., 1984. Allelopathy, second ed. Academic Press Inc., Orlando, FL.

Ricketts, T.H., Regetz, J., Steffan-Dewenter, I., Cunningham, S.A., Kremen, C., Bogdanski, A., Gemmill-Herren, B., Greenleaf, S.S., Klein, A.M., Mayfield, M.M., Morandin, L.A., Ochieng', A., Potts, S.G., Viana, B.F., 2008. Landscape effects on crop pollination services: are there general patterns? Ecol. Lett. 11, 499–515. http://dx.doi. org/10.1111/j.1461-0248.2008.01157.x.

Rohr, R.P., Saavedra, S., Bascompte, J., 2014. On the structural stability of mutualistic systems. Science 345, 416–425.

Rollin, O., Decourtye, A., 2015. Etudes des variations temporelles des populations d'abeilles mellifères et des communautés d'abeilles sauvages selon divers contextes paysagers—Test de l'hypothèse d'une interaction compétitive entre l'espèce *Apis mellifera* et les abeilles sauvages non-Apis (Compte rendu). ITSAP-Institut de l'Abeille.

Rollin, O., Bretagnolle, V., Decourtye, A., Aptel, J., Michel, N., Vaissière, B.E., Henry, M., 2013. Differences of floral resource use between honeybees and wild bees in an intensive farming system. Agric. Ecosyst. Environ. 179, 78–86. http://dx.doi.org/10.1016/ j.agee.2013.07.007.

Rollin, O., Bretagnolle, V., Fortel, L., Guilbaud, L., Henry, M., 2015. Habitat, spatial and temporal drivers of diversity patterns in a wild bee assemblage. Biodivers. Conserv. 24, 1195. http://dx.doi.org/10.1007/s10531-014-0852-x.

Roubi, D.W., Villanueva-Guttiérez, R., 2009. Invasive Africanized honeybee impact on native solitary bees: a pollen resource and trap nest analysis. Biol. J. Linn. Soc. 98, 152–160.

Roubik, D.W., 1978. Competitive interactions between neotropical pollinators and africanized honeybees. Science 201, 1030–1032. http://dx.doi.org/10.1126/science.201. 4360.1030.

Ruttner, F., 1988. Biogeography and Taxonomy of Honeybees. Springer, Berlin, Germany, p. 284.

Saavedra, S., Stouffer, D., Uzzi, B., Bascompte, J., 2011. Strong contributors to network persistence are the most vulnerable to extinction. Nature 478, 233–235.

Sáez, A., Morales, C.L., Ramos, L., Aizen, M.A., 2014. Extremely frequent bee visits increase pollen deposition but reduce drupelet set in raspberry. J. Appl. Ecol. 51, 1603–1612. http://dx.doi.org/10.1111/1365-2664.12325.

Sanguinetti, A., Singer, R.B., 2014. Invasive bees promote high reproductive success in Andean orchids. Biol. Conserv. 175, 10–20. http://dx.doi.org/10.1016/j.biocon.2014.04.011.

Santos, G.M., Aguiar, C.M., Genini, J., Martins, C.F., Zanella, F.C., Mello, M.A., 2012. Invasive Africanized honeybees change the structure of native pollination networks in Brazil. Biol. Inv. 14, 2369–2378.

Sauve, A.M.C., Thébault, E., Pocock, M.J., Fontaine, C., 2016. How plants connect pollination and herbivory networks and their contribution to community stability. Ecology 97, 908–917.

Sax, D.F., Gaines, S.D., 2008. Species invasions and extinction: the future of native biodiversity on islands. Proc. Natl. Acad. Sci. U.S.A. 105, 11 490–11 497. http://dx.doi.org/ 10.1073/pnas.0710824105.

Schaffer, W.M., Zeh, D.W., Buchmann, S.L., Kleinhans, S., Valentine Schaffer, M., Antrim, J., 1983. Competition for nectar between introduced honeybees and native North American bees and ants. Ecology 64, 564–577.

Schmid-Hempel, R., Eckhardt, M., Goulson, D., Heinzmann, D., Lange, C., Plischuk, S., Escudero, L.R., Salathé, R., Scriven, J.J., Schmid-Hempel, P., 2014. The invasion of southern South America by imported bumblebees and associated parasites. J. Anim. Ecol. 83, 823–837. http://dx.doi.org/10.1111/1365-2656.12185.

Seeley, T.D., 1985. Honeybee Ecology. Princeton University Press, Princeton, NJ.

Senapathi, D., Carvalheiro, L.G., Biesmeijer, J.C., Dodson, C., Evans, R.L., McKerchar, M., Morton, R.D., Moss, E.D., Roberts, S.P.M., Kunin, W.E., Potts, S.G., 2015. The impact of over 80 years of land cover changes on bee and wasp pollinator communities in England. Proc. R. Soc. B Biol. Sci. 282, 20150294. http://dx.doi.org/10.1098/rspb.2015.0294.

Shavit, O., Dafni, A., Ne'eman, G., 2009. Competition between honeybees (*Apis mellifera*) and native solitary bees in the Mediterranean region of Israel—implications for conservation. Isr. J. Plant Sci. 57, 171–183. http://dx.doi.org/10.1560/IJPS.57.3.171.

Sheffield, C.S., 2014. Pollination, seed set and fruit quality in apple: studies with *Osmia lignaria* (Hymenoptera: Megachilidae) in the Annapolis valley, Nova Scotia, Canada. J. Pollinat. Ecol. 12, 120–128.

Sheppard, W.S., Meixner, M.D., 2003. *Apis mellifera pomonella*, a new honeybee subspecies from Central Asia. Apidologie 34, 367–375.

Slaa, E.J., Sánchez Chaves, L.A., Malagodi-Braga, K.S., Hofstede, F.E., 2006. Stingless bees in applied pollination: practice and perspectives. Apidologie 37, 293–315. http://dx.doi.org/10.1051/apido:2006022.

Spiesman, B.J., Gratton, C., 2016. Flexible foraging shapes the topology of plant-pollinator interaction networks. Ecology 97, 1431–1441. http://dx.doi.org/10.1890/15-1735.1.

Squires, S.E., Hermanutz, L., Dixon, P.L., 2009. Agricultural insect pest compromises survival of two endemic Braya (Brassicaceae). Biol. Conserv. 142, 203–211.

Stanley, D.A., 2013. Pollinators and pollination in changing agricultural landscapes; investigating the impacts of bioenergy crops. Thesis, Trinity College Dublin.

Stanley, D.A., Stout, J.C., 2013. Quantifying the impacts of bioenergy crops on pollinating insect abundance and diversity: a field-scale evaluation reveals taxon-specific responses. J. Appl. Ecol. 50, 335–344.

Stanley, D.A., Stout, J.C., 2014. Pollinator sharing between mass-flowering oilseed rape and co-flowering wild plants: implications for wild plant pollination. Plant Ecol. 215, 315–325.

Steffan-Dewenter, I., Kuhn, A., 2003. Honeybee foraging in differentially structured landscapes. Proc. R. Soc. Lond. B 270, 569–575.

Steffan-Dewenter, I., Schiele, S., 2008. Do resources or natural enemies drive bee population dynamics in fragmented habitats? Ecology 89, 1375–1387.

Steffan-dewenter, I., Tscharntke, T., 2000. Resource overlap and possible competition between honeybees and wild bees in central Europe. Oecologia 122, 288–296.

Steffan-Dewenter, I., Tscharntke, T., 2001. Succession of bee communities on fallows. Ecography 24, 83–93.

Stelzer, R.J., Chittka, L., 2010. Bumblebee foraging rhythms under the midnight sun measured with radiofrequency identification. BMC Biol. 8, 93. http://dx.doi.org/10.1186/1741-7007-8-93.

Stern, R.A., Goldway, M., Zisovich, A.H., Shafir, S., Dag, A., 2004. Sequential introduction of honeybee colonies increases cross-pollination, fruit-set and yield of "Spadona" pear (*Pyrus communis* L.). J. Hortic. Sci. Biotechnol. 79, 652–658.

Stouffer, D.B., Cirtwill, A.R., Bascompte, J., 2014. How exotic plants integrate into pollination networks. J. Ecol. 102, 1442–1450. http://dx.doi.org/10.1111/1365-2745.12310.

Stout, J.C., Morales, C.L., 2009. Ecological impacts of invasive alien species on bees*. Apidologie 40, 388–409. http://dx.doi.org/10.1051/apido/2009023.

Stout, J.C., Kells, A.R., Goulson, D., 2002. Pollination of the invasive exotic shrub *Lupinus arboreus* (Fabaceae) by introduced bees in Tasmania. Biol. Conserv. 106, 425–434. http://dx.doi.org/10.1016/S0006-3207(02)00046-0.

Sugiura, S., 2016. Impacts of introduced species on the biota of an oceanic archipelago: the relative importance of competitive and trophic interactions. Ecol. Res. 31, 155–164. http://dx.doi.org/10.1007/s11284-016-1336-0.

Thébault, E., Fontaine, C., 2010. Stability of ecological communities and the architecture of mutualistic and trophic networks. Science 329, 853–856. http://dx.doi.org/10.1126/science.1188321.

Thomson, D., 2004. Competitive interactions between the invasive European honeybee and native bumble bees. Ecology 85, 458–470. http://dx.doi.org/10.1890/02-0626.

Thomson, D.M., 2016. Local bumble bee decline linked to recovery of honey bees, drought effects on floral resources. Ecol. Lett. 19, 1247–1255. http://dx.doi.org/10.1111/ele.12659.

Thorp, R.W., 1996. Resource overlap among native and introduced bees in California. In: Matheson, A., Buchmann, S.L., O'Toole, C., Westrich, P., Williams, I.H. (Eds.), The Conservation of Bees. Academic Press, London, pp. 134–152.

Tiedeken, E.J., Stout, J.C., 2015. Insect-flower interaction network structure is resilient to a temporary pulse of floral resources from invasive *Rhododendron ponticum*. PLoS One 10, e0119733. http://dx.doi.org/10.1371/journal.pone.0119733.

Torné-Noguera, A., Rodrigo, A., Osorio, S., Bosch, J., 2015. Collateral effects of beekeeping: impacts on pollen-nectar resources and wild bee communities. Basic Appl. Ecol. 17, 199–209. http://dx.doi.org/10.1016/j.baae.2015.11.004.

Traveset, A., Richardson, D.M., 2006. Biological invasions as disruptors of plant reproductive mutualisms. Trends Ecol. Evol. 21, 208–216. http://dx.doi.org/10.1016/j.tree.2006.01.006.

Traveset, A., Richardson, D.M., 2011. Mutualisms: key drivers of invasions ... key casualties of invasions. In: Richardson, D.M. (Ed.), Fifty Years of Invasion Ecology: The Legacy of Charles Elton. Wiley-Blackwell, Oxford, UK, pp. 143–160. http://dx.doi.org/10.1002/9781444329988.ch12.

Traveset, A., Richardson, D.M., 2014. Mutualistic interactions and biological invasions. Annu. Rev. Ecol. Evol. Syst. 45, 89–113. http://dx.doi.org/10.1146/annurev-ecolsys-120213-091857.

Traveset, A., Heleno, R., Chamorro, S., Vargas, P., McMullen, C.K., Castro-Urgal, R., Nogales, M., Herrera, H.W., Olesen, J.M., 2013. Invaders of pollination networks in the Galapagos Islands: emergence of novel communities. Proc. R. Soc. B Biol. Sci. 280, 20123040. http://dx.doi.org/10.1098/rspb.2012.3040.

Tscharntke, T., Klein, A.M., Kruess, A., Steffan-Dewenter, I., Thies, C., 2005. Landscape perspectives on agricultural intensification and biodiversity—ecosystem service management. Ecol. Lett. 8, 857–874. http://dx.doi.org/10.1111/j.1461-0248.2005.00782.x.

Tschumi, M., Albrecht, M., Collatz, J., Dubsky, V., Entling, M.H., Najar-Rodriguez, A.J., Jacot, K., 2016. Tailored flower strips promote natural enemy biodiversity and pest control in potato crops. J. Appl. Ecol. 53, 1169–1176. http://dx.doi.org/10.1111/1365-2664.12653.

Turk, M.A., Tawaha, A.M., 2003. Allelopathic effect of black mustard (*Brassica nigra* L.) on germination and growth of wild oat (*Avena fatua* L.). Crop Prot. 22, 673–677. http://dx.doi.org/10.1016/S0261-2194(02)00241-7.

Tylianakis, J.M., 2008. Understanding the web of life: the birds, the bees, and sex with aliens. PLoS Biol. 6, 0224–0228. http://dx.doi.org/10.1371/journal.pbio.0060047.

Tylianakis, J.M., Didham, R.K., Bascompte, J., Wardle, D.A., 2008. Global change and species interactions in terrestrial ecosystems. Ecol. Lett. 11, 1351–1363. http://dx.doi.org/10.1111/j.1461-0248.2008.01250.x.

Tylianakis, J.M., Laliberté, E., Nielsen, A., Bascompte, J., 2010. Conservation of species interaction networks. Biol. Conserv. 143, 2270–2279. http://dx.doi.org/10.1016/j.biocon.2009.12.004.

Valdovinos, F.S., Moisset De Espanes, P., Flores, J.D., Ramos-Jiliberto, R., 2013. Adaptive foraging allows the maintenance of biodiversity of pollination networks. Oikos 122, 907–917.

Van Engelsdorp, D., Meixner, M.D., 2010. A historical review of managed honeybee populations in Europe and the United States and the factors that may affect them. J. Invertebr. Pathol. 103, S80–S95. http://dx.doi.org/10.1016/j.jip.2009.06.011.

Vanbergen, A.J., 2013. Threats to an ecosystem service: pressures on pollinators. Front. Ecol. Environ. 11, 251–259. http://dx.doi.org/10.1890/120126.

Vaughton, G., 1996. Pollination disruption by European honeybees in the Australian bird-pollinated shrub Grevillea barklyana (Proteaceae). Plant Syst. Evol. 200, 89–100.

Vázquez, D.P., Aizen, M.A., 2004. Asymmetric specialization: a pervasive feature of plant-pollinator interactions. Ecology 85, 1251–1257.

Velthuis, H.H.W., Van Doorn, A., 2006. A century of advances in bumblebee domestication and the economic and environmental aspects of its commercialization for pollination. Apidologie 37, 421–451. http://dx.doi.org/10.1051/apido.

Vereecken, N.J., Dufrêne, E., Aubert, M., 2015. Sur la coexistence entre l'abeille domestique et les abeilles sauvages. Rapport de synthèse sur les risques liés à l'introduction de ruches de l'abeille domestique (Apis mellifera) vis-à-vis des abeilles sauvages et de la flore. Observatoire des Abeilles (OA).

Vilà, M., Bartomeus, I., Dietzsch, A.C., Petanidou, T., Steffan-Dewenter, I., Stout, J.C., Tscheulin, T., 2009. Invasive plant integration into native plant-pollinator networks across Europe. Proc. Biol. Sci. 276, 3887–3893. http://dx.doi.org/10.1098/rspb.2009.1076.

Villemant, C., Haxaire, J., Streito, J.C., 2006. Premier bilan de l'invasion de Vespa velutina Lepeletier en France (Hymenoptera, Vespidae). Bull. Soc. Entomol. Fr. 111, 535–538.

Villemant, C., Barbet-Massin, M., Perrard, A., Muller, F., Gargominy, O., Jiguet, F., Rome, Q., 2011a. Predicting the invasion risk by the alien bee-hawking Yellow-legged hornet Vespa velutina nigrithorax across Europe and other continents with niche models. Biol. Conserv. 144, 2142–2150. http://dx.doi.org/10.1016/j.biocon.2011.04.009.

Villemant, C., Muller, F., Haubois, S., 2011b. Bilan des travaux (MNHN et IRBI) sur l'invasion en France de Vespa velutina, le frelon asiatique predateur d'abeilles. In: Barbançon, J.-M, L'Hostis, M. (Eds.), Journée Scientifique Apicole JSA, Arles, 11 février 2011. ONIRIS-FNOSAD, Nantes, pp. 3–12.

Walther-Hellwig, K., Fokul, G., Frankl, R., Buchler, R., Ekschmitt, K., Wolters, V., 2006. Increased density of honeybee colonies affects foraging bumblebees. Apidologie 37, 517–532. http://dx.doi.org/10.1051/apido:2006035.

Watts, S., Ovalle, D.H., Herrera, M.M., Ollerton, J., 2012. Pollinator effectiveness of native and non-native flower visitors to an apparently generalist Andean shrub, Duranta mandonii (Verbenaceae). Plant Species Biol. 27, 147–158. http://dx.doi.org/10.1111/j.1442-1984.2011.00337.x.

Westphal, C., Steffan-Dewenter, I., Tscharntke, T., 2003. Mass flowering crops enhance pollinator densities at a landscape scale. Ecol. Lett. 6, 961–965. http://dx.doi.org/10.1046/j.1461-0248.2003.00523.x.

Westphal, C., Steffan-Dewenter, I., Tscharntke, T., 2006. Foraging trip duration of bumblebees in relation to landscape-wide resource availability. Ecol. Entomol. 31, 389–394. http://dx.doi.org/10.1111/j.1365-2311.2006.00801.x.

Westphal, C., Steffan-Dewenter, I., Tscharntke, T., 2009. Mass flowering oilseed rape improves early colony growth but not sexual reproduction of bumblebees. J. Appl. Ecol. 46, 187–193. http://dx.doi.org/10.1111/j.1365-2664.2008.01580.x.

Williams, I.H., 1994. The dependence of crop production within the European Union on pollination by honeybees. Agric. Zool. Rev. 6, 229–257.

Williams, P.H., An, J., Brown, M.J.F., Carolan, J.C., Goulson, D., Huang, J., Ito, M., 2012. Cryptic bumblebee species: consequences for conservation and the trade in greenhouse pollinators. PLoS One 7, 1–8. http://dx.doi.org/10.1371/journal.pone.0032992.

Zhang, H., Huang, J., Williams, P.H., Vaissière, B.E., Zhou, Z., Gai, Q., Dong, J., An, J., 2015. Managed bumblebees outperform honeybees in increasing peach fruit set in China: different limiting processes with different pollinators. PLoS One 10, e0121143. http://dx.doi.org/10.1371/journal.pone.0121143.

Zurbuchen, A., Landert, L., Klaiber, J., Mueller, A., Hein, S., Dorn, S., 2010. Maximum foraging ranges in solitary bees: only few individuals have the capability to cover long foraging distances. Biol. Conserv. 143, 669–676.

CHAPTER FIVE

Invasions of Host-Associated Microbiome Networks

C.L. Murall*,1, J.L. Abbate*,†, M. Puelma Touzel‡, E. Allen-Vercoe§, S. Alizon*, R. Froissart*, K. McCann§

*Laboratoire MIVEGEC (UMR CNRS 5290, UR IRD 224, UM), Montpellier, France
†UMI UMMISCO (UMI 209 IRD, UPMC), Bondy, France
‡Laboratoire de physique théorique (UMR 8549) CNRS and École Normale Supérieure, Paris, France
§University of Guelph, Guelph, Canada
1Corresponding author: e-mail address: carmenlia.murall@outlook.com

Contents

Advances in Ecological Research, Volume 57
ISSN 0065-2504
http://dx.doi.org/10.1016/bs.aecr.2016.11.002

Abstract

The study of biological invasions of ecological systems has much to offer research on within–host (WH) systems, particularly for understanding infections and developing therapies using biological agents. Thanks to the ground-work established in other fields, such as community ecology and evolutionary biology, and to modern methods of measurement and quantification, the study of microbiomes has quickly become a field at the forefront of modern systems biology. Investigations of host-associated microbiomes (e.g. for studying human health) are often centred on measuring and explaining the structure, functions and stability of these communities. This momentum promises to rapidly advance our understanding of ecological networks and their stability, resilience and resistance to invasions. However, intrinsic properties of host-associated microbiomes that differ from those of free-living systems present challenges to the development of a WH invasion ecology framework. The elucidation of principles underlying the invasibility of WH networks will ultimately help in the development of medical applications and help shape our understanding of human health and disease.

1. INTRODUCTION

Never before has there been such widespread appreciation that ecology happens not just around us but inside us. The ongoing discovery of the complexity of the web of microorganisms in and on our bodies has led to a rapidly emerging, systems-level picture of our **within–host (WH)** ecosystem. Owing to the recent advancements of '-*omics*' techniques and increasingly sensitive methods to probe immunity and host cells, it is now possible to study the ecology and evolution happening inside the human body at an unprecedented level of resolution and quantification. Arising from this is a challenge similar to the one ecologists face for free-living ecosystems: how can we harness community (i.e. network) level knowledge in order to predict the consequences of disturbances? In particular, what makes a WH community susceptible to or protected against invasion?

There is a wide range of biological invasions that can enter our bodies, from various kinds of micro- to macroorganisms (e.g. bacteria, viruses, worms) to human-created invaders (e.g. live vaccines, probiotics, bacteriophage therapies). Consequently, knowledge about biological invasions of the WH ecosystem is scattered across the scientific literature. The introduction of a nonnative biological agent into a WH system is usually referred to as an **infection**. A species introduced into a WH system is considered a successful **WH invader** when it is able to establish and maintain a viable population in or on the host. While most research is interested in infections that cause disease, i.e. **pathogenic** invaders, successful WH invasion does not necessarily harm or kill the host. An invasive species in a WH system may be beneficial, i.e. it may share a symbiotic relationship with the host. Similarly, in free-living systems an invading species is considered to be **invasive** if it has a disruptive or negative impact on the resident community; while frequent, this is not always the case. Here, we will consider invasions generally and thus include discussions about biological agents invading WH ecosystems regardless of medical implications to the host or negative impacts on the resident community.

The extent of the interdisciplinarity of WH studies is a challenge. Many insights, concepts and discoveries are found in very distant fields with often little communication with one another (e.g. microbiology, immunology, various clinical fields, community ecology, evolutionary biology and invasion biology where insights are often found in the plant or conservation literature). Also, the discovery of the members of WH communities has created a picture of an interkingdom network, replete with bacteria, viruses, fungi and archaea (Horz, 2015; Huffnagle and Noverr, 2013; Lecuit and Eloit, 2013; Nobbs and Jenkinson, 2015). We believe that ecological and evolutionary theory can be a unifying framework from which patterns and principles can emerge.

There have been several attempts at synthesizing ecological methods and concepts to aid researchers studying the microbiome (e.g. reviews by Costello et al., 2012; Ley et al., 2006; Pepper and Rosenfeld, 2012). Here, our focus is more targeted. We wish to illustrate how insights from the ecology and evolution of biological invasions can help us understand the massive and ever-growing knowledge about host-associated microbiomes. We use **'microbiome'** to mean the collection of microorganisms in and on its host and this also encompasses the set of genes from these microorganisms, their products and environment (Boon et al., 2014). These, then, are host-associated 'biomes'. We also discuss when these ecosystems exhibit novel features distinct from free-living ecosystems. Given the immense variety of invading species that can enter host-associated ecosystems, we will focus our examples mainly on invasions of bacterial communities by other bacteria and bacteriophages. We complement this selection with a non-exhaustive set of other examples chosen either for their importance in the development of the field or because they exhibit unique features worthy of mention. We end by highlighting the various invasions and perturbations that humans actively induce (e.g. probiotics or antibiotics) and finally conclude with a perspective of future directions for those interested in applying a community ecology and network approach.

2. ECOLOGICAL AND EVOLUTIONARY CONCEPTS APPLIED TO MICROBIOME NETWORKS: IN THE FACE OF INVADERS

2.1 Invasion and Ecological Stability

Community ecology has a long, rich history of studying the stability of ecological systems (Ives and Carpenter, 2007; May, 1972; McCann, 2000; Namba, 2015). The application of this knowledge, theory and methods should play a central role in the understanding of the human WH ecosystem. Indeed, humans can be seen as an ecosystem with many interacting species, perturbations, spatial structure, environmental gradients and successional patterns. In addition, studying invasions of the microbiome can help elucidate the underlying mechanisms that stabilise and destabilise these ecological systems.

The meaning of a 'stable community' can differ depending on the definition of stability or what metric is used (for more on stability's multidimensionality and other details, see Donohue et al., 2013; Ives and Carpenter, 2007; McCann, 2011). Two important concepts, rooted in dynamical systems theory, arise. A **steady state** is one where, in the absence of perturbations, the behaviour of the system remains constant over time. When a steady state is stable, the natural dynamics of the system will take its state to this stable state, i.e. to an attractor, which can have fixed, oscillating, or even chaotic values. The **return time** refers to how quickly a system can return to a stable steady state that it is perturbed away from (this is also called **resilience** in ecology). Changing parameters can alter the stability of steady states, for example by making them suddenly unstable, whereupon the natural dynamics takes the state away to some other steady state that is stable. Most theoretical studies of ecological stability focus on stable states while empirical studies use temporal (or spatial) variability of abundance or biomass to measure stability, where communities with wildly fluctuating populations are considered less stable (Donohue et al., 2016).

The theoretical literature on invasion and its implication on stability are twofold. One, there exists a literature deriving largely from Elton (1958) that addresses **invasion resistance**, which is measured as a change in magnitude of abundance and biomass after an invasion, and it is related to **persistence**, where a community that is difficult to invade is considered persistent. For a summary of the various hypotheses to explain successful invasions, including Elton's **biotic resistance hypothesis** (see Inderjit et al., 2005; in particular their Table 1), and Section 2.2 for discussion of classic empirical studies. A second, far less developed area, considers the impacts of invasions on the stability of the preexisting whole system. One metric used for this is **robustness**, which is the number of extinctions after an invasion. Charles Elton with his enormous historical influence weighed in on both of these aspects of invasion and stability by broadly arguing that more diverse ecosystems tend to be more resistant to invasion and less affected (Elton, 1958).

Since Elton (1958), there has been a strong push for understanding the role of diversity on invasion resistance, usually through intra- or interspecific competition. Case (Case, 1990; Case and Bolger, 1991) and others (Drake, 1990; Law and Morton, 1996; Levine and D'Antonio, 1999) employed communities matrices capable of producing complex communities (May, 1973) to show that, on average, more diverse communities were more resistant to invasion. In an excellent synthetic single trophic level contribution, Tilman (1999) argued that much of diversity–stability theory agreed with Elton's premise that increased diversity tends to decrease invasion success. Specifically, Tilman (1999) argued that increasing diversity has the tendency, on average, to reduce equilibrium resource levels, R^*. Since lower resource levels inhibit successful invasion, it remains that diverse communities should be more resistant (Tilman, 1999). While interesting, this perspective is largely considered from a negative interaction perspective, and others (e.g. Lawlor, 1979; Levine and D'Antonio, 1999) have pointed out that species interactions that include facilitative effects can alter this result in such a way that increased diversity can actually decrease invasion resistance.

In the past 20 years, both the general ecological literature on diversity–stability (e.g. McCann, 2000) and the invasion resistance literature discussed earlier (e.g. Levine and D'Antonio, 1999) have developed these arguments to find that diversity, in and of itself, is likely not the greatest contributor to stability in the face of invasion. The diversity–stability literature, for example, has argued that the structure and positioning of interaction strengths can dramatically increase stability relative to the random community approaches (McCann, 2011). One finding is that weak interactions, in low- or high-diversity trophic webs, can act to mute out the destabilising impacts of strong, highly suppressive or oscillatory, interactions (Gellner and Mccann, 2016). Indeed, in the largest analysis of empirical food webs to date, nonrandom structures, such as a high frequency of weak interactions and their topology, were found to stabilise complex food webs (Jacquet et al., 2016). From an invasion perspective, these results strongly suggest that invasive species with rapid growth potential and low mortality (e.g. the invader has no resident predator in the newly invaded community) are extremely likely to not only invade, but also reach high densities, and act as potent destabilisers of even diverse communities (see single trophic level ideas that are similar by Tilman, 1999). Thus, both demographic attributes (high growth potential) and a species position within the food web contrive to yield massive destabilising impacts on the resident food web.

2.2 Insights From Empirical Invasion Ecology

Empirical investigations of invasions of ecological systems have given ecologists insights into how ecological communities assemble and evolve over time (see Amsellem et al., 2017; Jackson et al., 2017; Médoc et al., 2017, in this issue). While there are several classic lessons that invasions have taught us about the ecology and evolution of communities (e.g. rapid geographical range changes are possible, speciation can be sympatric), more recent studies have pushed this body of knowledge further (e.g. communities are rarely saturated with species, adaptive genetic change can be occur quickly; see Sax et al., 2007 for a review). It can be argued that in these early years of microbiome studies, invasions have also played an important role in helping probe these networks. We will review some examples of these microbiome studies in later sections. Invasions of WH communities are implicated in a number of human diseases, particularly those induced by infectious pathogens, as well as in determining the success of prophylactic and therapeutic biological control agents. Therefore, invasion ecology of the WH ecosystem will continue to not only be central to teaching us how these systems work but is also crucial for applied research. Here we present some insights from invasions of free-living systems that might be of use to those studying WH systems.

The field of invasion ecology is focused primarily on what conditions determine either success at or susceptibility to invasion by a novel species into a resident community. From the perspective of the invader, this action is broken down into a set of successive processes that must be executed: (1) **dispersal** out of the native environment, (2) **colonisation** of a suitable new environment, (3) successful **establishment** in the new environment, and (4) **spread** across the novel landscape (Theoharides and Dukes, 2007). Experimental studies have explored the traits that favour or hinder these processes, and how each step contributes to the success and impact of the invasion. No fewer than 29 different hypotheses have been posited in the literature to explain either what determines a successful alien invader, or what factors might render a resident community vulnerable to invasion (recently reviewed in Jeschke, 2014). Some of the most pervasive and pertinent to invasions of the microbiome, discussed or alluded to later in this chapter, include the *biotic resistance hypothesis* (as mentioned earlier), the **adaptive evolution hypothesis** (the ability of a species to respond to selective pressures, see Keller and Taylor, 2008 for a synthesis), and the **invasional facilitation hypothesis** or '**invasional meltdown**', which posits that successive invasions happen through time due to facilitative interactions between invaders (Inderjit et al., 2005; Simberloff and Holle, 1999).

Empirical investigations into some of these hypotheses have found a number of biological properties that contribute to invasion success. Here, we describe a few well-supported properties, such as dispersal, adaptation, and phylogenetic relatedness.

Dispersal out of the native environment (invasion step 1) and colonisation into the new environment (invasion step 2) are processes that can be likened to epidemiological transmission of infectious agents (regardless of pathogenicity). If there is a low probability of success for each agent's propagule to disperse (make contact) and establish (cause infection) in a new host, then this can be offset by increasing the number of propagules leaving the original infected host (if the agent is pathogenic then this is equivalent to increasing its virulence) or, otherwise, by increasing the amount of contact events with the prospective new host (Alizon and Michalakis, 2015; Gilligan and Van Den Bosch, 2008). Indeed, in free-living communities, species with higher fecundity were more often a priori classified as invasive (Burns et al., 2013). The intuitive finding that species with life-history strategies that divert more energy to reproduction were more successful invaders than those that put more energy into survival (e.g. having higher overwintering success) is consistent with the growing evidence that suggests **propagule pressure** (i.e. the frequency and abundance of newly arriving, hopeful invaders) is one of the most important indicators of establishment in a new range (Lockwood et al., 2005). High propagule pressure can help an invading species achieve threshold densities of individuals necessary for population viability (e.g. allee effects; Hufbauer et al., 2013). This feature of invasibility is likely to be shared with WH systems, as many pathogens are thought to only be successful after achieving a threshold of *infection pressure* (Simberloff, 2009), minimum infectious dose (Yezli and Otter, 2011), or infection *spillover* events between species that often precede full-blown host shifts (Lloyd-Smith et al., 2009). Hartfield and Alizon (2014) laid out a mechanistic process by which epidemiological feedbacks may allow for this to happen, and this theory could be of use to help understand the determinants of successful establishment for an invader, particularly if the invader has arrived from a very different environment (e.g. not from direct transmission between hosts). However, too much propagule pressure can actually limit the success of an invasion, because colonisers with advantageous mutations in the new environment may be out-competed (in terms of opportunity for establishment) due to their rarity among other propagules that are more adapted, on average, to the home range (and not the new range). This is termed **gene swamping** and has been shown to hinder the ability to maintain novel adaptations to the new environment in small peripheral populations (Antonovics, 1968, 1976). While an organism that can reproduce without a partner is not limited by finding a suitable mate in the new range, clonality, self-fertilisation or even inbreeding within large homogeneous founder events can limit the probability of successful invasion (e.g. Hufbauer et al., 2013).

On the other hand, **prior adaptation**, i.e. selection in the native habitat before invasion that produces traits conferring an evolutionary advantage in the introduced habitat (also sometimes termed '*preadaptation*'; see Hufbauer et al., 2012 for a detailed description), can be important for biological invasions. For example, a comparative study among wildflower species in the family Caryophyllaceae showed that traits that improved home range size were tightly correlated with successful invasion in the new range (Jenkins and Keller, 2011). In cases such as these, one would expect different success rates by an invader coming from a completely different environment (e.g. abiotic/environmental transmission) vs from inside another host (e.g. direct contact or ingestion) or displacement within the same host (e.g. auto-inoculation). Along these lines, Tilman (1999) also pointed out that invaders from an entirely separate ecosystem may often evolve under different conditions giving them traits (e.g. low edibility) that produce a suite of particularly successful advantages in another differently evolved ecosystem.

The determination of successful invasion is likely to involve the interaction between several factors. Gallien et al. (2014) used community vegetation data from the French Alps to illustrate the insight offered by invasive species distribution models that incorporate biotic interactions with environmental and dispersal components. Whether through shared resources or escape from predators, invasion steps 3 (establishment) and 4 (spread), in particular, must be viewed in the context of the resident species in the new environment. Rather than intrinsic traits or a general rule, the advantageous characteristics of an invader depend on the functional and phylogenetic similarities shared with taxa in the resident community, and these relationships can change depending on the scale (Carboni et al., 2016) or environmental conditions (Burns, 2006) under which they are measured. However, there is a debate concerning the positive or negative impacts of relatedness on an invasion, as related resident taxa may either prime the environment to facilitate similar taxa (or simply indicate a permissive environment for similarly adaptive traits, termed the **preadaptation hypothesis**) or serve as exclusive competitors already occupying the shared 'niche' making it harder to invade, i.e. **Darwin's naturalisation hypothesis** (this is discussed in greater detail by Pantel et al., 2017, in this issue). In the context of the WH microbiome, where the sheer number and diversity of species interactions can be overwhelming, consideration of functional trait similarities may prove more informative than seemingly everchanging taxonomic classifications of microbiota (Martiny et al., 2015). This is important to establish, because a recent metaanalysis on the generality of relatedness effects on invasion success and impact showed that broad generalities appear to exist across a wide range of plant and animal studies (Ma et al., 2016; see also Box 1 for a discussion on relatedness and invasion success). Furthermore, experimental results from a longitudinal study of plant communities showed that the invader can quickly drive the native relative to extinction (Li et al., 2015), thus increasing the chances that it is missed by simple association surveys. This may be of particular importance for disease studies, in which the microbiota is typically assessed in a snapshot fashion after the invasion has occurred.

It is also helpful to view the biological properties of the resident community as dynamic rather than static. Thus, it matters *when* an invader enters along the evolutionary or ecological trajectory of the resident community. This invokes yet another major axis of fundamental ecology, **ecological succession**, or the progression of biological community assembly through time. In forest ecology, from which this literature was born, the pervasive idea is that there is a relatively predictable progression from a virgin environment devoid of life to a stable community of species best adapted to the habitat. Theory, and decades of empirical support from free-living species, posits that while the most successful of initial colonisers to a new community will be those with **r-selected** lifestyles (investing in a large number of fast-growing and wide-spreading offspring), those with more **K-selected** lifestyles (investing in fewer, larger, slower-growing offspring) that arrive later tend to be better competitors in the recently colonised environment (Duyck et al., 2004, 2006; Facon et al., 2005, 2008). The speed and timing

BOX 1 Does Relatedness Help or Hurt a Species' Ability to Invade?

Ecological studies have shown us that there can be both positive and negative consequences to sexual reproduction of invaders with taxa in the new environment (see Section 2), and similarly, functional or genetic relatedness between WH invaders and the resident microbiome can affect invasibility success, establishment and spread. Cross-immunity (or cross-reactivity) is a major evolutionary force that affects pathogen diversity (i.e. it drives viruses and microbes to be as distinct as possible from one another in order to avoid immunity detection, memory recognition and clearance). Mechanisms to escape cross-reactive immune responses include recombination (i.e. microbe sex), rapid generation times and substitution rates, and antigenic shedding. Paradoxically, interactions with related resident taxa can both help (e.g. provide beneficial genes conferring higher fitness, via recombination or gene transfer) or hinder (e.g. introducing genes conferring lower fitness) the invader. Finally, related resident taxa can both prime the environment to help invaders but can also be competitive within a shared 'niche', making it harder to invade (i.e. *Darwin's naturalisation hypothesis*). Whether relatedness of the invader to resident taxa indicates a permissive environment (termed the *preadaptation hypothesis*) or a more challenging battle with these native competitors is the subject of much debate. A recent metaanalysis across mostly plant and animal invasion studies found that the answer to this 'conundrum' appears to be dependent on the scale at which it is measured, but the patterns are fairly consistent across wide variation in taxa and ecosystems (Ma et al., 2016). Thus, determining whether relatedness can help or hurt invasion success in the microbiome will require similar studies and metaanalysis across various kinds of invaders (viruses, microbes, etc.) in order to find out whether these apparently global patterns also apply to WH microbial systems.

of this succession can therefore influence what taxa, and traits, are those considered successful for invasion. In the case of food webs, trophic level and generalism play such a role, i.e. islands are first colonised by producers, then primary consumers, and so forth (see Massol et al., 2017, in this issue). This is particularly applicable to how our bodies are colonised after birth. We are born essentially germfree, and the development from infant to childhood to adulthood is a long complex series of colonisation events (Koenig et al., 2011). While the concept also helps generate expectations for recolonisation of the microflora after perturbation (such as antibiotic exposure), we will discuss in later sections how succession has been important in experimental studies of the microbiome, which are often conducted on axenic in vivo or in vitro systems for tractability.

Finally, a classic insight from invasion ecology is that, counterintuitively, resident species are not always optimally adapted to their environment (Sax et al., 2007). The potential that invaders are better adapted to the host environment than resident species is underappreciated in microbiome studies (Cho and Blaser, 2012). It is tempting to view a resident WH species (commensal or mutualistic) as being highly adapted to their environment and host (e.g. Ley et al., 2006), and the expectation that symbiotic bacteria must be highly adapted to very specialised host niches (Danielsson et al., 2011) is found in the microbiome literature. However, the fact that some invaders can successfully invade and outperform local species in some functions means that the local strategy may not be the true optimal strategy, but rather is the best that the local species can do given its own evolutionary history or constraints. Like in free-living systems, resident bacterial taxa will have some traits that are more highly specialised than others. Similarly, community composition is often not due to ancient coevolutionary histories with their host but rather communities can be formed uniquely by species interactions (via *fitting* and *sorting*), and as a consequence, communities dominated by exotic species can form and persist (Sax et al., 2007).

2.3 Competitive Interactions and Invasions

Competitive species–species interactions are often found to destabilise ecological community dynamics (Ives and Carpenter, 2007) but yet competitive interactions in light of invasions can help to maintain stability of the resident system. Indeed, in WH systems, competitive interactions of resident species can help prevent invasions. For instance, gut microbiota species in leeches have been shown to prevent invasions by production of antibiotics (Graf et al., 2006; Tasiemski et al., 2015). Also, experimental evolution studies in a nematode model showed how a resident bacteria rapidly evolved increased antimicrobial superoxide production as a defence against *Staphylococcus aureus* infections (King et al., 2016). In a human example, lactic acid producing bacterial species lower the pH of the vaginal microbiome environment, which subsequently prevents the establishment of not only bacteria but also viruses (Danielsson et al., 2011). Importantly, resident bacterial species produce compounds (e.g. bacteriocins) to directly inhibit growth of invading bacteria thus using interference competition as a mechanism to prevent invasion (Dethlefsen et al., 2007). In a recent study, a cell-to-cell contact-dependent antagonism (where strains of Bacteroidetes inject toxic effectors into their neighbours) was described in detail (Wexler et al., 2016).

This phylum is very common in the human microbiome and thus, by demonstrating that this antagonistic behaviour affects the overall community composition and dynamics, they suggest that this interaction could be significantly shaping human microbiome communities. Locally, this antagonism allows these strains to limit their direct competition for resources but still benefit from diffusible compounds from other bacteria nearby (though not directly touching).

Interestingly, competition and not cooperative interactions appear to be more common in cultured communities (Foster and Bell, 2012). Similarly a study, that fit mathematical (ecologically based) models to time series data from the guts of mice, found that competitive interactions were most common, and in fact, parasitic/predatory and amensalistic interactions were also more common than cooperative interactions (Marino et al., 2014). Furthermore, they found that intraphylum interactions between all Firmicutes were competitive. These experimental results appear to help support a recent theoretical result that showed that microbial communities with more competitive interactions than cooperative interactions are more stable, with the caveat that too many competitive interactions can also have a destabilising effect (Coyte et al., 2015).

In free-living systems, invasions by competitors rarely cause complete extinction of endemic species (Sax et al., 2007), thus, suggesting that using a competitive strategy alone does not necessarily make invaders effective at disrupting the resident community. However, the same cannot be said for invading predators and pathogens which are more successful strategies for invasion (Sax et al., 2007), thus implying that these negative interactions can be more effective at destabilising the system. Insights such as these are particularly important for studies that wish to eradicate an unwanted resistant species using invaders as biological control (such as bacteriophage therapies) because using pathogens or possibly combining predators and competitors could be potentially more effective than solely introducing competitors.

There are other factors that matter as to whether competitive interactions are stabilising or destabilising, namely, the strength of the interactions and the topological context. For example, competitive interactions that form *transitive* (hierarchical) or *intransitive* (cyclic) networks have been shown to affect not only stability (e.g. networks of competitive interactions are stabilising; Allesina and Levine, 2011) but also promote or hinder diversity. For instance, Kerr et al. (2002) showed that *Escherichia coli* strains coexist when interactions are spatially explicit, thus echoing theoretical work on the coexistence of species in structured communities (Hofbauer and

Schreiber, 2010; Mouquet and Loreau, 2002). Indeed, experiments (in vivo or in vitro) of microbiome invasions, particularly those that combine them with ecological mathematical modelling (e.g. Marino et al., 2014; Stein et al., 2013; Wexler et al., 2016), could help further probe the properties of the WH networks that make competitive interactions either effective or ineffective at preventing invasions.

2.4 Evolution of Cooperation and Mutualism

2.4.1 Interactions Within Microbial Species

Altruistic strategies consist in conferring a fitness benefit to another individual at a personal cost. By definition, these are at the mercy of defecting strategies that do not pay the helping cost (but accept the benefits). There are two ways to explain how altruistic strategies can out-compete cheating strategies. First, individuals may have means to identify altruists, which is known as **kin discrimination** (Hamilton, 1964). One caveat, however, is that a defecting individual may end up bearing the same discrimination tag as altruists. Interestingly, cooperation can endure if one allows for enough different tag values and for mutation in this trait (Jansen and van Baalen, 2006); an idea which has been applied to siderophore production (Lee et al., 2012). The second process, also introduced by Hamilton (1964), through which altruists can out-compete defectors is known as **kin selection** and depends on the 'viscosity' of the environment. If the environment is spatially structured and if individuals are born locally and interact locally, then on average altruists should interact more with fellow, related altruists. As a group, altruists are then by definition fitter than defectors such that clusters of altruists can out-compete clusters of defectors. Much of the microbiota is found on epithelial surfaces and therefore dwell in highly structured environments (e.g. the gut or the skin), which means kin selection effects should be strong: even if a parasitic strain manages to create a small cluster, it is likely to be outcompeted by clusters of cooperators that are more productive.

2.4.2 Interactions Between Microbial Species

Positive reciprocal interactions between microbiota members are often cited as important protectors of WH invasions. Yet, the mechanisms for this are less clear. For example, a recent theoretical study has challenged this idea by showing that mutualistic interactions, in fact, weaken communities by introducing feedbacks that lessen the stability of the network (Coyte et al., 2015). That mutualisms themselves can be destabilising is something already known in general community ecology theory (Namba, 2015) and so considering other features that are linked to mutualisms, such as nestedness or evolution need to be considered. For instance, another theoretical study (not specific to the microbiome) shows that the evolutionary process itself is stabilising, and that, for low-diversity networks, evolution of mutualistic interactions are also stabilising, but this changes for larger networks (Loeuille, 2010). However, Rohr et al. (2014) argued that mutualistic systems have architectures that allow them to widen the parameter region where species coexistence is possible and thus are more persistent than often appreciated. In addition, adding a third mutualist to a pair of cooperators can increase the cost of cooperation (Harcombe et al., 2016), thus suggesting that increasing the number of competitors may weaken a network. Overall, then, more work is needed to understand the role microbe–microbe cooperative interactions have on stability and resilience of the whole community.

Earlier modelling has identified three factors that may lead to mutualistic interactions, namely (i) high benefit to cost ratio, (ii) high within-species relatedness and (iii) high between-species fidelity (Foster and Wenseleers, 2006). One challenge to model microbe evolution is that of **horizontal gene transfer**. Indeed, if the trait of interest, i.e. the one used to define the level of cooperation/mutualism, is carried on a mobile genetic element, the classic relatedness definition, which is based on identity by descent, becomes obsolete and requires adjustments (Rankin et al., 2011; Smith, 2001).

Horizontal gene transfer can play a striking role in the adaptation of microbial communities to fluctuating host environments. Interestingly, recent evidence suggests that the invasion of parasitic bacteria is itself a fuel for horizontal gene transfer (Stecher et al., 2013). Indeed, **dysbiosis** (a microbiome community that is pathogenic to the host, and is often the consequence of a successful invasion) is characterised by the growth of bacteria that were either rare or absent before the perturbation. The proliferation of these invaders represents an opportunity for the resident microbiota to acquire new genes via horizontal transfer. There is, of course, a risk that this blooming of rare bacteria destabilises the whole microbial community. However, the benefits might also outweigh the costs because of a tendency for the loss of genes that are redundant in a community. Morris et al. (2012) name it the **Black queen** effect and their idea is that at the genotype level, there is a selective pressure to get rid of nonessential genes that are present in other species in the community. The name of the process originates from the fact that, at some point, one genotype bears the last copy of the gene and cannot get rid of it without destabilising the whole community. This whole black queen effect is likely to destabilise the community (since gene redundancy has a stabilising effect at the community level), which is why dysbiosis could in the end prove somehow useful at the community level. Note that this is also a clear illustration of how, perhaps contrarily to free-living ecosystems, evolutionary and ecological processes regularly overlap in WH communities.

2.4.3 Interaction Between Host and Microbes

It is commonly said that hosts have coevolved with their microbiome to have a strong mutualistic relationship. However, specific human–microbe coevolution examples are difficult to find, possibly because most of the evidence relies on cophylogenetics (i.e. comparing the phylogeny of the hosts to that of its bacteria) and because it is extremely difficult to disentangle human evolution from phenotypic plasticity or development (but see Moeller et al., 2016 for an example on hominid–microbe coevolution). For instance, the gut microbiota is strongly affected by the host's diet and changes at the host population level (e.g. due to migration or external changes in the environment) do not require human evolution. In fact, it is easy to have a fixed view of interactions. As is also found in free-living systems, the mutualistic or parasitic nature of a resident microbe can very well depend on the environment and thus can shift without evolutionary changes. For instance, GB virus C is usually a commensal virus in humans but it becomes beneficial if there is a coinfection (i.e. simultaneous infection) with HIV (Bhattarai and Stapleton, 2012; Tillmann et al., 2001; Xiang et al., 2001). Envisaging human–microbiome interactions as 'dangerous liaisons' (i.e. allowing for dynamic partnerships and also for environmental feedbacks; van Baalen and Jansen, 2001), is perhaps key to understanding the ecology and evolution of this relationship. Epidemiological modelling can also help understand interactions between hosts and microbial species. It has been known for a long time that infections can have pleotropic effects on their host (Michalakis et al., 1992). The challenge, however, is that this effect on the host likely depends on the WH parasite community and few epidemiological models allow for coinfections by more than two parasite species (but see Sofonea et al., 2015) and even fewer allow for host evolution. Incorporating diverse microbial communities into a detailed epidemiological setting remains an open challenge.

3. QUANTIFYING INVASIONS OF THE MICROBIOME: DATA, MODELLING AND THEORY

Progress in our understanding of invasions into microbiomes will likely require parallel developments in data acquisition, model inference techniques and theoretical investigations. In this section, we begin with a discussion of the formulation of a microbiome that is amenable to quantitative analysis. We then summarise ecologically motivated approaches from each of the measurement, inference and mathematical theory of microbiomes relevant to invasion analyses.

3.1 Quantitative Formulation of a Microbiome

Despite the additional challenge posed by definitions of the species concept for microbial organisms, an ecological community as an interacting network of species populations remains a useful concept for microbiome research. Attempts to quantify this network face many of the same challenges found in quantifying of those of free-living communities. The microbiome exhibits a diversity that parallels that of free-living systems. We are finding many more phyla than expected and there is a great deal of interhost variability (e.g. not more, and likely less, than 30% of faecal bacteria species are shared across individuals; Faith et al., 2013). Thus, determining and then understanding even a specific microbiome network of even a single region of a single individual's whole WH ecosystem appears daunting (e.g. see Nobbs and Jenkinson, 2015, which reviews the diverse players of the interkingdom network of the oral ecosystem). The many players include bacteria, host cells (as the physical environment and as resources), bacteriophages, yeast, fungi, viruses and, not the least, immunity cells and compounds. Nevertheless, we are rapidly discovering and characterising what fungi (i.e. the mycobiota, Huffnagle and Noverr, 2013; Underhill and Iliev, 2014), viruses (i.e. the **virome**; Delwart, 2013; Foxman and Iwasaki, 2011; Minot et al., 2011, 2013; Wylie et al., 2013) and archaea are present in human bodies (Lurie-Weinberger and Gophna, 2015). Much of this progress comes from the attractive advantages that microbiomes offer to ecology over those of free-living systems: shorter generation times, larger samples and high-throughput sequencing methods, which together open up the possibility of measuring abundances of a significant fraction of microbiome species over timescales at which relevant community dynamics occurs.

Given the diversity of the microbiome, the scope of study is often narrowed with the research question onto only a subset of players. This carries the risk of neglecting important contributors. At the very least, however, the currently studied subnetworks provide a practical starting point from which to incorporate more players. As the field matures and the methods achieve more depth with less measurement issues, we are likely to access significant fractions of most regional microbiome networks.

Just as with free-living ecosystems, directed links between nodes in a microbiome network signify significant asymmetric energy transfer, termed interaction strength. Defining and measuring interaction strength is important (Berlow et al., 2004; McCann, 2000; O'Gorman et al., 2010), especially given the array of interactions WH ecological systems engage in. Capturing multiple forms of species interactions is also of interest to ecologists and recent work has proposed 'inclusive networks' which capture multiple kinds of interactions more mechanistically (e.g. feeding and nonfeeding; Kéfi et al., 2012).

At a deeper resolution, and in contrast to the taxonomy of free-living ecosystems, current bacterial species definitions are often constructed by applying user-defined (and thus subjective) sequence-similarity thresholds to cluster the data into so-called **operational taxonomic units (OTUs)**. However, despite the high genetic variability within bacterial species, in part due to their ability to pass genes between individuals, this clustering is in principle informative because it is typically performed on regions of 16S ribosomal RNA (rRNA), which contain highly variable regions that are nonetheless highly conserved within a bacterial species. Longer sequencing reads now lower the sensitivity of the clustering on the choice of this subjective threshold (Franzén et al., 2015), and attempts at more robust threshold-based and nonthreshold-based definitions are being proposed (Puillandre et al., 2012; Rossi-Tamisier et al., 2015).

As studies in some other complex biological networks have suggested (Marder and Taylor, 2011), such a detailed dissection of the system might obscure rather than clarify the principles at work. Indeed, a healthy microbiome network might be functionally redundant or degenerate regarding the set of functions that it provides for the host. A redundant community is one in which multiple species play similar roles. These communities are in principle robust up to perturbations that knock out all but one of the members within one such functional group. A similarity of function can also exist between two species in two instances of the same community across a pair of hosts in which case the community is degenerate. By particular life histories/contingency, etc., different host individuals may then have different bacterial actors performing in a functionally equivalent community. This degeneracy would lead to highly variable species-level measurements across individuals, hiding the underlying functional organisation. A current direction of research regarding a microbiome definition is then the determination of functional classes of microbiome members. Members of the same group would serve the same function (Boon et al., 2014; Shafquat et al., 2014), and it is hoped that this grouping would better capture the community's topology and dynamics. One popular option is to group taxa based on their metabolic activities, e.g. grouping by catabolisers of polysaccharides or proteins or by creating respiratory guilds (for a review of such models, see Song et al., 2014). A particular grouping, *Clusters of Orthologous Groups* (COGs), classes proteins into gross functions of the cell such as energy production and conversion, cell cycle control and mitosis, and replication and repair (Tatusov et al., 2003). While seemingly natural, such characterisations have so far been subjective and the need to make them less so remains. In particular, many species fulfil the same basic functions with similar pathways, and so distinguishing species via these functions is not likely meaningful. At the other extreme, many species exhibit unique functions that make comparisons ill-defined. Choosing the appropriate 'depth' in the degree of variation of functions across the community and knowing which functions are active and important for community dynamics, both present subtlety and require careful consideration. The singular focus of these genetic and molecular approaches on genes and their expression make them fundamentally different from traditional ecological approaches, such as trophic-level and guild methods of network reconstruction. Attempts to carry over techniques or make comparisons to these traditional methods may be fruitful in some cases and senseless in others.

Also important is the question of how resolved the temporal measurements need to be in order to capture the relevant WH dynamics of the microbiome. This is again a feature likely best tailored to the question. However, the choice may be more constrained by practical limitations than for the case of free-living communities. For instance, only very recently did a study carry out sufficient temporal sampling to uncover diurnal oscillations of expression patterns in mice microbiomes (Leone et al., 2015). In humans, daily sampling suggested rapid turnover of the community (Caporaso et al., 2011), yet recently, more reliable sequencing showed that 60% of species persist on the timescale of years (Faith et al., 2013). This follow-up study has reinvigorated the investigation of stable communities and serves as a lesson that the reliability of the experimental techniques should not yet be taken for granted. The majority of human studies has acquired samples at only a single, or a few, widely spaced time points. Depending on the microbiome under study and the sampling method, frequent sampling can also significantly disturb a microbiome, adding unwanted variability to the measurements. As the methods are refined to be less invasive, however, these undesirable interference effects can hopefully be made minimal.

Another important feature of the microbiome is spatial structure. It is not clear how spatially structured niches will skew the effective interactions between species, but some microbiome habitats are highly spatially structured and insights from ecology suggests these features can play an important role (Hastings, 1978; van Nes and Scheffer, 2005). The regions of the body are usually studied separately from one another (see Pedersen and Fenton, 2007 for an exception) and spatial features differ a great deal between them. Consequently, comparisons between microbiome networks across regions of the body show they can be quite different in their composition and structure (see Fig. 1). For instance, vaginal and skin communities tend to have lower alpha-diversity than gastrointestinal-related communities, and vaginal communities in particular are heavily dominated by only a few genera (see Fig. 1; Jordán et al., 2015; Zhou et al., 2014). Currently, the application of spatial ecology to studying WH systems and infections is very uncommon, even though it is very clear that some invasions require a spatial ecology approach. There is also growing evidence for the fact that resident taxa move around between local body sites, i.e. **translocate**, much more than originally thought (Balzan et al., 2007). In addition, spatial structure may also play an important role in structuring the microbial community within a given body site. For example, patch dynamics are likely to affect the outcomes of human papillomavirus infections (Murall et al., 2014) and possibly most skin or mucosal infections. Also, continuum models seem appropriate for sites containing biofilms, but there are few examples of this approach (though see D'Acunto et al., 2015). Complementary to sequencing approaches for providing both high temporal and spatial resolution, in vivo imaging of the microbiome is coming available (Geva-Zatorsky et al., 2015). Currently limited to animal models, and to a small handful of microbial species, this approach can reveal the effects of an infection in great temporal and spatial detail. Overall then, there is ample room for spatial ecologists to help study how this important phenomenon affects WH communities and their dynamics.

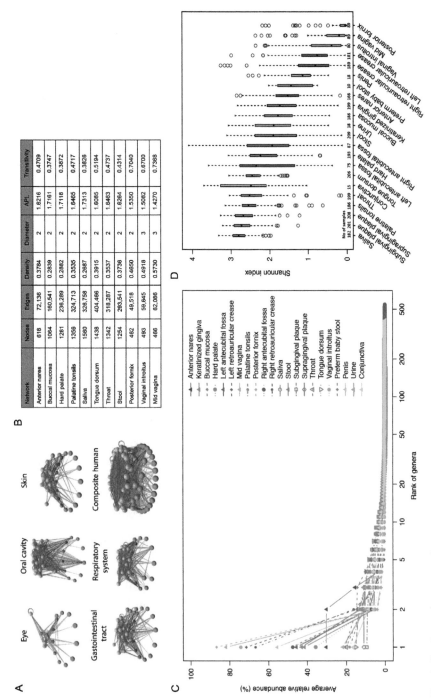

Fig. 1 See legend on opposite page.

In analogy with free–living ecosystems, the notion of a baseline (existing prior to the invasion) and its measurement need to be established in order to quantify the effects of an invasion on the resident network. A multistability of the community dynamics (known as **alternative stable states** in ecology) hypothesizes that the structure of the community dynamics induces multiple basins of attraction in the space of all possible community states. In the absence of strong external perturbations, communities found in a given basin tend in time towards the corresponding stable state as the local attractor of the dynamics. In support of the stable community hypothesis, studies have shown that the communities in healthy individuals indeed cluster into a handful of types (e.g. vaginal microbiome states, Gajer et al., 2012; and enterotypes of the gut, Arumugam et al., 2011; that persist over time, Faith et al., 2013). Nevertheless, there are reports (Gajer et al., 2012) of community dynamics with fast-fluctuating species abundances such that the identities of the prominent species undergo quick turnover. This kind of short correlation time dynamics, reminiscent of chaos but not necessarily chaotic, was nevertheless found only in healthy individuals. This is evidence against the hypothesis that such dynamics is pathological for the host and suggests that long-term maintenance of a given set of species may not be necessary for normal, healthy functioning. Regardless of the relevant timescale of microbiome dynamics, sampling and sequencing the microbiome at multiple time points provides the means to establish a baseline from which the effects of an invasion can hopefully be observed. We note

---

Fig. 1 Different body-site communities have heterogeneous network properties and diversity. (A) Community interaction networks from different human body sites. *Red nodes* are source-specific nutrients, while *orange* and *yellow* nodes are microbial genera. *Links* are either consumer–resource interactions or facilitative interactions. (B) Network properties of 11 human habitats. Nodes are OTUs; density is the number of links divided by the maximum number of links; APL is average path length and transitivity is the clustering coefficient. (C and D) Heterogeneous diversity across body-site specific microbiomes. (C) Rank abundance curves of bacterial genera from 22 human habitats. The female reproductive habitats (*light blue curves*) have few genera dominating their communities, while oral habitats (*red curves*) have fewer dominant genera. (D) Shannon diversity box plots of the same 22 human habitats. According to this index, the abundances of oral communities (*left*) are the most uniform, while those of vaginal communities (*right*) are the least uniform. *(A) Image modified with permission from Chapter 2 of Ritchie, M.L., 2011. Structure and Function of the Human Microbiome. Dalhousie University. (B) Table from Jordán, F. et al., 2015. Diversity of key players in the microbial ecosystems of the human body. Sci. Rep. 5, 15920. (C and D) Used with permission from Zhou, Y. et al., 2013. Biogeography of the ecosystems of the healthy human body. Genome Biol. 14(1), R1.*

that other kinds of measurements, for example of environmental properties such as pH, can also be informative across an invasion event (e.g. Gajer et al., 2012).

3.2 Measuring the Microbiome

The maturing revolution in sequencing techniques has focused attention on the characteristics of the microbiome accessible by methods from the three -*omics*: genomics, transcriptomics and proteomics, providing information on DNA, RNA and proteins, respectively. Respecting the terminology, the prefix 'meta-' should be prepended to these terms since, in microbiome research, they are applied to samples containing multiple species. DNA sequencing is often used to identify the species present. The number of reads of a given sequence can also be used in principle to infer cell counts. Primers can amplify different genes at different rates, and initial primer concentration should be carefully controlled so as not to introduce additional bias (Brooks et al., 2015; Kennedy et al., 2014; Tremblay et al., 2015). Which proteins are being expressed can in principle be obtained from RNA sequencing. It has become apparent over the last decade, however, that expression level does not reliably report the actual amount of protein present (Rogers et al., 2008). For this, protein purification using antibody assays are often used. Moreover, since an altered state of a protein often signals the activation of its function, current proteomic research is focusing on methods for identifying alternate, e.g. phosphorylated states of the same protein complex (Aebersold and Mann, 2016). More advanced sequencing methods seek to account (via, for example, molecular barcoding) for artefacts due to the amplification, such as copy error and primer bias (Best et al., 2015). Reliable or at least soon-to-be reliable estimates of abundances can be obtained with these methods. The abundances are nevertheless often taken relative to the sample size so that data from multiple samples can be compiled. This operation can lead to spurious correlations between species abundances that can, and should, be corrected (Berry and Widder, 2014).

3.3 Inferring the Interaction Network

The inference of human–microbiome interaction networks, in particular from metagenomic sequencing data, has become a preoccupation of many in the field. Doing so accurately will likely be required before understanding the effects of invasions on the microbiome community. These methods employ the abundance data from samples taken either from the same community at multiple time points, or from a single sample from the same body-site across various hosts, or in the best case both.

In a first approach, researchers analyse pairwise correlation matrices of taxa abundance, drawing a link between pairs of taxa whose pairwise correlation surpassed some threshold value. For instance, **cooccurrence networks** (also called **association** or **correlation** or **coexisting networks**; Vacher et al., 2016) are usually built on statistically significant copresence or coabsence of pairs of taxa. Taking such networks as indicative of ecological interactions is problematic, however, as coexistence patterns, such as presence and absence, are not a direct result of pairwise ecological interactions, such as mutualism and competition, respectively. They are shaped in nontrivial ways by the type of underlying interactions, habitat use and abiotic factors. Analysis of the approach has nevertheless framed the technical difficulties of the problem (see Faust and Raes, 2012 for a review and Marino et al., 2014 for a recent example). Recent studies show that these correlation-based networks perform poorly at recovering the actual interaction network from synthetically produced data (Berry and Widder, 2014; Fisher and Mehta, 2014a; Kurtz et al., 2015).

In performing analysis on sequence data, the often widely varying sample sizes and primer bias require first transforming the data. The most common transformation is simply to take relative abundance, but this introduces a sum-constraint making inadmissible many statistical inference approaches. The appropriate statistical theory for such compositional data was communicated to the field in Kurtz et al. (2015), in which the data are transformed to centred log ratios. Kurtz et al. (2015) go on to infer a network of associations by fitting a probabilistic graphical model to this transformed data. Such models are based on conditional dependencies and so can in principle distinguish coexistence patterns due to direct ecological interaction and those due to spurious correlation. An added regularisation term to select for sparse solutions allows for the problem to be solved efficiently, and they test the method on data generated in silico. The method's performance depends on some network properties, performing less well for larger maximum in-degree (the number of links going into a given node; see next section for a description of other network properties), and relies on the number of connections not scaling faster than the number of taxa (i.e. it may fail for densely linked microbiomes).

Another more ambitious class of approaches to inferring interaction matrices seeks to leverage the temporal correlations in the data. Such methods must handle both the deterministic and stochastic contributions to the observed time series. One subset of methods in this class use causality measures computed on time series to determine the presence and direction of a causal interaction between two species. Sugihara et al. (2012) is an example combining the stochastic prediction of Granger causality with the deterministic embedding of attractor reconstruction. The high dimensionality of sequence data can apparently help overcome the finite sample limitations of the latter (Ye and Sugihara, 2016). A more standard approach is to use a noise model for the observed abundances at each time point to extract a mean abundance, whose time series is then fit by deterministic models of the underlying community dynamics. The hope is that the temporal correlations in the dynamics provides additional information about interaction strengths allowing some of the noise issues in the measurements to be overcome and give more reliable results. Moreover, if the chosen model of the dynamics is sufficiently faithful to the real system, then the values of the inferred parameters (e.g. growth or attack rates) actually reflect the ecology of the system. **Generalised Lotka–Volterra (gLV) models** are often used with an additional species specific, time-dependent term representing disturbances. This type of approach has been pursued in a number of recent publications and there are a growing number of software packages available online, most notably MDSINE (Bucci et al., 2016; also see Eq. (1) and Fig. 2). A detailed comparison of these packages (by performance, accessibility, versatility, etc.) goes beyond the scope of this review, but would be a welcome contribution to this growing body of work. Here, we briefly discuss some of the major issues and proposed solutions.

The MDSINE package is a suite of methods. At each time point, they infer the mean of a probabilistic model of read counts to the abundance data, connecting means across time points using splined functions of time. The unbound growth, interaction and disturbance parameters of a gLV model are inferred on this splined representation of the data using gradient matching and also with a regularisation to overcome the undersampling. Relevant to invasion analysis, the quality of the inference is increased when applied to datasets obtained over periods of stepwise disturbances. Such cases allow the dynamics to exhibit its timescales making the data more informative to parameter estimation, and ultimately network reconstruction. The authors state that there was no clear relationship between the reconstruction performance and the in–degree. This benefit over the inference methods of Kurtz et al. (2015) and the literature cited therein could reflect the utility of dynamical information in this inference problem. More work is needed to settle the origin of this discrepancy.

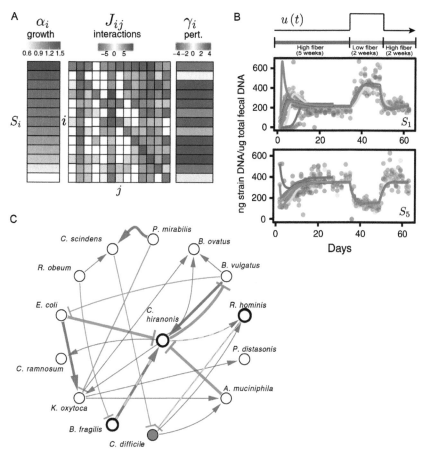

Fig. 2 Inferring microbiome interaction networks using community dynamics. The growth, interaction and perturbation parameters of community dynamics of the form of Eq. (1) (see Section 3) and shown in (A) are inferred from time series of expression of all species, two of which are shown in (B), measured over a window within which a step-like environmental change is made (here high- to low-fibre, and back again). (C) A similar procedure is applied to a different network observed over a window in which *Clostridium difficile* (*green circle*) is introduced. *Blue arrows* denote positive interaction strengths, while *red* denotes negative ones. Thickness denotes interaction strength. The authors find that, of all nodes, three (*Bacillus fragilis*, *Clostridium hiranonis* and *Roseburia hominis*, *thick circles*) are necessary and sufficient to effectively repel the invasion. In this network representation, a particular pathway (*green dashes*) involving these three can indeed inhibit *C. difficile*. *Modified images from Bucci, V. et al., 2016. MDSINE: microbial dynamical systems inference engine for microbiome time-series analyses. Genome Biol. 17(1), 121.*

Many studies test these methods on synthetic data. For example, the inferred matrices of Fisher and Mehta (2014a) have correlations of 0.8 with the original interaction matrix used to simulate the data. How this number depends on the form of the community dynamics was not investigated however, and different functional responses (i.e. types I, II and III; Holling, 1959) in the community dynamics equations may play a role in the strength of the correlation between the ground truth and the inferred network. Unlike in nervous systems where the ground truth synaptic connectivity between neurons can at least in principle be obtained empirically, there is currently no obvious definition or way of obtaining ground-truth connectivity in a real microbiome. Nevertheless, similar methods to infer parameters of the underlying network using dynamical models are being applied to other microbial networks, such as the bacteriophage–bacteria networks (Jover et al., 2013). With inference algorithms and machine learning developing at a fast pace, more work is needed to find more reliable and general methods for inferring microbiome interactions.

Once links are reliably inferred, interaction networks can be analysed with graph theoretic techniques. Simple measures can help distinguish nodes with atypical properties. Highly connected nodes, such as *Bacteroides stercoris* and *Bacteroides fragilis* (Fisher and Mehta, 2014a), may indicate the species on which the community's stability depends, in analogy with the *keystone species* concept from ecology (discussed more explicitly in Section 5).

Beyond single species, but before considering the community as a whole, systems of a handful of species nodes can be studied. When connected in particular ways and with classic interacting population dynamics, these are called **modules** in ecology (e.g. the diamond module; Holt et al., 1994), and each provides characteristic dynamical behaviours (reviewed in McCann and Gellner, 2012). Ecological modules have been identified and studied as a first step to understanding larger community networks that contain them as subnetworks. Ecological modules have been found to also appear in WH systems, particularly in pathogen–host interactions (see Fig. 3). Similarly, in the context of the microbiome, Stein et al. (2013) performed inference with a dynamical model and found particular circuits of species that are involved in successfully repelling an infection by *Clostridium difficile*.

For large, or many replicates of, interaction matrices, one can employ some of the many well-established graph statistics (Newman, 2003). These include how dense the network is (connectance); how many links 'wide' the network is (graph diameter); the degree of modularity (not to be confused

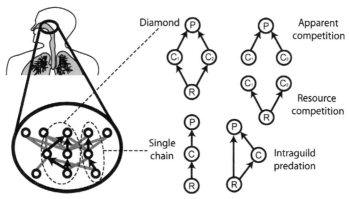

Fig. 3 Interaction networks can be decomposed into ecological modules. (*Left*) A schematic caricature of a sinus microbiome interaction network. A diamond and single chain module are highlighted by the *dashed ovals*. (*Right*) A set of typical modules, labelled with their name and showing the resource, R, the consumers, C and the predator, P. *Modified with permission from Murall, C.L., McCann, K.S., Bauch, C.T., 2012. Food webs in the human body: linking ecological theory to viral dynamics. PLoS ONE 7(11), e48812.*

with ecological modules), i.e. link prevalence with in clusters compared to link prevalence between clusters; the degree of clusteredness, i.e. how often neighbouring species with one species are also neighbours with each other (see Table 1 and David et al. (2017), in this issue, for a review of network properties and invasions). One can also calculate **motif statistics**, i.e. the over- or underrepresentation relative to chance of small groups of nodes connected in particular ways (but agnostic to their dynamics). On the other hand, one can incorporate information extrinsic to the graph, and for instance intrinsic to the nodes. For example, OTU assortativity is the propensity of connections between nodes with similar 16S rRNA profiles. The inferred gut networks in Kurtz et al. (2015) displayed a high degree of OTU assortativity. However, overall, very few studies have directly investigated motifs in host-associated microbiomes and most of what has been done is in metabolic network studies (Table 1).

Assessing the over- or underrepresentation of motifs relative to the prevalence expected by chance can be used to analyse microbiome abundance time series. With datasets sampled at a high enough temporal resolution, the possibility to classify their trajectories arises. Once sufficiently long time series can be measured, comparisons of cross correlations of the abundances from microbiome communities across body sites could provide signatures of leaders and laggards in the whole body microbiome dynamics.

Table 1 Comparison Between Microbiome Ecological Networks and Free-Living Ecological Networks, With Suggested Implications for the Stability and Invasibility of the Microbiome Community

Property	Known Properties of:		Implications for Invasibility
	Microbiome Networks	Free-Living Ecological Networks	
Network size (total number nodes, i.e. diversity)	Dysbiotic communities (which are usually lower in diversity than healthy communities) are more susceptible to invasions (e.g. in murine experiments artificially low-diversity communities are more susceptible to infection by *Salmonella enterica* but that protection can be restored by raising these mice with other mice having a diverse microbiota (Stecher et al., 2010)) Conversely, the vaginal microbiome is significantly less diverse than other communities (such as the gut) and yet, it is very stable and resistant to invasions (Jordán et al., 2015)	The positive relationship between high diversity and resistance to invasion is found frequently in theoretical literature, leading to the 'diversity resistance hypothesis'; suggesting that highly diverse systems have more competitive links and can thus resist invasion more readily. However, the mechanisms are not well understood (Stecher et al., 2010) and this is still not experimentally well founded (Levine and D'Antonio, 1999). For example, positive interactions may alter this theoretical hypothesis	Diversity alone probably does not determine the susceptibility to invasion; other features are important and underlie the mechanisms behind the diversity–stability relationship (see diversity–stability debate, review by McCann, 2000)
Motifs (and ecological modules)	Motifs found: Four motifs in healthy patients vs three motifs in alcoholics, and some motifs are more abundant in healthy than in alcoholics (Naqvi et al., 2010) Network inference of gut data found the network was a composite of several types: scale-free, band-like and clusters of subnetworks (Kurtz et al., 2015)	Ecological systems have been found to contain several overrepresented motifs: food chain, omnivory, generalist and apparent competition motif (Milo et al., 2002) Energy channels, i.e. highly interacting species that derive most of their energy or biomass from the same basal species are common structures in food webs (Rooney and McCann, 2012) and the pattern of top predators coupling different energy channels is an ecosystem scale structure that has been found repeatedly (McCann and Rooney,	Certain motifs are likely to affect the stability of the system and thus the robustness to invasion

Modularity	Invasion experiments into simulated food webs found lower modularity made the community more susceptible to invasion (Lurgi et al., 2014)	Obese and inflammatory bowel disease microbiomes, viewed through their metabolic networks, were found to have reduced modularity (Greenblum et al., 2012) / Modularity of cecum (gut) from metagenomics and metabolomics (McHardy et al., 2013)	Reduced modularity in microbiome communities is posited to be associated to lower variation in the environment (Greenblum et al., 2012). These disturbed systems are also less diverse and more susceptible to invasions
Node–degree distribution	Are generally not scale-free (Zhou et al., 2010)	Are often scale-free (i.e. the node–degree distribution is described by a power-law) but not always (Faust and Raes, 2012)	Scale-free are generally less sensitive to random node removal except when a hub node is removed, which implies that if an invader affects a hub node then it is more likely to establish
Connectance (connectedness)	Higher connectance has been found to decrease invasion success (Dunne et al., 2002; Romanuk et al., 2009); however, in some cases, larger food webs with low connectance (and high modularity) can be robust to invasion (Lurgi et al., 2014). Generally, though, it is thought that higher connectance is more resistant to invasion (Smith-Ramesh et al., 2016)	Host-associated microbiome networks (across several body sites and constructed with positive and negative interactions) showed higher connectedness (and network density) than soil networks (Faust et al., 2015)	Higher connectedness is believed to help prevent invasions
Average shortest path length (AL)	Food webs have been found to have short-path lengths (Dunne, 2006; Montoya et al., 2002)	Mostly have small AL (Faust and Raes, 2012)	Small AL increases the networks' responsiveness to perturbations, i.e. faster return (Faust and Raes, 2012; Zhou et al., 2010)

Continued

Table 1 Comparison Between Microbiome Ecological Networks and Free-Living Ecological Networks, With Suggested Implications for the Stability and Invasibility of the Microbiome Community—cont'd

Property	Known Properties of:		Implications for Invasibility
	Microbiome Networks	Free-Living Ecological Networks	
Hub nodes (core taxa, keystone and foundational species)	Identification of keystone and foundational species using network inference and time series metagenomics data (Fisher and Mehta, 2014a; Trosvik and de Muinck, 2015)	Removal of keystone species and fragmentation of food webs can destabilise food webs (Proulx et al., 2005; Solé and Montoya, 2001)	If an invader removes a keystone species or a foundational species this can have strong effects on the community
	Experimental example: A low-abundance biofilm species orchestrates periodontal disease (Hajishengallis et al., 2011)		Note: The trophic position into which the invader enters matters to the initial phases of the introduction (Romanuk et al., 2009)
Weak vs strong interactions (links)	We did not find any direct investigations of this; however, it is mentioned in Foster et al. (2008)	Weak interactions are generally stabilising while strong interactions are generally destabilising (Gellner and Mccann, 2016). Trade-offs (e.g. competition vs edibility) can readily produce these strong–weak associations. High frequency of weak interactions is stabilising (Jacquet et al., 2016)	The new weak interactions (links) created by an invader can affect stability of the resident community
Distribution of positive and competitive interactions	Negative interaction networks have been found to be less connected than the positive interactions of the same networks. Also, host-associated networks contained more positive edges than soil microbial communities (Faust et al., 2015; Friedman and Alm, 2012)	Invasion resistance arises in strongly interacting species-rich model competition communities (Case, 1990)	Invaders that compete with resident species can create different structures and whether they create transitive or intransitive structures could affect their invasion success
		Transitive (hierarchical) competitive interactions are less stable than intransitive (cyclical) networks of competitive interactions (Petraitis, 1979)	Predatory or parasitic strategies are likely to lead to more successful invasions than competitive strategies
		Invading species that feed on resident competitors are more effective at entering a new system than species that only compete	

Finally, we mention that network reconstruction competitions based on benchmark datasets can be a fruitful way to attract high-performing algorithms (e.g. Orlandi et al., 2014).

3.4 Theoretical Microbiome Invasion Dynamics

There is a wealth of theoretical ecology literature using dynamical systems theory to study population and community dynamics. The vast majority of existing models and theories that have taken this approach to WH invasions are found in the pathogen literature, more specifically the literature related to multiple infections (Read and Taylor, 2001). These works normally do not include commensal phyla (for an exception, see Lysenko et al., 2010). Generally, in this WH disease modelling framework, the host is abstracted as populations of resources (usually cells) and immunity cells. Resident species in this case are acute or chronic infections already established in the system. Invasion analysis is done to see if a second invader with new traits, or another species altogether, can take hold. These models have been very successful in helping study viral–immunity interactions (Perelson, 2002), studying viral or bacterial WH evolution (Luciani and Alizon, 2009) and aiding in antiviral and antibiotic treatment studies (Koizumi and Iwami, 2014). Attempts to capture microbiota interactions mathematically have not focused on feeding relationships, as is the case for pathogen–immunity models and free-living ecological systems. Microbiome network models, in contrast, define links and their weights by effective cooperative and competitive interactions between species (see Eq. 1).

An important ecological insight that has been found using the dynamical systems approach is that many ecological modules found in free-living systems are also found inside the body (Fenton and Perkins, 2010; Holt and Dobson, 2006; Murall et al., 2012; see also Fig. 3). With the inference of networks from sequencing data, we are only now beginning to see attempts at finding what modules are important to microbiome communities (Table 1), and particularly how they can help explain community stability in face of invasions (see analogous work for free-living systems and biological control; Holt and Hochberg, 2001).

Given that the field has been rightly preoccupied with inferring networks and interactions, it is not surprising that despite a considerable amount of discussion about microbiome community stability (e.g. Bäckhed et al., 2012; Hunt et al., 2011; Lozupone et al., 2012; Relman, 2012; Romero et al., 2014), the majority of human microbiome models have not yet focused on stability and resilience to invasions. A recent exception for stability analysis was presented by Coyte et al. (2015) who used a classic theoretical community ecology approach to study microbial community stability. This approach focuses on the stability properties of interaction matrices arising from community dynamics models such as a system of Lotka–Volterra-like population dynamics, a phenomenological population network model that can incorporate multiple types of interactions. Such models can incorporate invasions to the system by the addition of the appropriate equation for the invader's population dynamics. The interaction strengths with the rest of the community can be turned on at the desired time of invasion. Another kind of community perturbation is a disturbance incorporated as a time-dependent change in the affected parameter(s). The dynamics of a community of N species, S_i, $i = 1, \ldots, N$, interacting via a type I functional response, and which, for the purposes of example, experience a time-dependent disturbance in their growth rates can be written as:

$$\frac{\mathrm{d}S_i(t)}{\mathrm{d}t} = (\alpha_i + \gamma_i u(t)) S_i(t) + \sum_{j=1}^{N} J_{ij} S_i(t) S_j(t) \tag{1}$$

Here, J_{ij} is a matrix, the elements of which are real numbers corresponding to the interaction strength with which species j affects species i; α_i is the basic growth rate of species i; and, in a simple example of a disturbance, γ_i is the constant amount of change in the growth rate induced by the binary-valued function, $u(t)$, denoting whether the disturbance is present or absent at time t. Such a system has been effectively used to infer interactions (Fig. 2; Bucci et al., 2016). Mathematical systems of this form allow for pairwise interactions, such as competition ($-/-$), mutualism ($+/+$), predation, herbivory or parasitism ($-/+$), amensalism ($0/-$), commensalism ($0/+$) and no interaction ($0/0$). These equations in principle even allow for trophic levels, such as predation (e.g. protists or the immune system) or parasitism (bacteriophages), through particular block structures in the matrix, J_{ij} (see Johnson et al., 2014 for such an example using the matrix approach outlined next).

The large number of interactions in a diverse ecosystem makes the determination of elements of J_{ij} impractical. An alternative approach, the so-called **matrix approach** pioneered by Sir Robert May, specifies the interaction matrix only through the statistics of its elements. Results from random matrix theory can then be utilised to determine the average stability properties of the stationary states of the corresponding dynamics as a function of the remaining parameters in the model (May, 1972). Applying this approach to microbiome systems, Coyte et al. (2015) found results consistent with previous ecology research, namely that synergistic interactions are destabilising and weak interactions are stabilising. Gibson et al. (2016) employ a similar model to study the effects of one-time perturbations that knock the community between some small number of putatively stable states.

Reconstructing interaction matrices before and after an invasion may not be necessary if only the resilience of the network is of interest. For example, a recent work shows that a particular reduction of high dimensional dynamics onto a single effective dimension highlights a resilience function (how abundance decays as a function of some kind of loss) depending only on the form of the equations for the dynamics and not on the values of the system parameters or perturbations (Gao et al., 2016). In a bipartite network of plants and pollinators, they show that the form of the decay of the average population sizes as a function of the number of nodes or links lost, or the reduction in interaction weight, all collapse onto the same bifurcation diagram. This diagram shows how changing an effective parameter beyond a critical value leads to a change in the steady state from one in which species have relatively high abundance to one where they have low abundance.

Despite the seeming ease with which ecological theory can be transferred over to the microbiome, there are caveats. Creating a consumer–resource modelling framework for the microbiome is less straightforward than in free-living systems, since microbes interact in unique ways such as cross feeding and have complex nutrient use, and thus energy flow (Foster et al., 2008). We know from free-living systems that the destabilising effect of positive feedback in a community's dynamics can be controlled by the stabilising effects of predation, whose exploitative interactions then maintain the community's stability. The presence of microbial eukaryotes (i.e. protists; Parfrey et al., 2011) or predatory bacteria (even if they appear in low abundances) in host-associated environments are then likely important, but for now are understudied. Multitrophic interactions, for example, are much more investigated in other, e.g. terrestrial, microbial systems (see Saleem, 2015 book for review).

There are few other studies on perturbations in microbial systems, so we briefly review some of them here. Allison and Martiny (2008) discuss microbial community stability to disturbances; D'Acunto et al. (2015) model multispecies biofilms and their invisibility; and in a recent study,

Peura et al. (2015) numerically test the response of microbial networks to the removal of taxa. Indeed, some important properties of ecological networks, such as the pervasiveness of weak interactions instead of strong interactions (McCann, 2000) remain unstudied in human microbiome communities. Yet, the increasingly network-oriented approach to studying the microbiome lends itself to investigating these sorts of properties and how they mediate stability and resistance to invasion (Foster et al., 2008). In Table 1, we highlight some key findings of microbiome network properties and how they may help understand community stability.

Finally, we mention there are even fewer examples of attempts to create full networks of interacting immune cells, pathogens and resources. Griffiths et al. (2014) assemble a network of such interactions, drawing links between pairs of species on which some association has been published. Analysing the resulting three-trophic layer network, they argue that WH pathogen interactions tend to be predominantly indirect; that the network has several identifiable and repeated motifs; that most nodes have less than 10 links (nodes are connected to few other nodes); and that there is **bottom–up control** (the community dynamics is driven mostly by nutrient availability Griffiths et al., 2014). Unfortunately, this analysis did not include resident taxa (e.g. microbiome) which may change these results; nor did they study how the model responds to perturbations.

In conclusion, there is a large body of work in theoretical ecology that can likely be leveraged in investigations of the microbiome. Nevertheless, there are important distinctions and at this early stage, it would be wise to keep model development close to data. Fruitful future research will likely come as WH disease modelling begins to include more of the resident community (or at least important properties of these communities) and as microbiome modelling begins to apply invasion analysis. An exciting step possible when these other directions have matured will be to include the host resource and immune cells to see what level of control they can enact. With all this accumulated knowledge, we expect that models of the stability of WH systems will eventually inform the design of perturbative experiments of the microbiome, not only in microbiome science but also for human health.

4. UNIQUE FEATURES OF WH COMMUNITIES

While there are many similarities between WH and free-living ecosystems, there are several key features unique to WH biological invasions that affect their interaction with the host microbiome. One of the most important distinctive features, with regards to invasions, is the immune system, which is itself an interacting network of cell populations thought to be primarily specialised for defending the host (and by proxy its contained microbiome) against pathogens. Not surprisingly then, it also plays a role in managing the WH community composition and dynamics. We also discuss two additional unique features of WH communities in this section, namely invaders that are fatally virulent to the host, and interactions among resident species and with their host that can prevent invasions.

4.1 The Immune System

Research establishing the degree of trophic separation in WH microbial systems has identified the immune system as an important trophic level of top predators. In fact, a recent hypothesis on the origins of the immune system points out the similarities between immunity (e.g. phagocytosis and enzymes) and digestive properties and functions (Broderick, 2015). While the population dynamics of ciliates and predatory bacteria are likely to have many similarities to free-living predators (e.g. similar feeding functional forms), the immune system has several features that depart from this analogy. For instance, the set of components of the immune system active in the response to a pathogen invasion is itself a network made up of populations of cells that communicate with each other, via populations of helper cells and signalling compounds (e.g. cytokines). This degree of cooperation between predators, a consequence of a shared genetic trajectory, is not something that is observed in free-living systems. While there are parallels of generalists and specialists to innate effectors and adaptive effector cells, respectively, these effectors engage in collective behaviours that go beyond free-living systems, which we discuss in the following two paragraphs. Studying the population dynamics and evolution of this system is itself a growing field to which ecology and evolutionary biology have been applied (Alizon and van Baalen, 2008; Mayer et al., 2016; Murall et al., 2012; Pilyugin and Antia, 2000; Sadd and Schmid-Hempel, 2009; Wodarz, 2006). To close the section, we discuss the ways in which the WH community coordinates with the immune system in countering a pathogen invasion.

4.1.1 Host Barriers and the Innate Immune Response

The protection provided by physical barriers in the body is one of the fundamental forces shaping microbial communities, as they dictate which microbes interact. Hosts have constitutive defence systems, such as the skin barrier (e.g. cornified cells, surface fluids; Sanford and Gallo, 2013) that are fundamental in keeping unwanted invaders out or trying to prevent movement around the body. Bacteria that have the ability to translocate out of their primary niche to invade new body sites often produce pathogenesis (Ribet and Cossart, 2015). Similarly, pathogenic invaders regularly have traits that help them navigate through these barriers in order to spread to various target niches in the body. Overall, then, host barriers are not enough to keep out invasions.

The vertebrate immune system is characterised by two distinct but interacting suites of mechanisms: innate and adaptive responses. Innate responses use general mechanisms that do not target-specific pathogens (i.e. the generalist predators), whereas the adaptive responses are triggered by identification of specific antigens (i.e. the specialist predators). The innate response is primarily a first responder system that recognises nonself and aims to keep down early growth of an unwanted invader. The adaptive response is typically most effective at clearing an infection, and importantly, responsible for rapid response to an attempt to reinfect. The innate response actively interacts with the resident microbial community (more below) and important microbicides are secreted throughout the body in order to help prevent infections (e.g. angiogenins in the gut; Hooper et al., 2003). This level of community-wide 'border control' is truly a WH phenomenon.

4.1.2 Rapidly Evolving Predation

Investigations into infectious disease and immunity interactions have found that the immune response can cause dynamics similar to consumer–resource interactions. For instance, coinfecting pathogens that are antigenically similar (e.g. different strains of the same species) experience cross reactivity, or in ecological terms, **apparent competition** (Brown and Grenfell, 2001; Holt and Barfield, 2006; Smith and Holt, 1996). Similarly, the stimulation of responses by one pathogen can have a range of **top-down** impacts on other microbiota in the body (McSorley and Maizels, 2012). For instance, helminth infection can downregulate responses to intracellular pathogens, such as *Plasmodium* spp. (Graham, 2008), and as found in another study, helminth infection can promote a type 2 immunity response that helps prevent *Bacteroides* colonisation (Ramanan et al., 2016).

Nevertheless, unlike free-living consumer–resource interactions, the predator is evolving rapidly during the course of the invasion, and thus the evolutionary and ecological timescales are simultaneous. Historically, free-living predators were considered to be 'prudent' because they did not evolve to be so highly efficient as to drive their prey extinct (Abrams, 2000). Clearly, however, the immune system is able to evolve fast enough to drive invaders to extinction (i.e. clearance). Furthermore, if the invader is also able to rapidly evolve during the course of the invasion, then a **coevolutionary arms race** can take place.

The adaptive immune response consists of two main effector populations involved in clearing infections (i.e. ousting an invader): B and T cells that are generated with antigen receptors that are highly diverse. Once triggered by specific antigen, they are made effective through an amplification process by which the population of those cells with receptors that bind strongly to circulating antigen undergo accelerated division. Once amplified, they can rapidly deplete a pathogen invasion population. Even more striking is that B cells undergo evolution-like rounds of mutation and positive selection which results in hyperspecialist predation with orders of magnitude higher attack rates. Another unique feature is that once an adaptive immune response has mounted and has infiltrated the infection site, the response will often decouple its growth from the presence and density of the pathogen (**antigen-independent proliferation programme**; briefly reviewed in Bevan and Fink, 2001). This independent growth allows these potent immune effectors to swamp the area and completely clear the pathogen. Finally, the immune system, builds a library of memory cells in order to aid its recognition of, and speed up its response to, a similar invasion in the future. Overall, WH communities have in the immune system a highly specialised and effective defence against invasions, the main purpose of which is to prevent invaders from colonising, establishing and spreading within the body.

4.1.3 Immunity Controls Community Composition and Dynamics

An important part of the host immune system's ability to recognise self vs nonself is due to the family of proteins called the major histocompatibility complex (MHC) and the immunity cells that use them to create a response (e.g. T-helper cells). As the most diverse region of the entire genome, the multiple genes that code for these proteins dictate the immunological repertoire of what the host can recognise. The diversity of this complex plays a central role in the immune system of vertebrates, whose slow life histories might otherwise leave them at a severe disadvantage in the coevolutionary arms race between microbial invaders with short-generation times and large population sizes (Hamilton, 1980; Jaenlke, 1978; Lively, 2010; Sommer, 2005). Interestingly, animals, including humans and nonhuman primates, have been shown to choose mates based on cues that signal high MHC diversity; however, they do not always choose mates with dissimilar repertoires (which would, in theory, provide the benefit of increased immunological repertoire for their offspring; reviewed in Winternitz and Abbate, 2015), thus suggesting that optimal MHC diversity is not always maximal MHC diversity (Milinski, 2006; Wegner et al., 2003). One speculation, as to why this may be, is that too much MHC diversity may disrupt positive associations between the host and its microbiota. An experiment in three-spined sticklebacks has demonstrated that fish with higher MHC diversity have lower diversity in their gut flora, and that different MHC allele repertoires are associated with shifts in the relative abundance of some bacterial functional groups (Bolnick et al., 2014). Recently, Kubinak et al. (2015) found that the role of MHC in regulating gut microbial community composition is also applicable to mice, in which specific MHC genotypes reduce the fitness of resident microbiota, favouring systemic infection.

Evidence that the immune response affects resident microbiota composition also comes from the examination of cases where immune function is compromised. From studies of infectious diseases where immune-deficient patients are often unable to limit the number of infections, one could think that a less effective immune response would allow a resident microbial community to be more diverse. However, studies like Oh et al. (2013) show otherwise. Patients with various immunodeficiencies were found to have decreased skin microbial diversity as well as different relative abundances of taxa in comparison to their healthy counterparts. Interestingly, these altered microbiota compositions were also associated with increased susceptibility to infectious diseases or opportunistic pathogenic fungi. This correlation is also likely to be causal, as Taur et al. (2012) showed that temporary pretransplant removal of circulating granulocytes and monocytes in allogeneic haematopoietic stem cell transplantation patients leads to reduced gut microbiota diversity, followed by invasion and establishment of pathogenic species that compromise the patients' clinical outcomes. Of course the elevated incidence of pathogenic bacteria or fungi in these communities would need to be teased apart from the impact of the immunodeficiency itself; however, the diminished resident bacterial diversity is likely to also play a role. In a telling example, nonfunctional immune response to the flesh-eating protozoan parasite *Leishmania major* in germ-free mice is rescued by the addition of the skin commensal *Staphylococcus epidermidis*, suggesting that this resident bacteria aids local inflammation and thus boosts T lymphocyte function (Naik et al., 2012). These studies suggest that the immune system plays a complex role in engineering the WH community allowing or actively promoting a more diverse community composition (which is potentially more stable and less invasible).

4.2 Resident Species Actively Prevent Invasions

The direct thwarting of a pathogen invasion by members of the resident microbiota, termed **colonisation resistance**, is a phenomenon seen in various regions of the hosts' body. This relationship is similar to symbiotic relationships seen in plants, e.g. ants on acacia plants, and is thus a common host defence strategy. However, nonmicrobial free-living ecological networks typically do not have these kinds of invasion prevention mechanisms that are community-wide.

Resident bacteria can interact directly with host cells to help prevent or clear an infection. For instance, the skin resident species *S. epidermidis* interacts directly with host keratinocytes to inhibit infection of pathogenic *S. aureus* and group A *Streptococcus* species by triggering keratinocyte expression of antimicrobial peptides (AMPs). This resident species also produces phenol-soluble modulins that work with the AMPs to improve killing efficacy, and in addition they engage in signalling that helps keratinocytes cells repair and survive during infections (Grice and Segre, 2011). In fact, AMPs are produced throughout the body and so these host–resident interactions are probably more common than may be currently appreciated. The gut microbiota is also known to engage in similar relationships with host endothelium to help maintain homeostasis necessary for organ functioning. Notably, gut disease states such as inflammatory bowel diseases (e.g. Crohn's disease) are linked to reduced production of antimicrobial defences (reviewed in Buffie and Pamer, 2013; Ostaff et al., 2013).

In addition, resident bacteria directly interact with one another to prevent invasions. For instance, healthy female genital tracts are colonised by lactic acid bacteria, mostly *Lactobacillus* spp. These bacterial species are ubiquitous and found in various hosts (as diverse as mammals and birds) and in various organs (gut, mouth, female genital tract, breast milk). Numerous *Lactobacillus* spp. have been described in healthy human genital tracts and the most frequent and abundant species are *Lactobacillus crispatus*, *Lactobacillus gasseri*, *Lactobacillus jensenii* and *Lactobacillus iners* (Gajer et al., 2012; Ravel et al., 2011). The key protective roles of *Lactobacillus* spp. are usually attributed to (i) their lowering of environmental pH through lactic acid production, (ii) their production of bacteriocins and other bacteriocidal compounds such as biosurfactant or hydrogen peroxide (H_2O_2), and (iii) their competitive exclusion of other bacterial species (Boskey et al., 2001; Fayol-Messaoudi et al., 2005; Kaewsrichan et al., 2006; Ling et al., 2013; Ravel et al., 2011). Similarly, a study on the persistence of human-associated Bacteroidetes in the gut (discussed in Section 2.3) showed that some bacterial species can exclude an invader by producing bactericidal effectors triggered by direct physical contact, and that this activity is mediated by resistance evolution and relative population densities between species (Wexler et al., 2016). Finally, how our commensal or mutualist viruses interact with each other or the microbiome in order to prevent invasions is significantly less understood. An example reviewed in Clemente et al. (2012) showed that infecting mice with a latent gammaherpesvirus (a virus highly related to common childhood infections in humans, such as Epstein–Barr virus and human cytomegalovirus) seems to increase resistance to bacterial infections, at least for a period of time (Yager et al., 2009). More research into the human virome and its effects on invasions is greatly needed.

To conclude, it is tempting to anthropomorphise the mutualistic interaction between humans and their microbiota with the simplification that we provide food and shelter while microbiota prevents invasions. We caution that, in fact, invasion prevention mechanisms are often not for our direct benefit, since some may be secondary or indirect effects. Indeed, many protective effects are actually indirect effects via the immune response, and, in addition, the mutualistic effect of a resident can come with costs. For instance, chronic infections by *Helicobacter pylori* seems to confer some protection against tuberculosis (Perry et al., 2010), but chronic infection of *H. pylori* is also associated to other pathogenic effects (e.g. increased incidence of gastric cancer). Likewise, several viruses and bacteria are mainly commensal, but persistent carriage of them can lead to pathogenic effects (e.g. the oncoviruses such as Merkel cell polyomavirus, Epstein–Barr virus and high-risk human papillomaviruses). Some recent, interesting microbiome results are similar to earlier results on multiple infections showing that pathogens can have pleiotropic effects by protecting their hosts against other (more virulent) pathogens (Michalakis et al., 1992). As a counterexample, a study by Stecher et al. (2010) found that the presence of related bacteria in the resident microbiota can facilitate bacterial invasion. Overall, then, we do not have the full picture of this commensal-pathogenic continuum. As such, we echo the suggestion by Mushegian and Ebert (2016) that the term 'mutualism' be used more sparingly, i.e. only when there is a clear pairwise interaction or when mutual reciprocity has been directly investigated and demonstrated.

4.3 Fatally Virulent Invaders

Virulent invaders have negative health consequences on their hosts (e.g. morbidity or mortality), and some WH invaders can be so pathogenic that they kill their host. This would be equivalent to a free-living invader completely wiping out all biological life in the area it invades. In free-living systems, biological invasions can have very strong destabilising effects but rarely do they directly eliminate the entire community. There is clearly a cost to the invader to completely kill the ecosystem it grows and lives in, and thus there is a great deal of literature dedicated to understanding the evolutionary forces that maintain this costly trait.

In general, pathogen virulence is expected to be counterselected because killing the host decreases the duration of the infection and, hence, the epidemiological fitness of the pathogen. This led to the **avirulence hypothesis**, which postulates that given enough time, host–pathogen interactions should evolve to be benign (Méthot, 2012). The same intuition applies to the human microbiota and its invaders: virulence should be counterselected. Over the years, it has been realised that the avirulence hypothesis was at odds with the data (see, for example Ball, 1943). There are several explanations as to why virulent strains can persist and they broadly divide into nonadaptive reasons and adaptive reasons (Alizon and Michalakis, 2015). The former reasons imply that the host is not involved in the transmission route of the pathogen or, more generally, that virulence does not affect pathogen fitness so there is no selection against it. Virulence can persist if it confers an advantage in the between-host life-cycle of the pathogen. The classic perspective is that more virulent strains have a higher transmission rate. A more appropriate approach consists in measuring the pathogen between-host fitness as a function of its virulence (Alizon and Michalakis, 2015).

Another explanation for virulence, that is arguably more pertinent in a microbiota context, stems from WH interactions. Microbes compete for resources inside their host (Mideo, 2009; Smith and Holt, 1996), either via direct competition for nutrients (such as iron) or indirect competition (such as the apparent competition via the immune response). It is usually expected that the most virulent strains are the most competitive, as proven experimentally in the case of *Plasmodium chabaudi* (de Roode et al., 2005). However, this result strongly depends on the type of WH interactions (Alizon et al., 2013). For instance, if pathogens interact by competing for public goods they produce, then the less virulent strain wins the competition, as demonstrated by Pollitt et al. (2014) using quorum sensing in bacteria. Overall, this means that a more or a less virulent invading strain can be favoured depending on the nature of the interactions in the micro-biota (Lysenko et al., 2010). Furthermore, predicting the direction of evolution in multiple infections caused by different species is even more difficult given that it requires coevolution models. In addition, models of zoonotic or opportunistic infections—which is often the case for pathogens invading the microbiome—must consider the possibility that the host is not part of the parasite's life cycle, which means that there will be no direct selective pressure on the traits expressed in this host type as long as it remains an evolutionary dead end (Brown et al., 2012).

A final adaptive explanation for the persistence of virulent strain is WH evolution. In other words, these strains can be completely maladapted to the between host level (to the point that they are rarely transmitted), but they keep evolving de novo over the course of an infection because they are well adapted to the WH level and evolution is shortsighted (Levin and Bull, 1994). One of the best illustrations of this is provided by experiments using *Haemophilus influenzae* in mice (Margolis and Levin, 2007). These bacteria are usually commensals harboured in the nasopharynx but in some cases they can cause an invasive disease by colonising the bloodstream. By using tagged bacteria isolated from the blood and the nasopharynx of infected rats, the authors showed that the bacteria's diversity is lower in the blood than in the nasopharynx, suggesting a bottleneck occurred during the colonisation of the blood. Using these remaining bacteria to reinfect new mice, they found they could induce the disease. This example of WH evolution suggests a context dependence of a bacteria's virulence, in which context may well include other bacteria. The WH evolution literature, which dates back at least to the 1970s with articles on cancers as evolutionary processes (Nowell, 1976), is interesting as a starting point to import evolution ideas into the microbiota context because it may offer an adaptive explanation to virulence without involving fitness at the between host level, which is complicated to measure due to the difficulty in determining the life cycle of the microbe and due to epidemiological feedbacks (i.e. the fact that the outcome of an infection may be affected by processes occurring at the epidemiological level, such as a new parasite invading the host). Of course, there is always a risk that between-host selective pressures act against WH selective pressures.

In the context of microbiota invasions, there is a natural parallel to make with emerging infectious diseases because typically, when a pathogen first emerges, it will face a set of 'patches' (i.e. hosts) that are colonised by resident pathogens. In epidemiology, there is a long-lasting debate regarding the virulence of emerging infectious diseases. Some argue that these pathogens should be more virulent than resident pathogens, while others argue the opposite. By definition, emerging pathogens are less likely to be adapted to the host than resident pathogens. This does not, however, predict their virulence since maladaptation can lead to high virulence (e.g. due to immunopathological reactions of the host) or low virulence (e.g. rapid clearance by the immune system). However, if stochasticity is taken into account, things change. Indeed, for two pathogens that cause the same number of secondary infections (and should therefore be equally fit in a deterministic setting), the less virulent is more likely to emerge because it causes longer infections, which minimises stochasticity (André and Day, 2005).

Adopting a metacommunity approach to study epidemiology is tempting as hosts can be seen as communities of microbes (see Seabloom et al., 2015 for an interesting attempt). However, beyond the analogy, it is not clear how each field can help each other, especially in terms of virulence. In the earliest metapopulation/community models, patch extinction was usually a catastrophic event occurring at a constant rate. This has changed and models now allow this property to depend on patch density or on the status of neighbouring patches (e.g. Eaton et al., 2014). In epidemiology, there has been a longer focus on the idea that virulence is a property that originates from the WH (i.e. within patch) dynamics (Sasaki and Iwasa, 1991). Both in ecology or epidemiology, the challenge raised by including the host microbiota into the picture is the same: how to track WH diversity while allowing for dynamics to take place at a higher level (Sofonea et al., 2015).

5. EXPERIMENTAL TESTS OF MICROBIOME INVASIBILITY

Culturing microbial communities has played an important role in ecology for several decades (Faust and Raes, 2012). However, with more sophisticated cultivation methods and technology advances that allow measurements to be made in unprecedented detail, it is now possible to probe more complex microbial communities and test their stability. Several animal models have been developed in order to study microbe–host interactions and in particular the symbioses between them. Invertebrates (e.g. squid, nematode and *Drosophila melanogaster*) are best suited for investigating more binary interactions, while vertebrate model systems (e.g. mouse and zebrafish) are the best for investigating more complex microbial communities and their interactions with the host (Rader and Guilemin, 2013).

Indeed, colonisation of germ-free organisms with human microbial communities (e.g. colonisation of zebrafish guts with human-derived anaerobic bacteria, Toh et al., 2013) allows for controlled studies of these communities in lab settings. Of particular importance are animal models such as *Drosophila* or mice that can be manipulated using very precise genetic tools with the aim of defining host–microbe interactions and mechanisms that stabilise communities (reviewed in Goodman, 2014; Lee and Brey, 2013).

While there is a long history of infecting animal models with pathogens, these studies have usually aimed for molecular, immunological, genetic, evolutionary or pharmacological insights of host–pathogen interactions and often do not take the presence of a resident microbiota within a given host into consideration. In contrast, emerging microbiome-focused studies have provided a more holistic appreciation of community ecology, in which the focus is how resident microbes can help prevent infection, or how microbial species can be used to restore a dysfunctional ecosystem after perturbation, through assessment of factors such as community stability, resistance and resilience to invasion. The study of faecal ecosystem changes following faecal bacteriotherapy (stool transplants) in human patients with *C. difficile* infections represents a good example of microbial ecology applied to a medical problem, where the goal is to understand how the presence of certain microbial species can correlate with stability and hence resistance to disease caused by *C. difficile* (Seekatz et al., 2014; Shahinas et al., 2012; Shankar et al., 2014). In a more targeted approach, a 33-strain defined ecosystem (MET-1) derived from a single healthy human donor was used to treat recurrent *C. difficile* infections in two patients, tracking ecosystem changes with time after administration of this therapeutic ecosystem (Petrof et al., 2013). Both approaches clearly demonstrated a correlation between a lack of microbial diversity with the disease state, and a subsequent input of diversity with transplantation that correlated with restitution of the diseased state. The same approach may also be applied for the treatment of other diseases and infections; the 33-strain therapeutic ecosystem used to treat *C. difficile* infection (earlier) was tested for efficacy against *Salmonella enterica* infection in a murine model (Martz et al., 2015). *S. enterica* was chosen in this case because the infection is mediated by bacterial invasion whereas *C. difficile*-associated disease is mediated by exotoxin production and the bacterium itself is not invasive. While protection against disease was seen in MET-1-treated animals dosed with *S. enterica*, the mechanism of protection was thought to be a direct result of strengthening of host barrier functions thereby preventing pathogen entry. Thus, microbial ecosystem therapeutics as an approach is likely to have multifactorial effects on host and microbiota, reflective of the complex interactions between these two factors. There remains much to be learned from such studies, integrating microbial communities with host phenotypes and genotypes to determine how both factors interact together during pathogenesis.

An example of a true invasion ecology experiment was conducted by Seedorf et al. (2014) in which they colonised gastrointestinal tracts of germ-free mice with microbial communities obtained from several very different sources (humans, zebrafish, termites, soil and aquatic mats); as such, this can be considered as an invasion island biogeography experiment (Delong, 2014). The fact that all communities (albeit from very different origins) were able to colonise the mice and form stable and reproducible populations highlights that there are community assembly rules as well as common properties across communities that allow their members to invade new habitats. In particular, the taxa that were most successful in colonising mice were those with greater capacities to perform certain functions, such as metabolism of host nutrients and bile acids. This study also found that there were repeatable patterns of invading successions, i.e. some taxa invaded first, then declined in abundance, allowing other taxa to follow (for free-living examples of this, see Schreiber and Rittenhouse, 2004). In addition, they found that Firmicutes species (regardless of origin) were fundamental to the initial colonisation of the gut. These findings support the invasional facilitation hypothesis discussed in Section 2.2, which seeks to explain that the succession of a community is predictable because particular species facilitate invasion by other taxa. These findings also support the premise of **keystone species** (Paine, 1966, 1969), which carry out critical functions required to prime an ecosystem for colonisation and that allow community assembly. Keystone species have been found in several other microbiome studies as well (Allen-Vercoe, 2013; Ze et al., 2013).

In vitro systems allow for even more control of microbial communities but come with the downside that the contributions of the host to ecosystem parameters can only be measured superficially (e.g. by the addition of host proteins, such as mucins; Allen-Vercoe, 2013; McDonald et al., 2013). Culturing microbes from the mammalian gut can be challenging because many are fastidious in their nutritional needs and require cooperation with other species for optimal growth (Allen-Vercoe, 2013). Faecal samples offer a convenient starting material from which to culture gut microbial ecosystems, and in vitro bioreactor-based models, with parameters set to mimic conditions of the gut, can be used to support these communities for several

weeks at a time, allowing a steady-state equilibrium to be achieved that can
serve as a baseline for experimental perturbations (Fig. 4). Interestingly, in
human studies, the ecosystem at steady state tends to diverge from the faecal
inoculum sample (McDonald et al., 2013, 2015) and this may reflect the fact

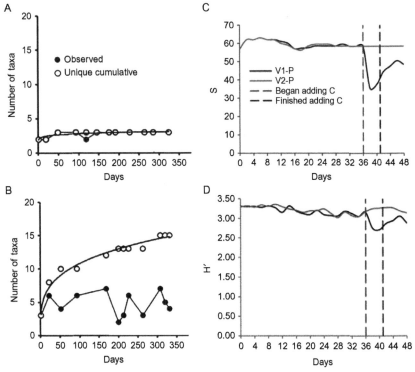

Fig. 4 Time series of microbiome communities. (A and B) Community of lung microbial communities of two cystic fibrosis patients over a period of 1 year, measured using
terminal restriction fragment length polymorphism. Diversity, or the number of
taxa, is shown as *filled circles*, while the unique cumulative diversity, or the number
of unique taxa seen up to that time, is shown as *open circles*. The community in
(A) shows little turnover and appears to persist in time, while that in (B) shows high
turnover with the numbers of taxa fluctuating in time. (C and D) Response of an
in vitro community to antibiotic treatment. Denaturing gradient gel electrophoresis
analysis of the planktonic communities from two chemostat vessels (V1-P, *black line*
and V2-P, *grey line*) modelling the human distal gut prior to, during, and following a
clindamycin treatment period (demarked by *vertical dashed lines*). From the number
of bands taken as species, S (shown in C), and their relative intensities taken as relative
abundances, the adjusted Shannon diversity index, H′ was calculated and shown in (D).
*(A and B) Modified from Stressmann, F.A. et al., 2012. Long-term cultivation-independent
microbial diversity analysis demonstrates that bacterial communities infecting the adult
cystic fibrosis lung show stability and resilience. Thorax 67(10), 867–873. (C and D)
Figure from McDonald, J.A.K. et al., 2015. Simulating distal gut mucosal and luminal
communities using packed-column biofilm reactors and an in vitro chemostat model. J.
Microbiol. Methods 108, 36–44.*

that faeces contains both autochthonous (resident) and allochthonous (nonresident) microbe representatives of the luminal content. However, a small number of mucosal microbes will be contained within faeces and thus attainment of steady state may represent the loss of allochthonous species (which cannot be supported by the ecosystem) as well as the bloom of lower abundance, mucosally associated microbes. In other words, the in vitro steady-state faecal ecosystem may represent the core microbiota of the human gut, although this remains to be tested.

Several in vitro systems for the culture of human gut microbiota, as well as oral microbiota using patient- or volunteer-derived samples, have been described and reviewed elsewhere (Edlund et al., 2013; Venema and van den Abbeele, 2013). While it is highly valuable to study complex communities such as these, there can be problems with reproducibility of these models because inoculum samples can differ if taken at separate time points. The high level of functional redundancy present within a natural microbial ecosystem does, however, lend itself to a partial solution to this problem. While not all of the microbes present within a human-derived ecosystem are easily cultured, it is possible to construct subset communities of these parent ecosystems which retain some functional similarities (Petrof et al., 2013; Yen et al., 2015). Such ecosystems, although not a complete facsimile of the natural situation, have the distinct advantage of being wholly reproducible, and furthermore, individual taxa can be removed or added precisely in order to determine their effects on the ecosystem. Such insights may then be extrapolated back to the natural ecosystem to help build hypotheses for testing.

Model in vitro ecosystems, both derived from complex or defined communities, are excellent platforms on which to model ecosystem perturbation with only minimal ethical consideration. Such platforms have been extensively used to study host microbiota/pathogen interactions, as well as the effects of dietary substrates, supplements and antibiotics on the microbial ecosystem (Aguirre et al., 2014, 2015; Baines et al., 2013; Crowther et al., 2016; Wissenbach et al., 2016). Perturbations can be assessed by measuring phylogenetic, metagenomic, transcriptomic or metabolomic outputs, and, in fact, the platform outperforms animal models in this respect since samples can be obtained from bioreactor vessels very easily and are not altered by absorption or transformation by the host. The platform may be particularly useful for studying microbial succession following invasion because of the controllability of the parameters.

A potentially very fruitful direction for future research would be to combine both in vitro and in vivo experiments with mathematical models incorporating ecological theory in order to help determine the drivers of community stability. Several groups have outlined frameworks for such models (Bucci and Xavier, 2014; Moorthy et al., 2015), and a recent and an elegant study by Buffie et al. (2015), successfully combined mathematical modelling with analysis of a mouse model of antibiotic induced *C. difficile* infection, by specifically determining how microbiota changes associated with antibiotic introduction could be correlated to disease status. The authors were able to extrapolate their findings to human patients and this work indicated that a specific bacterial species, *Clostridium scindens*, greatly contributes to resistance to *C. difficile* infection (Buffie et al., 2015). Approaches such as this may also be applied to other disease states in order to guide rational design of future therapeutic strategies.

Since it is becoming increasingly clear that the diversity of microbial communities significantly contributes to their stability, resilience and resistance (Allen-Vercoe, 2013; Lozupone et al., 2012), and that lack of microbial diversity seems to be associated with a growing number of human diseases including obesity, type 2 diabetes and inflammatory bowel diseases (Matsuoka and Kanai, 2015; Ridaura et al., 2013; Vrieze et al., 2012), it is important that a fuller understanding of ecosystem dynamics is found, which incorporates the roles of such ecological measures as network interactions and the significance of keystone species.

6. TYPES OF INVADERS

As laid out in Section 2.2, successful invasion by free-living species into new habitats requires completing four steps: transmission out of the home environment, introduction into the new environment, establishment and spread. WH invaders face these same challenges and they require many traits to complete these stages. Here we discuss invasion strategies by different kinds of invaders. In Box 2 we discuss specific features of invaders that are not bacteria.

6.1 Single and Polymicrobial Invasions

Given the tradition of using Koch's postulate approaches to find specific etiological agents for diseases, WH systems have traditionally been studied from the perspective of the invader. Important killers, both historically and presently, such as *Mycobacterium tuberculosis*, *Yersinia pestis* (plague) and *Streptococcus pneumoniae*, have been studied separately from the microbiome they encounter when colonising a new host. Nevertheless, community thinking is becoming more common. In particular, there is active research into how enteric infections, such as *Salmonella* spp., *Clostridium perfringens*,

BOX 2 Non-Bacterial Invaders

Viruses (Other Than Bacteriophages)

While viral infections are usually studied on their own without consideration for the microbial communities into which they land, there are examples of how resident microbes can protect the host from viral infections. For instance, *Wolbachia*, a Gram-negative alpha-proteobacteria that infects insects has been shown to protect *Drosophila* or mosquitos from RNA viruses (e.g. Dengue or Chikungunya virus). It uses mechanisms such as competition for cholesterol and regulating host gene expression, and thus inhibits virus replication (Hamilton and Perlman, 2013).

Some viruses, like human immunodeficiency virus (HIV), can have catastrophic effects on the entire WH ecosystem because they target the cells of the immune system, allowing themselves to be captured only to then destroy the cell (Carrington et al., 1999). By disabling the immune system, the virus is free to proliferate, as are all other pathogens (and the microbiome) that had previously been regulated by the immune system. This manipulation of the host immune defences can also be specific, such as in antibody-dependent enhancement of dengue virus. Here, the initial infection provokes a normal antibody response targeting the virus. However, upon reinfection with a sufficiently heterologous dengue virus, those antibodies still bind the virus, but trigger enlargement of the infected cell rather than neutralisation, allowing the virus to replicate unchecked and causing severe disease (Flipse et al., 2013).

With increased metagenomic approaches and microbiome research, we hope that human studies of viral infections will now investigate more direct interactions between the microbiota and viruses. These interactions may be playing a bigger role than is currently believed in preventing invasions, and affecting the immune system's efficacy, infection virulence or duration of infection.

Macroparasites

Macroparasites, such as intestinal worms, are likely to be involved in many trophic interactions in the gut, though studies are lacking. Helminths and the microbiome do appear to interact both directly and indirectly, and especially through the immune system (Zaiss and Harris, 2016). Theories of the association between helminth infection and anorexia are numerous and largely still unsubstantiated, but include the possibility of a behavioural response of the host to change amount or selectivity of food intake in order to enhance immune response to the infection (Kyriazakis et al., 1998). If this is the case, it is also likely to have impacts on the composition of host intestinal flora.

Invasive nontyphoidal *Salmonella enterica* is strongly associated to populations with high malaria transmission, and in addition to immunological potential mechanisms, malaria's sequestration of red blood cells in the intestine could free iron, which promotes *Salmonella* growth and the translocation of the bacteria through the gut barrier (Biggs et al., 2014).

Continued

BOX 2 Non-Bacterial Invaders—cont'd

Coinfections of macroparasite and microparasites affect the immune response in a unique way, whereby they trigger different, yet competing, immune
response pathways (Th2 and Th1, respectively). This can lead to longer durations
of infection or immunity escape (reviewed in Maizels and Yazdanbakhsh, 2003).
Whether and how this Th1/Th2 switch affects the dynamics of the WH system
(resident microbiota with the immune response) and the resulting epidemiology
is an interesting avenue to consider (e.g. Ezenwa and Jolles, 2015). For the microbiome in particular, both responses may be involved in WH interactions, because
intracellular bacteria trigger the Th1 response, whereas those that proliferate
outside of the cell trigger the competing Th2 responses.

Yeast and Fungi

Fungal species can form biofilms along with resident species and as such are
often implicated in polymicrobial biofilms and their associated diseases. A very
common example of a pathogenic yeast is *Candida albicans* which can have both
positive or negative interactions with resident bacterial species; for example, it
forms a scaffold base on which *Staphylococcus aureus* builds its biofilm (which
it does not readily do on its own) and it can facilitate the growth of *Staphylococcus
epidermidis*, yet *Candida* interacts antagonistically with *Lactobacillus* (Harriott and
Noverr, 2011).

Vibrio cholerae, *E. coli* and *Enterococcus faecalis*, interact with the resident community. Invading strains of bacteria face several challenges when entering a
new environment with resident bacteria, such as biofilm adhesion, availability of nutrients (residents can deplete nutrients), antibacterial effectors produced by residents (e.g. bacteriocins, metabolic by-products) and signalling
the immune system (Stecher and Hardt, 2011). Owing to these
bacteria-specific defences, many invading strains have evolved mechanisms
that make them more effective colonisers. For example, *V. cholerae* uses
quorum sensing (which is density–dependent chemical signalling allowing
perception of conspecific density) to increase the expression of its virulence
factors in the presence of conspecifics and to form biofilms when it is at low
density, thus giving it an advantage to colonise a gut even when low quantities are ingested (Hammer and Bassler, 2003). Indeed, many pathogenic
bacteria sense nutrient concentrations in their new environment and then
alter their virulence factors in order to better compete with resident taxa
(for review, see Rohmer et al., 2011). Finally, from a community
perspective, invading bacterial strains must also face natural predators and
bacteriophages already present in the WH environment (more on

bacteriophages in sections that follow). The degree to which these community members help prevent successful colonisation requires more research.

Historically, a single etiological agent was sought to explain a single disease. However, it is becoming clear that some diseases are caused by a group of microorganisms, i.e. **polymicrobial infections**. These invaders are often ineffective WH colonisers (or less virulent) unless they are together with other specific taxa. For instance, to test the hypothesis that periodontitis (gum disease) is caused by a *disbiotic* microbial community, mice were infected with four different bacteria species simultaneously and compared them to mono-infections (Chukkapalli et al., 2015). They found that polymicrobial infections were more virulent, formed plaques more readily and altered immunity functioning. Interestingly, it is hypothesized that *Porphyromonas gingivalis* plays a keystone species role in these polymicrobial infections by actively downregulating immune surveillance, which facilitates the growth of the other species in the biofilms, and these communities then become stable (Hajishengallis and Lamont, 2012). In fact, there are many examples of **polymicrobial synergy**, i.e. the interactive activity of two or more microbes which allows achievement of results otherwise not possible alone (reviewed in Murray et al., 2014). For instance, coinfections in the otitis media by *H. influenzae* and *Moraxella catarrhalis* appear to form stronger, more antibiotic-resistant biofilms, than when coinfecting with other bacteria, because of effective quorum sensing between them (Murray et al., 2014).

In another example, multispecies biofilms, grown in vitro beforehand, were placed onto wounds of mice and found that healing was impaired when compared to single-species infections (Dalton et al., 2011). Likewise, in a *Drosophila* model, Sibley et al. (2008) did several single vs coinfections of different strains (e.g. of *Staphylococcus* spp. and *Streptococcus* spp.) with *Pseudomonas aeruginosa* and found that in some cases, when infecting together, there is enhanced killing of the host even if some of the strains were not pathogenic on their own. Conversely, they also found that coinfection (i.e. simultaneous infection by two strains or species) can lead to faster clearance (Sibley et al., 2008). Indeed, in a rat model, the same group found that coinfection enhanced inflammation (Duan et al., 2003). Findings like these point to a potential disadvantage of multiple invaders entering at once: they are more conspicuous to the immune system. This could be because they bring more antigens into the site at once or because they cause more local damage; both of which will alert the innate immune response and cause more local inflammation. This reasoning is also used in combination vaccines in which combining vaccines into one shot leads to a stronger and more rapid local immune response. Overall, determining whether invading as a group or alone confers an advantage or disadvantage for the invaders will require more investigation.

Unsurprisingly, the community into which a polymicrobial infection lands matters, since much of the clinical focus of polymicrobial infections comes from studies of patients with some other underlying medical issue, such as cystic fibrosis (a genetic disorder; Rogers et al., 2010), diabetes (Mastropaolo et al., 2005) or immune system disease. The experiments mentioned earlier by Seedorf et al. (2014) are designed to identify what features of communities may favour establishment of an invader in a new host. They do so by looking at what affects establishment of an invader (i) into a naive or nearly naive habitat (e.g. in newborn infants) or (ii) into an established community. For example, see their stage four experiments in which native mice communities invaded stable, nonnative communities in other mice, only to quickly destabilise them and take over, thus demonstrating that communities that have longer evolutionary history with the host have an invasion advantage.

As we better characterise healthy and diseased WH ecosystems, we are discovering more examples of polymicrobial infections. Recently, **bacterial vaginosis (BV)** has been hypothesized to be initially triggered by the sexual transmission of a structured polymicrobial biofilm synthesized by *Gardnerella vaginalis*, which then subsequently displaces resident lactic acid bacteria (mainly *Lactobacillus* spp.) from their ecological niche. In fact, the first experiments aiming at testing *G. vaginalis* as the etiologic agent of BV did not fulfil Koch's postulates (Criswell et al., 1969; Gardner and Dukes, 1955). Moreover, *G. vaginalis* can be isolated from healthy genital tracts of adult and adolescent girls (Hickey et al., 2015; Ravel et al., 2013) and healthy genital tracts are free of structured biofilms contrary to genital tracts with BV. It has, thus, been concluded that 'there is no BV without *G. vaginalis*, but *G. vaginalis* is not BV' and a consensus now exists on the hypothesis that the transmission unit of BV is the structured polymicrobial biofilm (Swidsinski et al., 2014). Interestingly, several different bacterial species cooccur with BV. For instance, more than 90% of a biofilm identified on vaginal epithelial cells of women with BV was composed of *Atopobium vaginae* and *G. vaginalis* (Swidsinski et al., 2005). In another study, two of seven key species (*Prevotella* spp. and *A. vaginae*) were significantly associated with BV and *G. vaginalis* and *Prevotella* spp. defined the majority of BV clusters (Datcu et al., 2013). These latter results suggest a specific interaction that may lead to enhanced selective advantage of both bacteria. Such synergistic interactions among the BV-associated bacteria have been observed. On the one hand, the reproduction rate of planktonic (i.e. free living, nonbiofilm) *G. vaginalis* and biofilm growth are promoted by the presence of some additional species such as *A. vaginae, Prevotella bivia, Fusobacterium nucleatum* and *Mobiluncus mulieris*, irrespectively of which other pathogenic species are involved in particular cases of BV (Machado and Cerca, 2015; Machado et al., 2013). On the other hand, some tested BV-associated bacteria had their growth enhanced by the presence of *G. vaginalis*-derived biofilms, such as *P. bivia* that was able to increment its concentration within the biofilm, when added to a preformed *G. vaginalis* biofilm (Machado et al., 2013; Pybus and Onderdonk, 1997).

As discussed earlier, insights from infectious disease evolution research can help better understand polymicrobial invasions. Studies have shown that if invasion occurs by a group of propagules rather than by a single propagule, this can deeply affect the evolutionary dynamics. In evolutionary biology, this is called **budding dispersal** (Goodnight, 1992). Experimentally, it was shown that this process helps to maintain cooperative behaviour (producing of siderophores) in a bacterial population over generations (Kümmerli et al., 2009). In epidemiology, this is called cotransmission and mathematical modelling predicts that it can affect the evolution of virulence in directions that are counterintuitive with respect to classic coinfection models (Alizon, 2013).

6.2 Bacteriophages

As part of microbiome networks, **bacteriophages** (bacteria-infecting viruses) are major players affecting the function and structure of bacterial communities. Bacteriophages influence both the frequencies of bacterial genotypes (through modification of fitness and/or virulence) and thus the environmental conditions in which bacteria reproduce, through the promotion of some bacterial genotypes providing new structures (e.g. biofilm) or modification of the physiology of the host. Existing data on interactions between bacteriophages and bacteria in human- or animal-associated microbiota suggest a major—though underevaluated—role of bacteriophages in the dynamics and evolution of individual bacterial species/genotypes, but also the bacterial community as a whole.

Bacteriophages follow two major cycles of viral reproduction: (i) the *lytic* cycle consists of infection, replication and lysis of their bacterial hosts (and thus *virulent bacteriophages* are restricted to horizontal transmission) and (ii) the *lysogenic* cycle consists of integration of the bacteriophage genome—called a *prophage* once integrated—into its bacterial host genome. While *temperate bacteriophages* (those with a lysogenic cycle) are transmitted vertically between bacteria, external factors can trigger induction of the prophage, i.e. expression and encapsidation of the bacteriophage genome, lysis of the bacterial host cell and the release of viral particles into the environment which can then horizontally infect new susceptible hosts. Sometimes, after induction, temperate bacteriophages can carry genes that are not required for their own life cycle, but that modify the phenotype (and fitness) of their new bacterial host (i.e. *transduction* of bacterial genes). Bacteriophages are, thus, prime suspects in the adaptation of bacterial pathogens to new eukaryotic hosts and the emergence of new pathogens (Brüssow et al., 2004). Indeed, it has been shown that bacteriophages transduce antibiotic resistance genes (e.g. Enault et al., 2016; Schuch and Fischetti, 2006). Moreover, pathogenic strains of *Corynebacterium diphtheriae* (diphtheria), *S. enterica* (Salmonellosis), *Clostridium botulinum* (the leading cause of food poisoning), *Streptococcus pyogenes* (scarlet fever), *S. aureus* (causing a number of skin and blood diseases) and enterohemorrhagic *E. coli* (causing hemorrhagic colitis) all harbour toxins with bacteriophage origins. Specifically, some of these toxic proteins are involved in different steps of the bacterial life cycle, such as eukaryotic host immune evasion, intracellular and extracellular colonisation and proliferation (Boyd, 2012; Brüssow et al., 2004).

There are several ways bacteriophages interact with the microbiota (reviewed in De Paepe et al., 2014). One type of interaction, termed '*community shuffling*', may occur when an environmental stressor induces a lysogenic phage to kill its host. While this releases horizontally infectious viral particles that can infect other cells, which in turn may cause further lysis and potential extinction of the bacteriophage and its host. Moreover, the debris resulting from the simultaneous lysis of many bacteria is likely to stimulate a host response. When this happens, the composition of the microbiota will be affected and opens the opportunity for invasion by new bacterial genotypes or species. Another form of bacteriophage–microbiota interaction, termed '*kill the winner*', is one in which bacteriophages will only deplete a resident bacterial population that is above a critical density, meaning that only the dominant members of the ecosystem are directly affected. Conversely, bacteriophages can also indirectly affect microbiome networks by modifying phenotypes and/or fitness of specific bacterial species in the community. It has long been observed that temperate bacteriophages allow their host to outcompete their nonlysogenised competitors in environments in which nutrients are scarce, because lysogenic strains, counterintuitively, have higher metabolic rates, and thus reproduce faster than wild-type nonlysogenised strains (De Paepe et al., 2016; Duerkop et al., 2012; Edlin and Bitner, 1977; Edlin et al., 1975). Bacteriophages have also been hypothesized to be involved in bacterial capacity to adapt to adverse environments, for instance through biofilm formation. Indeed, considering that extracellular DNA (eDNA) is a key factor in the biofilm matrix, activation of *S. pneumoniae* prophage, resulting in host lysis, showed that localised release of eDNA favours biofilm formation by the remaining bacterial population (Carrolo et al., 2010).

Another interaction is the use of bacteriophages by bacteria as 'natural' biological weapons that they deploy to '*kill the relatives*' that do not harbour the same prophages (i.e. identifying which resident relatives are not part of the same *immunity-group*; Brown et al., 2006). This applies particularly in cases in which new bacterial genotypes invade new ecological niches and use bacteriophages as allelopathic agents that will replicate into, and lyse, their targets. In addition, due to the fact that some bacteriophages present a large host range, such as those able to infect bacteria belonging to different families (Meaden and Koskella, 2013), this interaction probably affects other susceptible bacterial species. However, this interaction can only explain short-term invasions. In fact, taking the example of enterohemorrhagic *E. coli*, it has also been shown in vivo in the murine gastrointestinal tract that resident susceptible *E. coli* were lysogenysed by Shiga toxin-converting prophages present in the invading enterohemorrhagic *E. coli* (Acheson et al., 1998). Thus, after horizontal transmission, temperate bacteriophages may establish themselves in the resident bacterial hosts and then confer immunity to bacteriophages. This property would then confer a disadvantage to the invading lysogenic bacteria. Moreover, the presence of new resident lysogenic genotypes will even accelerate as the bacteriophage population size grows because lysogeny rate has been shown to increase with increases in multiplicity of infection (Gama et al., 2013).

To date, no human disease has been strictly related to virulent bacteriophage invasion and depletion of bacterial communities. Still, several dysbioses have been hypothesized to have a direct bacteriophage origin such as Crohn's disease and ulcerative colitis, or bacterial vaginosis (Kiliç et al., 2001; Norman et al., 2015; Turovskiy et al., 2011). For instance, Crohn's disease and ulcerative colitis have been shown to be associated with a significant expansion of bacteriophages belonging to the *Caudovirales* family and a decrease in bacterial diversity (Norman et al., 2015). Moreover, bacteriophages were mostly virulent (T4 and T4-like) in individuals suffering from diarrhoea, whereas they were mostly temperate in samples from healthy controls (Chibani-chennou et al., 2004; Furuse et al., 1983).

As new technological tools continue to increase the precision with which we understand spatial and temporal compositions of bacteriophage–bacterial communities, it will help inform interpretations of clinical and biological data that will, in turn, allow the application of better and potentially new therapies such as bacteriophage therapy (see Section 7.1).

7. ACTIVELY ACQUIRED INVASIONS

As in free-living systems, humans attempt to engineer the ecological communities inside their bodies in many ways. Here we discuss examples of biological invaders that we use either to prevent disease (i.e. **prophylactics**) or as treatments. Prophylactic practises are often used to prevent other invasions, which can be done directly by adding a new resident to the community (e.g. adding a bacteriophage to guard against an unwanted infection) or indirectly by training the immune response to identify a pathogenic invader (e.g. using live vaccines to prime the adaptive immune response). These WH biological control agents can themselves be single or multiple agents administered into human bodies. Consequently, the way they subsequently interact with the WH system can be quite different (e.g. a single bacteriophage can cause predator–prey dynamics while cocktails of probiotics composed of several antagonist microorganisms such as commensal bacteria and bacteriophages can engage in many forms of interactions with resident community members). From a network perspective, these therapies can remove single nodes or multiple nodes of the resident community or even replace the entire system. Therefore, achieving success with these control methods challenges us to both understand WH invasions better but also to see where our gaps in ecology and evolution are in general.

7.1 Bacteriophage Therapies

Pioneered by the work of Twort and Felix d'Hérelle in the early part of the 20th century, bacteriophage therapy has a long history of study and application, particularly in Europe and the states of the former Soviet Union. Despite early enthusiasm, it was displaced by the advent of cheap, effective antibiotics in Western medicine in the 1940s. The promise of bacteriophage therapy as a weapon against bacterial infection is threefold (comprehensively reviewed in Keen, 2012; Loc-Carrillo and Abedon, 2011). First, in contrast to broad-spectrum antibiotics (Looft and Allen, 2012), targeted specificity for a single pathogen should have minimal impact on the microbiome community (aside from control of the pathogenic bacterium being targeted), and very few direct negative effects on the infected host. The second benefit is that bacteriophage will amplify itself as it destroys its target bacterium, reducing the need for repeated administration to the patient. The third intended benefit is that bacteriophage has the ability to adapt in response to the evolution of resistance in the targeted bacteria. It is thus touted as an evolution-proof alternative to the application of static chemical treatments. Though this may not always be the case (see Meaden and Koskella, 2013 for examples), bacterial resistance to bacteriophage is generally expected to occur through the loss of the receptors to which they bind on the bacterium's cell surface, which are often the very virulence factors causing the disease (Inal, 2003). While this may make the bacteriophage itself less able to propagate and spread, the goal of treating the disease would nonetheless be achieved through attenuation of pathogenicity. Among the few empirical examples to test this, it was demonstrated that bacteriophages selected to target *Y. pestis* virulence factors induced bacteriophage-resistant mutations which largely attenuated virulence, while the remaining few had no effect on host mortality (Filippov et al., 2011). Despite the ability of bacteria to become resistant to therapeutic bacteriophages, in vitro experimental coevolution can increase the efficiency of bacteriophage therapy to limit the evolution of resistance of bacteria targeted by the bacteriophages (Friman et al., 2016). These results thus call for a possible individual adjustment of therapeutic bacteriophages according to the bacterial genotypes infecting each patient. Moreover, such studies exemplify the fact that bacteriophage therapy is evolution-proof at a short and manageable scale contrary to other therapies (e.g. antibiotic or AMPs). Nevertheless, multibacteriophage cocktails can be used to reduce the chances of resistance (Torres-Barcelo and Hochberg, 2016) and combination with antibiotic treatments can also have synergistic impacts which improve treatment efficacy (Santos et al., 2009). Bacteriophage–antibiotic synergy is probably the result of several processes, both mechanistic and populational. One mechanism underlying this synergy could be the employment by bacteriophages of efflux pumps as receptors: if this was to be the case, bacterial resistance to bacteriophage would lead to decreased pump activity and consequently increased sensitivity to antibiotics (Chaturongakul and Ounjai, 2014). Synergy also arises due to the fact that bacteriophages decrease the bacterial population size and thus increase the efficiency of the chemical molecules targeting the remaining bacteria. As a concomitant effect, natural selection by bacteriophages significantly constrains the evolution of antibiotic resistance as it decreases the rate of appearance of bacteria resistant to antibiotics while also reducing the minimum concentration level of antibiotic necessary to inhibit the growth of the resistant bacteria (Torres-Barceló et al., 2016). In addition to the interest in bacteriophage therapy for overcoming established multidrug resistant infections, multiphage cocktails have also been shown to act as an effective prophylactics (e.g. in preventing dysentery in children Sulakvelidze, 2011).

While the victories of bacteriophage therapy are encouraging, there are some aspects that must be considered when designing a bacteriophage that will be able to invade and reach its target. First, like any foreign invader of the body, bacteriophage elicits both innate and adaptive immune responses upon entry into the host (Kaur et al., 2012). Bacteriophages must be selected or modified to escape innate immune mechanisms and to control the pathogen before neutralising antibacteriophage antibodies start to be produced. Furthermore, bacterial cell lysis by the bacteriophage can induce further responses from the immune system. For these reasons, cross–reactive adaptive immunity must be considered if further treatment against the same bacterium is required. Second, effective bacteriophage discovery relies heavily on understanding the pharmacokinetics, absorption and distribution, which, respectively, determine the ability of the phage to disseminate from the site of administration to the site of action and to enter the target bacterium (Abedon, 2014). Third, temperate phages—which, as discussed earlier, can result in transduction of antibiotic resistant genes or bacterial virulence factors—should clearly not be used, despite any benefits they may offer (Gama et al., 2013). Although bacteriophages in natural systems may run the risk of destabilising microbial communities and 'healthy' function, targeting of bacteriophage to a single pathogenic bacterium causing dysbiosis in a diseased patient should lessen the likelihood of disruption to the host's natural microbial community. While bacteriophage therapy seems to yield promising results, the population and evolutionary dynamics have yet to be thoroughly explored in the context of the human microbiome (Levin and Bull, 2004).

7.2 Probiotics and Food-Borne Invaders

Humans regularly ingest bacteria in food, especially from dairy products and fermented foods (Lang et al., 2014). The idea that these are 'good' bacteria is quite old. However, research into the human microbiome is revealing how these groups of potentially helpful invaders can affect our resident communities. Metagenomic analyses are being used to identify the various bacteria and even viruses in our foods (e.g. in Kimchi a fermented traditional Korean food Jung et al., 2011). Food–borne bacteria can interact with the resident microbiota in various ways, both directly and indirectly (e.g. supplying metabolites or stimulating mucin production, which would then affect resident mucosa-associated bacteria), and, thus, have effects on their fitness (Derrien and van Hylckama Vlieg, 2015). In fact, some argue that fermented foods could serve as experimentally tractable systems to study microbial community assembly, dynamics and invasibility (Wolfe and Dutton, 2015).

Probiotics are bacteria used to treat or prevent health disorders, and thus they must be nonpathogenic. There are many ways to introduce probiotics, such as food (e.g. yoghurts with *Bifidobacteria*), capsules filled with bacterial cultures (e.g. *Lactobacillus acidophilus* pills) and creams or suppositories (e.g. vaginal suppositories with Lactobacilli). Some desired effects are: secretion of inhibitory or beneficial substances, inhibition of growth of pathogens and positive interactions with the immune system (Macfarlane and Macfarlane, 2013). Probiotics are often marketed as helpful for maintaining a healthy, 'balanced' microbiome. However, it is far from understood what constitutes a healthy and stable microbiome. This is particularly challenging because there are many 'healthy' microbiomes, i.e. various stable communities with different taxonomic compositions, so how to define a healthy microbiome system is still under debate. Consequently, using a more ecological function approach, probiotics in the future may be designed for bolstering functional attributes of the resident community (Lemon et al., 2012).

The aim of treating or preventing obesity (i.e. improving patients' lipid profiles) has led to many studies into potential probiotic treatments. The main premise is that the ingestion of certain bacterial species (e.g. *Lactobacillus fermentum*, certain *Lactobacillus plantarum* strains, *Lactobacillus reuteri*) may lead to particular activities of these species, such as interactions with cholesterol, thereby reducing circulating levels of cholesterol in blood. There are several proposed mechanisms by which candidate bacterial species may produce this desired effect. For example, probiotic strains can bind cholesterol directly, incorporate it into their cell membranes, convert cholesterol into coprostanol which then gets directly excreted, induce enzymatic deconjugation of bile acids, etc. (reviewed in Ooi and Liong, 2010). However, the results of in vivo and human trials have been mixed. It could be that the interactions with the resident community are contributing to creating this bag of mixed results. For instance, probiotics that induce bile salt deconjugation subsequently liberate amino acids that can be used as sources of carbon and nitrogen by the invading species (Begley et al., 2006), but also by other species around it. Producing nutrient sources to local species should influence not just the likelihood of the probiotic species to establish in the community, but also how cholesterol levels are affected overall.

The successful establishment of an invader is, of course, fundamental to the success of probiotic treatments. For example, a cholesterol-reducing effect in mice, but long-term establishment of the probiotic bacteria was not obtained (Li et al., 1998). Indeed, a common failing of probiotic treatments and trials is that the probiotic bacteria are not able to colonise, i.e. establish in the resident community, or do so only transiently. In ecological terms, the invaders are not able to fully go through all steps in the invasion process (particularly, establishment and spread). Human gut communities are often found to be persistent when faced with food–borne or probiotic treatments. For example, Zhang et al. (2016) found that the structure of the gut community, and not its diversity, affected how quickly an invasive, food–borne bacteria were cleared from the system. Studies interested in this topic often focus on finding which specific bacterial phyla were correlated with persistent communities or with the resilience of the community. Perhaps, however, a network approach, such as inferring the preinvasion resident network and seeing how its properties change (e.g. types of interactions or other network statistics), would help improve inferring what properties of the communities make them permissive. In addition, the insights from these studies will help uncover what strategies WH invaders need (and what trade-offs shape them) in order to successfully establish. Hence, studies in probiotics and food–borne invasions could help both advance WH community studies and increase the efficacy of these medical therapies.

7.3 Other Therapies

Faecal bacteriotherapy follows the principle of probiotic use, but instead of a single or limited number of species, faecal bacteriotherapy aims to incorporate an entire healthy ecosystem into a sick ecosystem, either to integrate with, or to displace, the sick ecosystem. The process itself is fairly primitive; in general, a healthy donor is sought and then screened for the presence of pathogens in their stool, and provided these are not present, then a faecal sample is obtained from this donor, homogenised into a slurry with water or saline solution, and instilled directly into a patient (Allen-Vercoe et al., 2012). The approach has become fairly widespread in the treatment of recurrent *C. difficile* infection and is associated with a very high cure rate of over 90% (van Nood et al., 2013). However, the procedure is not without risk, particularly from the transfer of as-yet unknown pathogens from donor to patient, as well as the risk of 'unknown unknowns' such as long-term complications of gut microbiota modulation which have yet to be measured (Allen-Vercoe and Petrof, 2013; Allen-Vercoe et al., 2012). Despite this, the potential of the therapy for the treatment of other diseases with gut microbiota disturbance as a key feature, such as ulcerative colitis and type 2 diabetes, is currently being explored (Vrieze et al., 2012). The approach is also subject to improvement. A recent study, Li et al. (2016) examined the fate of applied and autochthonous communities for up to 3 months following faecal microbiota transplantation (FMT) for the treatment of metabolic syndrome and found that FMT-supplied strains of the same species as the recipient's microbiota tended to integrate themselves more readily than new species into the community after treatment. However, the patterns of microbial replacement/colonisation in each recipient were found to be quite distinct from each other, suggesting a large host-driven effect on gut microbiota colonisation. Several groups have been developing standardised or defined microbial ecosystems as treatment modalities (instead of faeces, which is inherently variable; Khanna et al., 2016; Petrof et al., 2013; Ray and Jones, 2016). Such standardised therapies will contribute to our understanding of how microbial ecosystem invasions affect health, and will shape this therapeutic approach further in the future.

In the face of globally rising antibiotic resistance, predatory bacteria are being considered as potential alternatives and have shown effectiveness against biofilms, which are particularly tough to break up even by antibiotics (Allen et al., 2014). Ecologically, this strategy is one that adds a trophic level and strong new links of predator–prey interactions. Thus, it is not surprising that these predatory bacteria therapies have the potential to effectively destabilise a pathogenic community. Nevertheless, as mentioned earlier (in bacteriophage therapies), unwanted effects on the resident bacterial communities of the host are a real danger and considerable research is needed to investigate their safety. Here, using an ecological network approach would help better understand how these predatory bacteria interact with their new community.

Other examples of human-induced biological invaders that are discussed less in the microbiome literature are live vaccines. Live vaccines are vaccines where the pathogenic microorganism (viruses or bacteria) are attenuated and these new strains with lost functions are used to prime immune responses to prevent later challenge infections by more pathogenic strains. Attenuated live vaccines are some of the most successful infection control in history (just one example: eradicating smallpox), and arguably some of the most successful biological control strategies across all ecosystems. These vaccines show the central importance the immune response plays in protecting the host (and its residents) from invasions. This is highlighted by the fact that it is recommended not to vaccinate immunosuppressed hosts with live vaccines since, in these hosts, the vaccine strains can cause infection and disease (Kew et al., 2005; Shearer et al., 2014).

Generally, any live vaccine strain has the potential to interact with the resident community. A viral example is that the oral poliovirus vaccine strains can interact with other resident, distantly related enteroviruses, recombine and regain their virulence (Riquet et al., 2008), causing disease and potentially spreading to other hosts. This example highlights the importance of the human virome, which is less understood than the microbiome. In spite of these potential resident–invader interactions, the fact that they are very rare speaks volumes to the efficacy of the immune system (i.e. stronger predation than in free-living systems). However, as we develop new live vaccines, particularly against bacterial pathogens (Girard et al., 2006), we need to be mindful of the resident community (Detmer and Glenting, 2006). Of particular worry are gene transfers and unexpected bacteria–bacteria interactions. Interestingly, vaccine designers are experimenting with answering similar ecological questions as discussed in this paper. Namely, what properties of the invader, the immune system or of the system itself make an invasion event unsuccessful? Understanding these mechanisms, which include ecological considerations, will be important for future successes of novel vaccines.

Likewise, live therapies (probiotics, transplants, predator introductions, etc.) can have long-term, potentially unwanted effects. Indeed our history of biological control in free-living ecological systems offers up many examples. Some issues to keep in mind are: unintended effects on the nontargeted species can be both direct and indirect, introduced invaders can disperse from the target site, relatives of the target species are likely to be affected and underlying environmental conditions affect the impact of the invader (Louda et al., 2003). A currently popular technique is 'inundative' biological control, whereby large numbers of the invader are introduced at once with the intention to have a high dose effect and not necessarily be a long-term pest management strategy. It is not clear that this control strategy is able to avoid the long-term issues that classic biological control has on nontarget species (Van Lenteren et al., 2003). However, dose effects should also play a key role in the success of invasions and thus live therapies. Quorum sensing is one important reason why this is the case (see polymicrobial infections in

Section 6.1). Therefore, applying this inundative approach may increase the chances of success, but maintain the risk of unwanted effects. Finally, a potential, and particularly unwanted, nontarget effect of live therapies is evolutionary effects, such as the evolution of virulence (Ford and King, 2016). With the aim of introducing stronger competitors, therapies may inadvertently select for higher virulence in the target pathogen, specifically by increasing the weapons it uses against the new residents (Ford and King, 2016; Vale et al., 2016). However, it has been suggested that if the therapy is sufficiently aggressive, i.e. high pathogen killing, then this evolutionary response could be avoided (McNally et al., 2015). Concomitantly, the new residents introduced by the therapy themselves can evolve novel weapons to increase their competitive ability, which is an idea coming from free-living invasion biology, namely the **novel weapons hypothesis** (reviewed in Callaway and Ridenour, 2004). Understanding potential long-term evolutionary outcomes will play an important role in ensuring efficacy of these live human–induced invasion treatments.

7.4 A Note on Nonliving Perturbations

Perturbations not due to biological invaders play a very important role in shaping WH communities. Abiotic disturbances can be seen through the analogy of environmental shifts such as changes in nutrient gradients, or where the concept of **catastrophic regime shifts** applies, i.e. when a disturbance is sufficiently large that the community shifts to an alternative stable state (Relman, 2012; Scheffer et al., 2001; Scheffer and Carpenter, 2003). Indeed, the concept of *alternative stable states* (or *multistability*; see Section 3.1) is one that has been embraced by microbiome researchers. An example an abiotic driver of changing states is a study of small bowl transplant patients that found that the oxygen brought in by an ileostomy changed the gut community away from a strictly anaerobe-dominated community, but then it went back to the original resident community composition after the ileostomy was closed (Hartman et al., 2009). Often abiotic perturbations can precede biological invasions and change the properties of the resident community to be more susceptible to invasions. The most common form of abiotic disturbance of the microbiome is antibiotic use. Antibiotics have the potential to directly affect the stability and resilience of the microbiome to biological invasions. Antibiotic administration represents a strong perturbation that inflicts heavy collateral damage to an ecosystem (see Fig. 4C and D; Antonopoulos et al., 2009; Heinsen et al., 2015; Jakobsson et al., 2010; Looft and Allen, 2012; Nobel et al., 2015;

Robinson and Young, 2010). It has been repeatedly shown that antibiotic use negatively affects microbial community diversity and hence resilience, potentially allowing microbiome communities to become more susceptible to invasions by pathogens such as *C. difficile* (Allen-Vercoe, 2013).

The nonliving components of diet, e.g. abiotic or biotic compounds that are ingested, are further key modulators of the gut microbiome. This is especially the case because different dietary substrates stimulate the growth of different gut microbes; for example, prebiotic substrates inulin and arabinoxylan have been shown to stimulate the growth of different groups of bifidobacteria (Van den Abbeele et al., 2013). Diet also greatly affects not only the abundance of microbial community members, but also the metabolic output of an ecosystem. An elegant example of this was demonstrated by switching the diets of healthy African American and rural South African men for a 2-week period, whereby the Americans, whose diets shifted from high fat, low-fibre to low fat, high fibre, experienced a reduction in detrimental faecal metabolites such as choline (a precursor of the atherogenic metabolite trimethylamine N-oxide), whereas the South Africans on the reciprocal diet experienced an increase in detrimental metabolites as well as increased markers of intestinal inflammation (O'Keefe et al., 2015). See Fig. 2B for an example of a community dynamic response to changes in diet. Finally, the western diet is also rich in food additives such as artificial sweeteners and emulsifiers, each of which has been clearly demonstrated to cause disturbances in the gut microbiota in animal models (Chassaing et al., 2015; Suez et al., 2014).

Drugs other than antibiotics also represent a further modulator of the gut microbiota, although the impact of many drugs on microbial ecosystems has yet to be evaluated. There is, however, clear evidence of the modulatory effects of particular drugs such as metformin on the gut microbiota (Cabreiro, 2016 and reviewed by Li et al., 2016). Overall, it is encouraging to see the application of ecological concepts in these applied research fields, and more interaction between ecologists and microbiome researchers promises to be fruitful.

8. PERSPECTIVES AND CONCLUSIONS

We have assembled work from various disciplines with the intention to understand biological invasions of WH systems from an ecological perspective. The emerging picture is a fast-moving field with ambitious goals and substantial challenges that need to be met in the near future. The host-associated microbiome communities measured so far vary widely in their species richness, have low connectance and high assortativity. Community dynamics can appear to exhibit alternative stable states and diversity does not seem to be strictly tied to stability (Table 1).

One of the benefits of working under an evolution and ecology framework is that it allows for finding patterns and general principles that might otherwise be seen as specifics of a particular infection or medical treatment. The theoretical framework of networks and community dynamics is naturally suited for this pursuit. It can help formulate hypotheses in mathematical terms, explain high-dimensional features of large, complex datasets and incorporate data obtained from various techniques.

Nevertheless, given that network inference and analysis methods are still under development, it is important to be mindful of what are the underlying assumptions and that they best represent our most up-to-date ecological understanding. There is a risk of oversimplifying ecological concepts, which can then lead to incorrectly inferred networks or erroneous predictions. Indeed, not all past examples of research communities embracing the conceptually attractive autonomously regulated complex network have turned out well (Eichmann, 2008). In a topic with so much complexity and with experimental techniques still under development, we would be wise then to keep hypotheses grounded in thoroughly vetted experimental conclusions obtained from techniques validated in multiple ways. This is certainly a challenge worth meeting because the payoff is substantial. We are living in a time when some of the most exciting community ecology experiments are happening inside labs trying to cure diarrheal infections. While microbiome research is a relatively new field, this excitement has rapidly attracted researchers that have brought with them a constellation of state-of-the-art approaches that together promise to build a vibrant understanding of this new, and fascinating biological domain. Greater communication across fields will facilitate the successful integration of this constellation and help accelerate our knowledge of not only WH invasions, but also biological invasions in general.

ACKNOWLEDGEMENTS

S.A. and C.L.M. would like to acknowledge funding from the European Research Council (ERC) under the European Union's Horizon 2020 research and innovation programme (grant agreement no. 648963). J.L.A. would like to acknowledge post-doctoral support provided by a grant from the French National Research Agency (ANR JC "STORY"). M.P.T. was supported in part by grant ERCStG no. 306312. We would also like to thank our reviewers for improving this review.

REFERENCES

van den Abbeele, P., et al., 2013. Prebiotics, faecal transplants and microbial network units to stimulate biodiversity of the human gut microbiome. J. Microbial. Biotechnol. 6 (4), 335–340.

Abedon, S.T., 2014. Phage therapy: eco-physiological pharmacology. Scientifica 2014, 1–30.

Abrams, P.A., 2000. The evolution of predator-prey interactions: theory and evidence. Annu. Rev. Ecol. Evol. Syst. 31, 79–105.

Acheson, D.W.K., et al., 1998. In vivo transduction with shiga toxin 1-encoding phage. Infect. Immun. 66 (9), 4496–4498.

Aebersold, R., Mann, M., 2016. Mass-spectrometric exploration of proteome structure and function. Nature 537 (7620), 347–355.

Aguirre, M., et al., 2015. Diet drives quick changes in the metabolic activity and composition of human gut microbiota in a validated in vitro gut model. Res. Microbiol. 167, 1–12.

Aguirre, M., et al., 2014. In vitro characterization of the impact of different substrates on metabolite production, energy extraction and composition of gut microbiota from lean and obese subjects. PLoS One 9 (11), 1–23.

Alizon, S., 2013. Parasite co-transmission and the evolutionary epidemiology of virulence. Evolution 67 (4), 921–933.

Alizon, S., van Baalen, M., 2008. Acute or chronic? Within-host models with immune dynamics, infection outcome, and parasite evolution. Am. Nat. 172 (6), E244–E256.

Alizon, S., Michalakis, Y., 2015. Adaptive virulence evolution: the good old fitness-based approach. Trends Ecol. Evol. 30 (5), 248–254.

Alizon, S., de Roode, J.C., Michalakis, Y., 2013. Multiple infections and the evolution of virulence. Ecol. Lett. 16, 556–567.

Allen, E.K., et al., 2014. Characterization of the nasopharyngeal microbiota in health and during rhinovirus challenge. Microbiome 2, 1–11.

Allen-Vercoe, E., et al., 2012. A Canadian Working Group report on fecal microbial therapy: microbial ecosystems therapeutics. Can. J. Gastroenterol. 26 (7), 457–462.

Allen-Vercoe, E., 2013. Bringing the gut microbiota into focus through microbial culture: recent progress and future perspective. Curr. Opin. Microbiol. 16 (5), 625–629.

Allen-Vercoe, E., Petrof, E.O., 2013. Artificial stool transplantation: progress towards a safer, more effective and acceptable alternative. Expert Rev. Gastroenterol. Hepatol. 7 (4), 291–293.

Allesina, S., Levine, J.M., 2011. A competitive network theory of species diversity. Proc. Natl. Acad. Sci. U.S.A. 108 (14), 5638–5642.

Allison, S.D., Martiny, J.B.H., 2008. Resistance, resilience, and redundancy in microbial communities. Proc. Natl. Acad. Sci. U.S.A. 105 (Suppl. 1), 11512–11519.

Amsellem, L., Brouat, C., Duron, O., Porter, S.S., Vilcinskas, A., Facon, B., 2017. Importance of microorganisms to macroorganisms invasions: is the essential invisible to the eye? Adv. Ecol. Res. 57, 99–146.

André, J.-B., Day, T., 2005. The effect of disease life history on the evolutionary emergence of novel pathogens. Proc. Biol. Sci. 272 (1575), 1949–1956.

Antonopoulos, D.A., et al., 2009. Reproducible community dynamics of the gastrointestinal microbiota following antibiotic perturbation. Infect. Immun. 77 (6), 2367–2375.

Antonovics, J., 1968. Evolution in closely adjacent plant populations. Heredity 23 (4), 507–524.

Antonovics, J., 1976. The nature of limits to natural selection. Ann. Mo. Bot. Gard. 63 (2), 224–247.

Arumugam, M., et al., 2011. Enterotypes of the human gut microbiome. Nature 473 (7346), 174–180.

van Baalen, M., Jansen, V.A.A., 2001. Dangerous liaisons: the ecology of private interest and common good. Oikos 95 (2), 211–224.

Bäckhed, F., et al., 2012. Defining a healthy human gut microbiome: current concepts, future directions, and clinical applications. Cell Host Microbe 12 (5), 611–622.

Baines, S.D., et al., 2013. Mixed infection by Clostridium difficile in an in vitro model of the human gut. J. Antimicrob. Chemother. 68 (5), 1139–1143.

Ball, G.H., 1943. Parasitism and evolution. Am. Nat. 77 (771), 345–364.

Balzan, S., et al., 2007. Bacterial translocation: overview of mechanisms and clinical impact. J. Gastroenterol. Hepatol. 22 (4), 464–471.

Begley, M., Hill, C., Gahan, C.G.M., 2006. Bile salt hydrolase activity in probiotics bile salt hydrolase activity in probiotics. Appl. Environ. Microbiol. 72 (3), 1729–1738.

Berlow, E.L., et al., 2004. Interaction strengths in food webs: issues and opportunities. J. Anim. Ecol. 73 (3), 585–598.

Berry, D., Widder, S., 2014. Deciphering microbial interactions and detecting keystone species with co-occurrence networks. Front. Microbiol. 5 (219), 1–14.

Best, K., et al., 2015. Computational analysis of stochastic heterogeneity in PCR amplification efficiency revealed by single molecule barcoding. Sci. Rep. 5, 14629.

Bevan, M.J., Fink, P.J., 2001. The CD8 response on autopilot. Nat. Immunol. 2 (5), 381–382.

Bhattarai, N., Stapleton, J.T., 2012. GB virus C: the good boy virus? Trends Microbiol. 20 (3), 124–130.

Biggs, H.M., et al., 2014. Invasive Salmonella infections in areas of high and low malaria transmission intensity in Tanzania. Clin. Infect. Dis. 58 (5), 638–647.

Bolnick, D.I., et al., 2014. Major histocompatibility complex class IIb polymorphism influences gut microbiota composition and diversity. Mol. Ecol. 23 (19), 4831–4845.

Boon, E., et al., 2014. Interactions in the microbiome: communities of organisms and communities of genes. FEMS Microbiol. Rev. 38 (1), 90–118.

Boskey, E.R., et al., 2001. Origins of vaginal acidity: high D/L lactate ratio is consistent with bacteria being the primary source. Hum. Reprod. 16 (9), 1809–1813.

Boyd, E.F., 2012. Bacteriophage-encoded bacterial virulence factors and phage-pathogenicity island interactions. Adv. Virus Res. 82, 91–118.

Broderick, N.A., 2015. A common origin for immunity and digestion. Front. Immunol. 6, 1–3.

Brooks, J.P., et al., 2015. The truth about metagenomics: quantifying and counteracting bias in 16S rRNA studies. BMC Microbiol. 15 (1), 66.

Brown, S.P., et al., 2006. Ecology of microbial invasions: amplification allows virus carriers to invade more rapidly when rare. Curr. Biol. 16 (20), 2048–2052.

Brown, S.P., Cornforth, D.M., Mideo, N., 2012. Evolution of virulence in opportunistic pathogens: generalism, plasticity, and control. Trends Microbiol. 20 (7), 336–342.

Brown, S.P., Grenfell, B.T., 2001. An unlikely partnership: parasites, concomitant immunity and host defence. Proc. Biol. Sci. 268, 2543–2549.

Brüssow, H., et al., 2004. Phages and the evolution of bacterial pathogens: from genomic rearrangements to lysogenic conversion. Microbiol. Mol. Biol. Rev. 68 (3), 560–602.

Bucci, V., et al., 2016. MDSINE: microbial dynamical systems INference Engine for microbiome time-series analyses. Genome Biol. 17 (1), 121.

Bucci, V., Xavier, J.B., 2014. Towards predictive models of the human gut microbiome. J. Mol. Biol. 426 (23), 3907–3916.

Buffie, C.G., et al., 2015. Precision microbiome reconstitution restores bile acid mediated resistance to Clostridium difficile. Nature 517 (7533), 205–208.

Buffie, C.G., Pamer, E.G., 2013. Microbiota-mediated colonization resistance against intestinal pathogens. Nat. Rev. Immunol. 13 (11), 790–801.

Burns, J.H., et al., 2013. Greater sexual reproduction contributes to differences in demography of invasive plants and their noninvasive relatives. Ecology 94 (5), 995–1004.

Burns, J.H., 2006. Relatedness and environment affect traits associated with invasive and noninvasive. Ecol. Appl. 16 (4), 1367–1376.

Cabreiro, F., 2016. Metformin joins forces with microbes. Cell Host Microbe 19 (1), 1.

Callaway, R.M., Ridenour, W.M., 2004. Novel weapons: invasive success and the evolution of increased competitive ability. Front. Ecol. Environ. 2 (8), 436–443.

Caporaso, J.G., et al., 2011. Moving pictures of the human microbiome. Genome Biol. 12 (5), R50.

Carboni, M., et al., 2016. What it takes to invade grassland ecosystems: traits, introduction history and filtering processes. Ecol. Lett. 19 (3), 219–229.

Carrington, M., et al., 1999. Genetics of HIV-1 infection: chemokine receptor CCR5 polymorphism and its consequences. Hum. Mol. Genet. 8 (10), 1939–1945.

Carrolo, M., et al., 2010. Prophage spontaneous activation promotes DNA release enhancing biofilm formation in Streptococcus pneumoniae. PLoS One 5 (12), e15678.

Case, T.J., 1990. Invasion resistance arises in strongly interacting species-rich model competition communities. Proc. Natl. Acad. Sci. U.S.A. 87 (24), 9610–9614.

Case, T.J., Bolger, D.T., 1991. The role of introduced species in shaping the abundance and distribution of island reptiles. Evol. Ecol. 5, 272–290.

Chassaing, B., et al., 2015. Dietary emulsifiers impact the mouse gut microbiota promoting colitis and metabolic syndrome. Nature 519 (7541), 92–96.

Chaturongakul, S., Ounjai, P., 2014. Phage-host interplay: examples from tailed phages and Gram-negative bacterial pathogens. Front. Microbiol. 5, 442.

Chibani-chennou, S., et al., 2004. In vitro and in vivo bacteriolytic activities of. Society 48 (7), 2558–2569.

Cho, I., Blaser, M.J., 2012. The human microbiome: at the interface of health and disease. Nat. Rev. Genet. 13 (4), 260–270.

Chukkapalli, S.S., et al., 2015. Polymicrobial oral infection with four periodontal bacteria orchestrates a distinct inflammatory response and atherosclerosis in ApoE null mice. PLoS One 10 (11), e0143291.

Clemente, J.C., et al., 2012. The impact of the gut microbiota on human health: an integrative view. Cell 148 (6), 1258–1270.

Costello, E.K., et al., 2012. The application of ecological theory toward an understanding of the human microbiome. Science 336 (6086), 1255–1262.

Coyte, K.Z., Schluter, J., Foster, K.R., 2015. The ecology of the microbiome: networks, competition, and stability. Science 350 (6261), 663–666.

Criswell, B.S., et al., 1969. Haemophilus vaginalis: vaginitis by inoculation from culture. Obstet. Gynecol. 33 (2), 195–199.

Crowther, G.S., et al., 2016. Efficacy of vancomycin extended-dosing regimens for treatment of simulated Clostridium difficile infection within an in vitro human gut model. J. Antimicrob. Chemother. 71 (4), 986–991.

D'Acunto, B., et al., 2015. Modeling multispecies biofilms including new bacterial species invasion. Math. Biosci. 259, 20–26.

Dalton, T., et al., 2011. An in vivo polymicrobial biofilm wound infection model to study interspecies interactions. PLoS One 6 (11), e27317.

Danielsson, D., Teigen, P.K., Moi, H., 2011. The genital econiche: focus on microbiota and bacterial vaginosis. Ann. N. Y. Acad. Sci. 1230, 48–58.

Datcu, R., et al., 2013. Vaginal microbiome in women from Greenland assessed by microscopy and quantitative PCR. BMC Infect. Dis. 13, 480.

David, P., Thébault, E., Anneville, O., Duyck, P.-F., Chapuis, E., Loeuille, N., 2017. Impacts of invasive species on food webs: a review of empirical data. Adv. Ecol. Res. 56, 1–60.

Delong, E.F., 2014. Alien invasions and gut island biogeography. Cell 159 (2), 233–235.

Delwart, E., 2013. A roadmap to the human virome. PLoS Pathog. 9 (2), e1003146.

Derrien, M., van Hylckama Vlieg, J.E.T., 2015. Fate, activity, and impact of ingested bacteria within the human gut microbiota. Trends Microbiol. 23 (6), 354–366.

Dethlefsen, L., McFall-Ngai, M., Relman, D.A., 2007. An ecological and evolutionary perspective on human–microbe mutualism and disease. Nature 449 (7164), 811–818.

Detmer, A., Glenting, J., 2006. Live bacterial vaccines—a review and identification of potential hazards. Microb. Cell Fact. 5, 23.

Donohue, I., et al., 2016. Navigating the complexity of ecological stability. Ecol. Lett. 19 (9), 1172–1185.

Donohue, I., et al., 2013. On the dimensionality of ecological stability. Ecol. Lett. 16 (4), 421–429.

Drake, J.A., 1990. Communities as assembled structures: do rules govern pattern? Trends Ecol. Evol. 5 (5), 159–164.

Duan, K., et al., 2003. Modulation of Pseudomonas aeruginosa gene expression by host microflora through interspecies communication. Mol. Microbiol. 50 (5), 1477–1491.

Duerkop, B.A., et al., 2012. A composite bacteriophage alters colonization by an intestinal commensal bacterium. Proc. Natl. Acad. Sci. U.S.A. 109 (43), 17621–17626.

Dunne, J., 2006. The network structure of food webs. In: Pascual, M., Dunne, J.A. (Eds.), Ecological Networks: Linking Structure to Dynamics in Food Webs. Oxford University Press, New York, NY, pp. 27–86.

Dunne, J.A., et al., 2002. Food-web structure and network theory: the role of connectance and size. Proc. Natl. Acad. Sci. U.S.A. 99 (20), 12917–12922.

Duyck, P.F., et al., 2006. Importance of competition mechanisms in successive invasions by polyphagous Tephritids in la reunion. Ecology 87 (7), 1770–1780.

Duyck, P.F., David, P., Quilici, S., 2004. A review of relationships between interspecific competition and invasions in fruit flies (Diptera: Tephritidae). Ecol. Entomol. 29 (5), 511–520.

Eaton, M.J., et al., 2014. Testing metapopulation concepts: effects of patch characteristics and neighborhood occupancy on the dynamics of an endangered lagomorph. Oikos 123 (6), 662–676.

Edlin, G., Bitner, R., 1977. Reproductive fitness of P1, P2 and Mu lysogens of Escherichia coli. J. Virol. 21 (2), 560–564.

Edlin, G., Lin, L., Kudrna, R., 1975. Gamma lysogens of E. coli reproduce more rapidly than non-lysogens. Nature 255, 735–737.

Edlund, A., et al., 2013. An in vitro biofilm model system maintaining a highly reproducible species and metabolic diversity approaching that of the human oral microbiome. Microbiome 1 (1), 25.

Eichmann, K., 2008. The Network Collective: Rise and Fall of a Scientific Paradigm. Birkhäuser, Basel.

Elton, C.S., 1958. The Ecology of Invasions by Animals and Plants. Methuen & Co., Chapman & Hall, Kluwer Academic Publishers B. V., London.

Enault, F., et al., 2016 (June 21). Phages rarely encode antibiotic resistance genes: a cautionary tale for virome analyses. ISME J.

Ezenwa, V.O., Jolles, A.E., 2015. Opposite effects of anthelmintic treatment on microbial infection at individual versus population scales. Science 347 (6218), 175–177.

Facon, B., et al., 2008. High genetic variance in life-history strategies within invasive populations by way of multiple introductions. Curr. Biol. 18 (5), 363–367.

Facon, B., et al., 2005. Hybridization and invasiveness in the freshwater snail Melanoides tuberculata: hybrid vigour is more important than increase in genetic variance. J. Evol. Biol. 18 (3), 524–535.

Faith, J.J., et al., 2013. The long-term stability of the human gut microbiota. Science 341 (6141), 1237439.

Faust, K., et al., 2015. Cross-biome comparison of microbial association networks. Front. Microbiol. 6, 1200.

Faust, K., Raes, J., 2012. Microbial interactions: from networks to models. Nat. Rev. Microbiol. 10 (8), 538–550.

Fayol-Messaoudi, D., et al., 2005. pH-, lactic acid-, and non-lactic acid-dependent activities of probiotic Lactobacilli against Salmonella enterica Serovar Typhimurium. Appl. Environ. Microbiol. 71 (10), 6008–6013.

Fenton, A., Perkins, S.E., 2010. Applying predator-prey theory to modelling immune-mediated, within-host interspecific parasite interactions. Parasitology 137 (6), 1027–1038.

Filippov, A.A., et al., 2011. Bacteriophage-resistant mutants in Yersinia pestis: Identification of phage receptors and attenuation for mice. PLoS One 6 (9), e25486.

Fisher, C.K., Mehta, P., 2014. Identifying keystone species in the human gut microbiome from metagenomic timeseries using sparse linear regression. PLoS One 9 (7), 1–17.

Flipse, J., Wilschut, J., Smit, J.M., 2013. Molecular mechanisms involved in antibody-dependent enhancement of dengue virus infection in humans. Traffic 14 (1), 25–35.

Ford, S.A., King, K.C., 2016. Harnessing the power of defensive microbes: evolutionary implications in nature and disease control. PLoS Pathog. 12 (4), 1–12.

Foster, J.A., Krone, S.M., Forney, L.J., 2008. Application of ecological network theory to the human microbiome. Interdiscip. Perspect. Infect. Dis. 2008, 839501.

Foster, K.R., Bell, T., 2012. Competition, not cooperation, dominates interactions among culturable microbial species. Curr. Biol. 22 (19), 1845–1850.

Foster, K.R., Wenseleers, T., 2006. A general model for the evolution of mutualisms. J. Evol. Biol. 19 (4), 1283–1293.

Foxman, E.F., Iwasaki, A., 2011. Genome-virome interactions: examining the role of common viral infections in complex disease. Nat. Rev. Microbiol. 9 (4), 254–264.

Franzén, O., et al., 2015. Improved OTU-picking using long-read 16S rRNA gene amplicon sequencing and generic hierarchical clustering. Microbiome 3, 43.

Friedman, J., Alm, E.J., 2012. Inferring correlation networks from genomic survey data. PLoS Comput. Biol. 8 (9), 1–11.

Friman, V.P., et al., 2016. Pre-adapting parasitic phages to a pathogen leads to increased pathogen clearance and lowered resistance evolution with Pseudomonas aeruginosa cystic fibrosis bacterial isolates. J. Evol. Biol. 29 (1), 188–198.

Furuse, K., et al., 1983. Distribution of RNA coliphages in Senegal, Ghana, and Madagascar. Microbiol. Immunol. 27 (4), 347–358.

Gajer, P., et al., 2012. Temporal dynamics of the human vaginal microbiota. Sci. Transl. Med. 4 (132), 132ra52.

Gallien, L., et al., 2014. Contrasting the effects of environment, dispersal and biotic interactions to explain the distribution of invasive plants in alpine communities. Biol. Invasions 17 (5), 1407–1423.

Gama, J.A., et al., 2013. Temperate bacterial viruses as double-edged swords in bacterial warfare. PLoS One 8 (3), e59043.

Gao, J., Barzel, B., Barabási, A.-L., 2016. Universal resilience patterns in complex networks. Nature 530 (7590), 307–312.

Gardner, H.L., Dukes, C.D., 1955. Haemophilus vaginalis vaginitis: a newly defined specific infection previously classified "nonspecific" vaginitis. Am. J. Obstet. Gynecol. 69 (5), 962–976.

Gellner, G., Mccann, K.S., 2016. The consistent role of weak and strong interactions in high and low-diversity food webs. Nat. Commun. 7 (11180), 1–7.

Geva-Zatorsky, N., et al., 2015. In vivo imaging and tracking of host–microbiota interactions via metabolic labeling of gut anaerobic bacteria. Nat. Med. 21 (9), 1091–1100.

Gibson, T.E., et al., 2016. On the origins and control of community types in the human microbiome. PLoS Comput. Biol. 12 (2), e1004688.

Gilligan, C.A., Van Den Bosch, F., 2008. Epidemiological models for invasion and persistence of pathogens. Annu. Rev. Phytopathol. 46, 385–418.

Girard, M.P., et al., 2006. A review of vaccine research and development: human enteric infections. Vaccine 24 (15), 2732–2750.

Goodman, A.L., 2014. In vivo and animal models of the human gut microbiome. In: Marchesi, J.R. (Ed.), The Human Microbiota and Microbiome. CABI Publishing, Wallingford, UK, p. 208.

Goodnight, K.F., 1992. The effect of stochastic variation on kin selection in a budding-viscous population. Am. Nat. 140 (6), 1028–1040.

Graf, J., Kikuchi, Y., Rio, R.V.M., 2006. Leeches and their microbiota: naturally simple symbiosis models. Trends Microbiol. 14 (8), 365–371.

Graham, A.L., 2008. Ecological rules governing helminth-microparasite coinfection. Proc. Natl. Acad. Sci. U.S.A. 105 (2), 566–570.

Greenblum, S., Turnbaugh, P.J., Borenstein, E., 2012. Metagenomic systems biology of the human gut microbiome reveals topological shifts associated with obesity and inflammatory bowel disease. Proc. Natl. Acad. Sci. U.S.A. 109 (2), 594–599.

Grice, E.A., Segre, J.A., 2011. The skin microbiome. Nat. Rev. Microbiol. 9 (4), 244–253.

Griffiths, E.C., et al., 2014. Analysis of a summary network of co-infection in humans reveals that parasites interact most via shared resources. Proc. Biol. Sci. 281, 20132286.

Hajishengallis, G., et al., 2011. Article low-abundance biofilm species orchestrates inflammatory periodontal disease through the commensal microbiota and complement. Cell Host Microbe 10 (5), 497–506.

Hajishengallis, G., Lamont, R.J., 2012. Beyond the red complex and into more complexity: the polymicrobial synergy and dysbiosis (PSD) model of periodontal disease etiology. Mol. Oral Microbiol. 27 (6), 409–419.

Hall, R.J., 2011. Eating the competition speeds up invasions. Biol. Lett. 7 (2), 307–311.

Hamilton, P.T., Perlman, S.J., 2013. Host defense via symbiosis in Drosophila. PLoS Pathog. 9 (12), 1–4.

Hamilton, W., 1980. Sex versus non-sex versus parasite. Oikos 35 (2), 282–290.

Hamilton, W.D., 1964. The genetical evolution of social behaviour. I. J. Theor. Biol. 7 (1), 1–16.

Hammer, B.K., Bassler, B.L., 2003. Quorum sensing controls biofilm formation in Vibrio cholerae. Mol. Microbiol. 50 (1), 101–114.

Harcombe, W.R., et al., 2016. Adding biotic complexity alters the metabolic benefits of mutualism. Evolution 70 (8), 1871–1881.

Harriott, M.M., Noverr, M.C., 2011. Importance of Candida-bacterial polymicrobial biofilms in disease. Trends Microbiol. 19 (11), 557–563.

Hartfield, M., Alizon, S., 2014. Epidemiological feedbacks affect evolutionary emergence of pathogens. Am. Nat. 183 (4), E105–E117.

Hartman, A.L., et al., 2009. Human gut microbiome adopts an alternative state following small bowel transplantation. Proc. Natl. Acad. Sci. U.S.A. 106 (40), 17187–17192.

Hastings, A., 1978. Spatial heterogeneity and the stability of predator-prey systems: predator-mediated coexistence. Theor. Popul. Biol. 14 (3), 380–395.

Heinsen, F.A., et al., 2015. Dynamic changes of the luminal and mucosaassociated gut microbiota during and after antibiotic therapy with paromomycin. Gut Microbes 6 (4), 243–254.

Hickey, R.J., et al., 2015. Vaginal microbiota of adolescent girls prior to the onset of menarche resemble those of reproductive-age women. MBio 6 (2), 1–14.

Hofbauer, J., Schreiber, S.J., 2010. Robust permanence for interacting structured populations. J. Differ. Eq. 248 (8), 1955–1971.

Holling, C.S., 1959. Some characteristics of simple types of predation and parasitism. Can. Entomol. 91 (7), 385–398.

Holt, R.D., Grover, J., Tilman, D.T., 1994. Simple rules for interspecific dominance in systems with exploitaitve and apparent competition. Am. Nat. 144 (5), 741–771.

Holt, R.D., Hochberg, M., 2001. Indirect interactions, community modules and biological control: a theoretical perspective. In: Wajnberg, G., Scott, J.K., Quimby, P.C. (Eds.), Evaluating Indirect Ecological Effects of Biological Control. CABI Publishing, Wallingford, UK, p. 261.

Holt, R.D., Barfield, M., 2006. Within-host pathogen dynamics: some ecological and evolutionary consequences of transients, dispersal mode, and within-host spatial heterogeneity. DIMACS Series in Discrete Mathematics and Theoretical Computer Science vol. 71. 45–66.

Holt, R.D., Dobson, A.P., 2006. Extending the principles of community ecology to address the epidemiology of host–pathogen systems. In: Collinge, S.K., Ray, C. (Eds.), Disease Ecology: Community Structure and Pathogen Dynamics. Oxford University Press, Oxford, pp. 6–27.

Hooper, L.V., et al., 2003. Angiogenins: a new class of microbicidal proteins involved in innate immunity. Nat. Immunol. 4 (3), 269–273.

Horz, H.-P., 2015. Archaeal lineages within the human microbiome: absent, rare or elusive? Life 5 (2), 1333–1345.

Hufbauer, R.A., et al., 2012. Anthropogenically induced adaptation to invade (AIAI): contemporary adaptation to human-altered habitats within the native range can promote invasions. Evol. Appl. 5 (1), 89–101.

Hufbauer, R.A., et al., 2013. Role of propagule pressure in colonization success: disentangling the relative importance of demographic, genetic and habitat effects. J. Evol. Biol. 26 (8), 1691–1699.

Huffnagle, G.B., Noverr, M.C., 2013. The emerging world of the fungal microbiome. Trends Microbiol. 21 (7), 334–341.

Hunt, K.M., et al., 2011. Characterization of the diversity and temporal stability of bacterial communities in human milk. PLoS One 6 (6), 1–8.

Inal, J.M., 2003. Phage therapy: a reappraisal of bacteriophages as antibiotics. Arch. Immunol. Ther. Exp. (Warsz.) 51, 237–244.

Inderjit, Cadotte, M.W., Colautti, R.I., 2005. The ecology of biological invasions: past, present and future. In: Birkhauser (Ed.), Invasive Plants: Ecological and Agricutural Aspects. Verlag, Switzerland, p. 19.

Ives, A.R., Carpenter, S.R., 2007. Stability and diversity of ecosystems. Science 317 (5834), 58–62.

Jackson, M.C., Wasserman, R.J., Grey, J., Ricciardi, A., Dick, J.T.A., Alexander, M.E., 2017. Novel and disrupted trophic links following invasion in freshwater ecosystems. Adv. Ecol. Res. 57, 55–97.

Jacquet, C., et al., 2016. The complexity–stability relationship in empirical ecosystems. Nat. Commun. 7, 12573.

Jaenlke, J., 1978. A hypothesis to account for the maintenance of sex within populations. Evol. Theory 94, 191–194.

Jakobsson, H.E., et al., 2010. Short-term antibiotic treatment has differing long-term impacts on the human throat and gut microbiome. PLoS One 5 (3), e9836.

Jansen, V.A.A., van Baalen, M., 2006. Altruism through beard chromodynamics. Nature 440 (7084), 663–666.

Jenkins, C., Keller, S.R., 2011. A phylogenetic comparative study of preadaptation for invasiveness in the genus silene (Caryophyllaceae). Biol. Invasions 13 (6), 1471–1486.

Jeschke, J.M., 2014. General hypotheses in invasion ecology. Divers. Distrib. 20 (11), 1229–1234.

Johnson, S., et al., 2014. Trophic coherence determines food-web stability. Proc. Natl. Acad. Sci. U.S.A. 1 (50), 1–6.

Jordán, F., et al., 2015. Diversity of key players in the microbial ecosystems of the human body. Sci. Rep. 5, 15920.

Jover, L.F., Cortez, M.H., Weitz, J.S., 2013. Mechanisms of multi-strain coexistence in host-phage systems with nested infection networks. J. Theor. Biol. 332, 65–77.

Jung, J.Y., et al., 2011. Metagenomic analysis of kimchi, a traditional Korean fermented food. Appl. Environ. Microbiol. 77 (7), 2264–2274.

Kaewsrichan, J., Peeyananjarassri, K., Kongprasertkit, J., 2006. Selection and identification of anaerobic lactobacilli producing inhibitory compounds against vaginal pathogens. FEMS Immunol. Med. Microbiol. 48 (1), 75–83.

Kaur, T., et al., 2012. Immunocompatibility of bacteriophages as nanomedicines. J. Nanotechnol. 2012, 13. Article ID 247427.

Keen, E.C., 2012. Phage therapy: concept to cure. Front. Microbiol. 3, 1–3.

Kéfi, S., et al., 2012. More than a meal… integrating non-feeding interactions into food webs. Ecol. Lett. 15 (4), 291–300.

Keller, S.R., Taylor, D.R., 2008. History, chance and adaptation during biological invasion: separating stochastic phenotypic evolution from response to selection. Ecol. Lett. 11 (8), 852–866.

Kennedy, K., et al., 2014. Evaluating bias of Illumina-based bacterial 16S rRNA gene profiles. Appl. Environ. Microbiol. 80 (18), 5717–5722.

Kerr, B., et al., 2002. Local dispersal promotes biodiversity in a real-life game of rock-paper-scissors. Nature 418 (6894), 171–174.

Kew, O.M., et al., 2005. Vaccine-derived polioviruses and the endgame strategy for global polio eradication. Annu. Rev. Microbiol. 59, 587–635.

Khanna, S., et al., 2016. A novel microbiome therapeutic increases gut microbial diversity and prevents recurrent Clostridium difficile infection. J. Infect. Dis. 214 (2), 173.

Kiliç, A.O., et al., 2001. Comparative study of vaginal Lactobacillus phages isolated from women in the United States and Turkey: prevalence, morphology, host range, and DNA homology. Clin. Diagn. Lab. Immunol. 8 (1), 31–39.

King, K.C., et al., 2016. Rapid evolution of microbe-mediated protection against pathogens in a worm host. ISME J. 10 (8), 1915–1924.

Koenig, J.E., et al., 2011. Succession of microbial consortia in the developing infant gut microbiome. Proc. Natl. Acad. Sci. U.S.A. 108 (Suppl. 1), 4578–4585.

Koizumi, Y., Iwami, S., 2014. Mathematical modeling of multi-drugs therapy: a challenge for determining the optimal combinations of antiviral drugs. Theor. Biol. Med. Model. 11 (1), 41.

Kubinak, J.L., et al., 2015. MHC variation sculpts individualized microbial communities that control susceptibility to enteric infection. Nat. Commun. 6, 8642.

Kümmerli, R., et al., 2009. Limited dispersal, budding dispersal, and cooperation: an experimental study. Evolution 63 (4), 939–949.

Kurtz, Z.D., et al., 2015. Sparse and compositionally robust inference of microbial ecological networks. PLoS Comput. Biol. 11 (5), 1–25.

Kyriazakis, I., Tolkamp, B., Hutchings, M., 1998. Towards a functional explanation for the occurrence of anorexia during parasitic infections. Anim. Behav. 56 (2), 265–274.

Lang, J.M., Eisen, J.A., Zivkovic, A.M., 2014. The microbes we eat: abundance and taxonomy of microbes consumed in a day's worth of meals for three diet types. PeerJ 2, e659.

Law, R., Morton, R.D., 1996. Permanence and the assembly of ecological communities. Ecology 77 (3), 762–775.

Lawlor, L.R., 1979. Direct and indirect effects of n-species competition. Oecologia 43 (3), 355–364.

Lecuit, M., Eloit, M., 2013. The human virome: new tools and concepts. Trends Microbiol. 21 (10), 510–515.

Lee, W., van Baalen, M., Jansen, V.A.A., 2012. An evolutionary mechanism for diversity in siderophore-producing bacteria. Ecol. Lett. 15 (2), 119–125.

Lee, W.-J., Brey, P.T., 2013. How microbiomes influence metazoan development: insights from history and Drosophila modeling of gut–microbe interactions. Annu. Rev. Cell Dev. Biol. 29 (1), 571–592.

Lemon, K.P., et al., 2012. Microbiota-targeted therapies: an ecological perspective. Sci. Transl. Med. 4 (137), RV5.

Van Lenteren, J.C., et al., 2003. Environmental risk assessment of exotic natural enemies used in inundative biological control. BioControl 48, 3–38.

Leone, V., et al., 2015. Effects of diurnal variation of gut microbes and high-fat feeding on host circadian clock function and metabolism. Cell Host Microbe 17 (5), 681–689.

Levin, B.R., Bull, J.J., 2004. Population and evolutionary dynamics of phage therapy. Nat. Rev. Microbiol. 2 (2), 166–173.

Levin, B.R., Bull, J.J., 1994. Short-sighted evolution and the virulence of pathogenic microorganisms. Trends Microbiol. 2 (3), 76–81.

Levine, J.M., D'Antonio, C.M., 1999. Elton revisited: a review of evidence linking diversity and invasibility. Oikos 87 (1), 15–26.

Ley, R.E., Peterson, D.A., Gordon, J.I., 2006. Ecological and evolutionary forces shaping microbial diversity in the human intestine. Cell 124 (4), 837–848.

Li, L., et al., 1998. Effect of feeding of a cholesterol-reducing bacterium, Eubacterium coprostanoligenes, to germ-free mice. Lab. Anim. Sci. 48 (3), 253–255.

Li, S.P., et al., 2015. The effects of phylogenetic relatedness on invasion success and impact: deconstructing Darwin's naturalisation conundrum. Ecol. Lett. 18 (12), 1285–1292.

Li, S.S., et al., 2016. Durable coexistence of donor and recipient strains after fecal microbiota transplantation. Science 352 (6285), 586–589.

Ling, Z., et al., 2013. The restoration of the vaginal microbiota after treatment for bacterial vaginosis with metronidazole or probiotics. Microb. Ecol. 65 (3), 773–780.

Lively, C.M., 2010. A review of red queen models for the persistence of obligate sexual reproduction. J. Hered. 101 (Suppl. 1), 13–20.

Lloyd-Smith, J.O., et al., 2009. Epidemic dynamics at the human–animal interface. Science 326, 1362–1367.

Loc-Carrillo, C., Abedon, S.T., 2011. Pros and cons of phage therapy. Bacteriophage 1 (2), 111–114.

Lockwood, J.L., Cassey, P., Blackburn, T., 2005. The role of propagule pressure in explaining species invasions. Trends Ecol. Evol. 20 (5), 223–228.

Loeuille, N., 2010. Influence of evolution on the stability of ecological communities. Ecol. Lett. 13 (12), 1536–1545.

Looft, T., Allen, H.K., 2012. Collateral effects of antibiotics on mammalian gut microbiomes. Gut microbes 3 (5), 463–467.

Louda, S.M., et al., 2003. Nontarget effects—the Achilles' heel of biological control? Retrospective analysis to reduce risk associated with biocontrol introductions. Annu. Rev. Entomol. 48 (1), 365–396.

Lozupone, C.A., et al., 2012. Diversity, stability and resilience of the human gut microbiota. Nature 489, 220.

Luciani, F., Alizon, S., 2009. The evolutionary dynamics of a rapidly mutating virus within and between hosts: the case of hepatitis C virus. PLoS Comput. Biol. 5 (11), e1000565.

Lurgi, M., et al., 2014. Network complexity and species traits mediate the effects of biological invasions on dynamic food webs. Front. Ecol. Evol. 2, 1–11.

Lurie-Weinberger, M.N., Gophna, U., 2015. Archaea in and on the human body: health implications and future directions. PLoS Pathog. 11 (6), 1–8.

Lysenko, E.S., et al., 2010. Within-host competition drives selection for the capsule virulence determinant of Streptococcus pneumoniae. Curr. Biol. 20 (13), 1222–1226.

Ma, C., et al., 2016. Different effects of invader–native phylogenetic relatedness on invasion success and impact: a meta-analysis of Darwin's naturalization hypothesis. Proc. Biol. Sci. 283 (1838), 20160663.

Macfarlane, G.T., Macfarlane, S., 2013. Manipulating the indigenous microbiota in humans: prebiotics, probiotics and synbiotics. In: Fredricks, D.N. (Ed.), The Human Microbiota: How Microbial Communities Affect Health and Disease. Wiley-Blackwell, Hoboken, NJ, p. 389.

Machado, A., Cerca, N., 2015. The influence of biofilm formation by Gardnerella vaginalis and other anaerobes on bacterial vaginosis. J. Infect. Dis. 212, 1–17.

Machado, A., Jefferson, K.K., Cerca, N., 2013. Interactions between Lactobacillus crispatus and bacterial vaginosis (BV)-associated bacterial species in initial attachment and biofilm formation. Int. J. Mol. Sci. 14 (6), 12004–12012.

Maizels, R.M., Yazdanbakhsh, M., 2003. Immune regulation by helminth parasites: cellular and molecular mechanisms. Nat. Rev. Immunol. 3 (9), 733–744.

Marder, E., Taylor, A.L., 2011. Multiple models to capture the variability in biological neurons and networks. Nat. Neurosci. 14 (2), 133–138.

Margolis, E., Levin, B.R., 2007. Within-host evolution for the invasiveness of commensal bacteria: an experimental study of bacteremias resulting from Haemophilus influenzae nasal carriage. J. Infect. Dis. 196 (7), 1068–1075.

Marino, S., et al., 2014. Mathematical modeling of primary succession of murine intestinal microbiota. Proc. Natl. Acad. Sci. U.S.A. 111 (1), 439–444.

Martiny, J.B.H., et al., 2015. Microbiomes in light of traits: a phylogenetic perspective. Science 350 (6261), aac9323.

Martz, S.-L.E., et al., 2015. Administration of defined microbiota is protective in a murine Salmonella infection model. Sci. Rep. 5, 16094.

Massol, F., Dubart, M., Calcagno, V., Cazelles, K., Jacquet, C., Kéfi, S., Gravel, D., 2017. Island biogeography of food webs. Adv. Ecol. Res. 56, 183–262.

Mastropaolo, M.D., et al., 2005. Synergy in polymicrobial infections in a mouse model of type 2 diabetes. Infect. Immun. 73 (9), 6055–6063.

Matsuoka, K., Kanai, T., 2015. The gut microbiota and inflammatory bowel disease. Semin. Immunopathol. 37 (1), 47–55.

May, R.M., 1973. Stability and Complexity in Model Ecosystems. Princeton University Press, Princeton, NJ.

May, R.M., 1972. Will a large complex system be stable? Nature 238 (5364), 413–414.

Mayer, A., et al., 2016. Diversity of immune strategies explained by adaptation to pathogen statistics. Proc. Natl. Acad. Sci. U.S.A. 113 (31), 8630–8635.

McCann, K.S., 2011. Food webs. In: Levin, S.A., Horn, H.S. (Eds.), Monographs in Population Biology. Princeton University Press, NJ.

McCann, K.S., 2000. The diversity–stability debate. Nature 405 (6783), 228–233.

McCann, K.S., Gellner, G., 2012. Food chains and food web modules. In: Hastings, A., Gross, L. (Eds.), Encyclopedia of Theoretical Ecology. University of California Press, Berkeley, CA, pp. 288–294.

McCann, K.S., Rooney, N., 2009. The more food webs change, the more they stay the same. Philos. Trans. R. Soc. Lond. B Biol. Sci. 364 (1524), 1789–1801.

McDonald, J.A.K., et al., 2013. Evaluation of microbial community reproducibility, stability and composition in a human distal gut chemostat model. J. Microbiol. Methods 95 (2), 167–174.

McDonald, J.A.K., et al., 2015. Simulating distal gut mucosal and luminal communities using packed-column biofilm reactors and an in vitro chemostat model. J. Microbiol. Methods 108, 36–44.

McHardy, I.H., et al., 2013. Integrative analysis of the microbiome and metabolome of the human intestinal mucosal surface reveals exquisite inter-relationships. Microbiome 1 (1), 17.

McNally, L., Vale, P.F., Brown, S.P., 2015. Microbiome engineering could select for more virulent pathogens. bioRxiv. http://dx.doi.org/10.1101/027854.

McSorley, H.J., Maizels, R.M., 2012. Helminth infections and host immune regulation. Clin. Microbiol. Rev. 25 (4), 585–608.

Meaden, S., Koskella, B., 2013. Exploring the risks of phage application in the environment. Front. Microbiol. 4 (11), 358.

Médoc, V., Firmat, C., Sheath, D.J., Pegg, J., Andreou, D., Britton, J.R., 2017. Parasites and biological invasions: predicting ecological alterations at levels from individual hosts to whole networks. Adv. Ecol. Res. 57, 1–54.

Méthot, P.-O., 2012. Why do parasites harm their host? On the origin and legacy of Theobald Smith's "law of declining virulence"—1900–1980. Hist. Philos. Life Sci. 34 (4), 561–601.

Michalakis, Y., et al., 1992. Pleiotropic action of parasites: how to be good for the host. Trends Ecol. Evol. 7 (2), 59–62.

Mideo, N., 2009. Parasite adaptations to within-host competition. Trends Parasitol. 25 (6), 261–268.

Milinski, M., 2006. The major histocompatibility complex, sexual selection, and mate choice. Annu. Rev. Ecol. Evol. Syst. 37, 159–186.

Milo, R., et al., 2002. Network motifs: simple building blocks of complex networks. Science 298 (5594), 824–827.

Minot, S., et al., 2013. Rapid evolution of the human gut virome. Proc. Natl. Acad. Sci. U.S.A. 110 (30), 12450–12455.

Minot, S., et al., 2011. The human gut virome: inter-individual variation and dynamic response to diet. Genome Res. 21 (10), 1616–1625.

Moeller, A., et al., 2016. Social behavior shapes the chimpanzee pan-microbiome. Sci. Adv. 2, e1500997.

Montoya, J.M., et al., 2002. Small world patterns in food webs. J. Theor. Biol. 214 (3), 405–412.

Moorthy, A.S., et al., 2015. A spatially continuous model of carbohydrate digestion and transport processes in the colon. PLoS One 10 (12), e0145309.

Morris, J., et al., 2012. The Black Queen hypothesis: evolution of dependencies through adaptive gene loss. mBio 3 (2), e00036-12.

Mouquet, N., Loreau, M., 2002. Coexistence in metacommunities: the regional similarity hypothesis. Am. Nat. 159 (4), 420–426.

Murall, C.L., McCann, K.S., Bauch, C.T., 2012. Food webs in the human body: linking ecological theory to viral dynamics. PLoS One 7 (11), e48812.

Murall, C.L., McCann, K.S., Bauch, C.T., 2014. Revising ecological assumptions about human Papillomavirus interactions and type replacement. J. Theor. Biol. 350, 98–109.

Murray, J.L., et al., 2014. Mechanisms of synergy in polymicrobial infections. J. Microbiol. 52 (3), 188–199.

Mushegian, A.A., Ebert, D., 2016. Rethinking "mutualism" in diverse host-symbiont communities. Bioessays 38 (1), 100–108.

Naik, S., et al., 2012. Compartmentalized control of skin immunity by resident commensals. Science 337, 1115–1120.

Namba, T., 2015. Multi-faceted approaches toward unravelling complex ecological networks. Popul. Ecol., 57 (1), 3–19.

Naqvi, A., et al., 2010. Network-based modeling of the human gut microbiome. Chem. Biodivers. 7 (5), 1040–1050.

van Nes, E.H., Scheffer, M., 2005. Implications of spatial heterogeneity for catastrophic regime shifts in ecosystems. Ecology 86 (7), 1797–1807.

Newman, M.E.J., 2003. The structure and function of complex networks. SIAM Rev. 45 (2), 167–256.

Nobbs, A.H., Jenkinson, H.F., 2015. Interkingdom networking within the oral microbiome. Microbes Infect. 17 (7), 484–492.

Nobel, Y.R., et al., 2015. Metabolic and metagenomic outcomes from early-life pulsed antibiotic treatment. Nat. Commun. 6 (5), 7486.

van Nood, E., et al., 2013. Duodenal infusion of donor feces for recurrent Clostridium difficile. N. Engl. J. Med. 368 (5), 407–415.

Norman, J.M., et al., 2015. Disease-specific alterations in the enteric virome in inflammatory bowel disease. Cell 160 (3), 447–460.

Nowell, P.C., 1976. The clonal evolution of tumor cell populations. Science 194, 23–28.

O'Gorman, E.J., et al., 2010. Interaction strength, food web topology and the relative importance of species in food webs. J. Anim. Ecol. 79 (3), 682–692.

O'Keefe, S.J.D., et al., 2015. Fat, fibre and cancer risk in African Americans and rural Africans. Nat. Commun. 6 (2014), 6342.

Oh, J., et al., 2013. The altered landscape of the human skin microbiome in patients with primary immunodeficiencies. Genome Res. 23, 2103–2114.

Ooi, L.G., Liong, M.T., 2010. Cholesterol-lowering effects of probiotics and prebiotics: a review of in vivo and in vitro findings. Int. J. Mol. Sci. 11 (6), 2499–2522.

Orlandi, J., et al., 2014. First connectomics challenge: from imaging to connectivity. In: JMLR: Workshop and Conference Proceedings, vol. 1, pp. 1–17.

Ostaff, M.J., Stange, E.F., Wehkamp, J., 2013. Antimicrobial peptides and gut microbiota in homeostasis and pathology. EMBO Mol. Med. 5 (10), 1465–1483.

De Paepe, M., et al., 2014. Bacteriophages: an underestimated role in human and animal health? Front. Cell. Infect. Microbiol. 4, 39.

De Paepe, M., et al., 2016. Carriage of λ latent virus is costly for its bacterial host due to frequent reactivation in monoxenic mouse intestine. PLoS Genet. 12 (2), 1–20.

Paine, R.T., 1969. A note on trophic complexity and community stability. Am. Nat. 103 (929), 91–93.

Paine, R.T., 1966. Food web complexity and species diversity. Am. Nat. 100, 65–75.

Pantel, J.H., Bohan, D.A., Calcagno, V., David, P., Duyck, P.-F., Kamenova, S., Loeuille, N., Mollot, G., Romanuk, T.N., Thébault, E., Tixier, P., Massol, F., 2017. 14 questions for invasion in ecological networks. Adv. Ecol. Res. 56, 293–340.

Parfrey, L.W., Walters, W.A., Knight, R., Parfrey, L.W., Walters, W.A., Knight, R., 2011. Microbial eukaryotes in the human microbiome: ecology, evolution, and future directions. Front. Microbiol. 2, 1–6.

Pedersen, A.B., Fenton, A., 2007. Emphasizing the ecology in parasite community ecology. Trends Ecol. Evol. 22 (3), 133–139.

Pepper, J.W., Rosenfeld, S., 2012. The emerging medical ecology of the human gut microbiome. Trends Ecol. Evol. 27 (7), 381–384.

Perelson, A.S., 2002. Modelling viral and immune system dynamics. Nat. Rev. Immunol. 2 (1), 28–36.

Perry, S., et al., 2010. Infection with Helicobacter pylori is associated with protection against tuberculosis. PLoS One 5 (1), e8804.

Petraitis, P.S., 1979. Competitive networks and measures of intransitivity. Am. Nat. 114 (6), 921–925.

Petrof, E.O., et al., 2013. Microbial ecosystems therapeutics: a new paradigm in medicine? Benef. Microbes 4 (1), 53–65.

Peura, S., et al., 2015. Resistant microbial co-occurrence patterns inferred by network topology. Appl. Environ. Microbiol. 81 (6), 2090–2097.

Pilyugin, S.S., Antia, R., 2000. Modeling immune responses with handling time. Bull. Math. Biol. 62 (5), 869–890.

Pollitt, E.J.G., et al., 2014. Cooperation, quorum sensing, and evolution of virulence in Staphylococcus aureus. Infect. Immun. 82 (3), 1045–1051.

Proulx, S.R., Promislow, D.E.L., Phillips, P.C., 2005. Network thinking in ecology and evolution. Trends Ecol. Evol. 20 (6 Special Issue), 345–353.

Puillandre, N., et al., 2012. ABGD, automatic barcode gap discovery for primary species delimitation. Mol. Ecol. 21, 1864–1877.

Pybus, V., Onderdonk, A.B., 1997. Evidence for a commensal, symbiotic relationship between Gardnerella vaginalis and Prevotella bivia involving ammonia: potential significance for bacterial vaginosis. J. Infect. Dis. 175, 406–413.

Rader, B.A., Guilemin, K., 2013. Insights into the human microbiome from animal models. In: Fredricks, D.N. (Ed.), The Human Microbiota: How Microbial Communities Affect Health and Disease. Wiley-Blackwell, Hoboken, NJ, p. 389.

Ramanan, D., et al., 2016. Helminth infection promotes colonization resistance via type 2 immunity. Science 352, 608–612.

Rankin, D.J., Rocha, E.P.C., Brown, S.P., 2011. What traits are carried on mobile genetic elements, and why? Heredity 106 (1), 1–10.

Ravel, J., et al., 2013. Daily temporal dynamics of vaginal microbiota before, during and after episodes of bacterial vaginosis. Microbiome 1 (1), 29.

Ravel, J., et al., 2011. Vaginal microbiome of reproductive-age women. Proc. Natl. Acad. Sci. U.S.A. 108 (Suppl. 1), 4680–4687.

Ray, A., Jones, C., 2016. Does the donor matter? Future Microbiol. 11 (5), 611–616.

Read, A.F., Taylor, L.H., 2001. The ecology of genetically diverse infections. Science 292 (5519), 1099–1102.

Relman, D.A., 2012. The human microbiome: ecosystem resilience and health. Nutr. Rev. 70 (Suppl. 1), S2–S9.

Ribet, D., Cossart, P., 2015. How bacterial pathogens colonize their hosts and invade deeper tissues. Microbes Infect. 17 (3), 173–183.

Ridaura, V.K., et al., 2013. Gut microbiota from twins discordant for obesity modulate metabolism in mice gut microbiota from twins metabolism in mice. Science 341, 1241214.

Riquet, F.B., et al., 2008. Impact of exogenous sequences on the characteristics of an epidemic type 2 recombinant vaccine-derived poliovirus. J. Virol. 82 (17), 8927–8932.

Robinson, C.J., Young, V.B., 2010. Antibiotic administration alters the community structure of the gastrointestinal microbiota. Gut Microbes 1 (4), 279–284.

Rogers, G.B., et al., 2010. Revealing the dynamics of polymicrobial infections: implications for antibiotic therapy. Trends Microbiol. 18 (8), 357–364.

Rogers, S., et al., 2008. Investigating the correspondence between transcriptomic and proteomic expression profiles using coupled cluster models. Bioinformatics 24 (24), 2894–2900.

Rohmer, L., Hocquet, D., Miller, S.I., 2011. Are pathogenic bacteria just looking for food? Metabolism and microbial pathogenesis. Trends Microbiol. 19 (7), 341–348.

Rohr, R.P., Saavedra, S., Bascompte, J., 2014. On the structural stability of mutualistic systems. Science 345 (6195), 1253497.

Romanuk, T.N., et al., 2009. Predicting invasion success in complex ecological networks. Philos. Trans. R. Soc. Lond. B Biol. Sci. 364 (1524), 1743–1754.

Romero, R., et al., 2014. The composition and stability of the vaginal microbiota of normal pregnant women is different from that of non-pregnant women. Microbiome 2 (1), 4.

de Roode, J.C., et al., 2005. Virulence and competitive ability in genetically diverse malaria infections. Proc. Natl. Acad. Sci. U.S.A. 102 (21), 7624–7628.

Rooney, N., et al., 2006. Structural asymmetry and the stability of diverse food webs. Nature 442, 265–269.

Rooney, N., McCann, K.S., 2012. Integrating food web diversity, structure and stability. Trends Ecol. Evol. 27 (1), 40–46.

Rossi-Tamisier, M., et al., 2015. Cautionary tale of using 16S rRNA gene sequence similarity values in identification of human-associated bacterial species. Int. J. Syst. Evol. Microbiol. 65 (6), 1929–1934.

Sadd, B.M., Schmid-Hempel, P., 2009. Principles of ecological immunology. Evol. Appl. 2 (1), 113–121.

Saleem, M., 2015. Microbiome community ecology: fundamentals and applications. In: Springer Briefs in Ecology. Springer International Publishing, Switzerland.

Sanford, J.A., Gallo, R.L., 2013. Seminars in immunology functions of the skin microbiota in health and disease. Semin. Immunol. 25 (5), 370–377.

Santos, S.B., et al., 2009. The use of antibiotics to improve phage detection and enumeration by the double-layer agar technique. BMC Microbiol. 9 (1), 148.

Sasaki, A., Iwasa, Y., 1991. Optimal growth switching schedule of pathogens within a host: between lytic and latent cycles. Theor. Popul. Biol. 39, 201–239.

Sax, D.F., et al., 2007. Ecological and evolutionary insights from species invasions. Trends Ecol. Evol. 22 (9), 465–471.

Scheffer, M., et al., 2001. Catastrophic shifts in ecosystems. Nature 413 (6856), 591–596.

Scheffer, M., Carpenter, S.R., 2003. Catastrophic regime shifts in ecosystems: linking theory to observation. Trends Ecol. Evol. 18 (12), 648–656.

Schreiber, S.J., Rittenhouse, S., 2004. From simple rules to cycling in community assembly. Oikos 105 (2), 349–358.

Schuch, R., Fischetti, V.A., 2006. Detailed genomic analysis of the wbeta and gamma phages infecting Bacillus anthracis: implications for evolution of environmental fitness and antibiotic resistance. Microbiology 188 (8), 3037–3051.

Seabloom, E.W., et al., 2015. The community ecology of pathogens: coinfection, coexistence and community composition. Ecol. Lett. 18, 401–415.

Seedorf, H., et al., 2014. Bacteria from diverse habitats colonize and compete in the mouse gut. Cell 159 (2), 253–266.

Seekatz, A.M., Aas, J., Gessert, C.E., 2014. Recovery of the gut microbiome following fecal microbiota transplantation. mBio 5 (3), 1–9.

Shafquat, A., et al., 2014. Functional and phylogenetic assembly of microbial communities in the human microbiome. Trends Microbiol. 22 (5), 261–266.

Shahinas, D., et al., 2012. Toward an understanding of changes in diversity. MBio 3 (5), e00338-12.

Shankar, V., et al., 2014. Species and genus level resolution analysis of gut microbiota in Clostridium difficile patients following fecal microbiota transplantation. Microbiome 2 (1), 13.

Shearer, W.T., et al., 2014. Recommendations for live viral and bacterial vaccines in immunodeficient patients and their close contacts. J. Allergy Clin. Immunol. 133 (4), 961–966.

Sibley, C.D., et al., 2008. Discerning the complexity of community interactions using a Drosophila model of polymicrobial infections. PLoS Pathog. 4 (10), e1000184.

Simberloff, D., 2009. The role of propagule pressure in biological invasions. Annu. Rev. Ecol. Evol. Syst. 40, 81–102.

Simberloff, D., Holle, B. Von, 1999. Positive interactions of nonindigenous species: invasional meltdown? Biol. Invasions 1 (1), 21–32.

Smith, J., 2001. The social evolution of bacterial pathogenesis. Proc. Biol. Sci. 268 (1462), 61–69.

Smith, V.H., Holt, R.D., 1996. Resource competition and within-host disease dynamics. Trends Ecol. Evol. 11 (9), 386–389.

Smith-Ramesh, L.M., Moore, A.C., Schmitz, O.J., 2016. Global synthesis suggests that food web connectance correlates to invasion resistance. Glob. Chang. Biol., 1–9.

Sofonea, M.T., Alizon, S., Michalakis, Y., 2015. From within-host interactions to epidemiological competition: a general model for multiple infections. Philos. Trans. R. Soc. B: Biol. Sci. 370 (1675), 20140303.

Solé, R.V., Montoya, J.M., 2001. Complexity and fragility in ecological networks. Proc. Biol. Sci. 268 (1480), 2039–2045.

Sommer, S., 2005. The importance of immune gene variability (MHC) in evolutionary ecology and conservation. Front. Zool. 2, 16.

Song, H.-S., et al., 2014. Mathematical modeling of microbial community dynamics: a methodological review. Processes 2 (4), 711–752.

Stecher, B., et al., 2010. Like will to like: abundances of closely related species can predict susceptibility to intestinal colonization by pathogenic and commensal bacteria. PLoS Pathog. 6 (1), e1000711.

Stecher, B., Hardt, W.-D., 2011. Mechanisms controlling pathogen colonization of the gut. Curr. Opin. Microbiol. 14 (1), 82–91.

Stecher, B., Maier, L., Hardt, W.-D., 2013. "Blooming" in the gut: how dysbiosis might contribute to pathogen evolution. Nat. Rev. Microbiol. 11 (4), 277–284.

Stein, R.R., et al., 2013. Ecological modeling from time-series inference: insight into dynamics and stability of intestinal microbiota. PLoS Comput. Biol. 9 (12), e1003388.

Suez, J., et al., 2014. Artificial sweeteners induce glucose intolerance by altering the gut microbiota. Nature 514 (7521), 181–186.

Sugihara, G., et al., 2012. Detecting causality in complex ecosystems. Science 338, 496–500.

Sulakvelidze, A., 2011. The challenges of bacteriophage therapy. Ind. Pharm. 45 (31), 14–18.

Swidsinski, A., et al., 2005. Adherent biofilms in bacterial vaginosis. Obstet. Gynecol. 106 (5 Pt. 1), 1013–1023.

Swidsinski, A., et al., 2014. Infection through structured polymicrobial Gardnerella biofilms (StPM-GB). Histol. Histopathol. 29 (5), 567–587.

Tasiemski, A., et al., 2015. Reciprocal immune benefit based on complementary production of antibiotics by the leech Hirudo verbana and its gut symbiont Aeromonas veronii. Sci. Rep. 5, 17498.

Tatusov, R.L., et al., 2003. The COG database: an updated version includes eukaryotes. BMC Bioinform. 4, 41.

Taur, Y., et al., 2012. Intestinal domination and the risk of bacteremia in patients undergoing allogeneic hematopoietic stem cell transplantation. Clin. Infect. Dis. 55 (7), 905–914.

Theoharides, K.A., Dukes, J.S., 2007. Plant invasion across space and time: factors affecting nonindigenous species success during four stage of invasion. New Phytol. 176 (2), 256–273.

Tillmann, H.L., et al., 2001. Infection with GB virus C and reduced mortality among HIV-infected patients. New Engl. J. Med. 345 (10), 715–724.

Tilman, D., 1999. The ecological consequences of changes in biodiversity: a search for general principles. Ecology 80 (5), 1455–1474.

Toh, M.C., et al., 2013. Colonizing the embryonic zebrafish gut with anaerobic bacteria derived from the human gastrointestinal tract. Zebrafish 10 (2), 194–198.

Torres-Barceló, C., et al., 2016. Long-term effects of single and combined introductions of antibiotics and bacteriophages on populations of Pseudomonas aeruginosa. Evol. Appl. 9 (4), 583–595.

Torres-Barcelo, C., Hochberg, M.E., 2016. Evolutionary rationale for phages as complements of antibiotics. Trends Microbiol. 24 (4), 249–256.

Tremblay, J., et al., 2015. Primer and platform effects on 16S rRNA tag sequencing. Front. Microbiol. 6, 1–15.

Trosvik, P., de Muinck, E.J., 2015. Ecology of bacteria in the human gastrointestinal tract—identification of keystone and foundation taxa. Microbiome 3 (1), 44.

Turovskiy, Y., et al., 2011. The etiology of bacterial vaginosis. J. Appl. Microbiol. 110 (5), 1105–1128.

Underhill, D.M., Iliev, I.D., 2014. The mycobiota: interactions between commensal fungi and the host immune system. Nat. Rev. Immunol. 14 (6), 405–416.

Vacher, C., et al., 2016. Learning ecological networks from next-generation sequencing data. Adv. Ecol. Res. 54, 1–39.

Vale, P.F., et al., 2016. Beyond killing: can we find ways to manage infection? Evol. Med. Public Health 1, 148–157.

Venema, K., van den Abbeele, P., 2013. Experimental models of the gut microbiome. Best Pract. Res. Clin. Gastroenterol. 27 (1), 115–126.

Vrieze, A., et al., 2012. Transfer of intestinal microbiota from lean donors increases insulin sensitivity in individuals with metabolic syndrome. Gastroenterology 143 (4), 913–916.e7.

Wegner, K.M., et al., 2003. Parasite selection for immunogenetic optimality. Science 301, 1343.

Wexler, A.G., et al., 2016. Human symbionts inject and neutralize antibacterial toxins to persist in the gut. Proc. Natl. Acad. Sci. U.S.A. 113 (13), 3639–3644.

Winternitz, J.C., Abbate, J.L., 2015. Examining the evidence for major histocompatibility complex-dependent mate selection in humans and nonhuman primates. Res. Rep. Biol. 6, 73–88.

Wissenbach, D.K., et al., 2016. Optimization of metabolomics of defined in vitro gut microbial ecosystems. Int. J. Med. Microbiol. 306 (5), 280–289.

Wodarz, D., 2006. Ecological and evolutionary principles in immunology. Ecol. Lett. 9 (6), 694–705.

Wolfe, B.E., Dutton, R.J., 2015. Fermented foods as experimentally tractable microbial ecosystems. Cell 161 (1), 49–55.

Wylie, K.M., Weinstock, G.M., Storch, G.A., 2013. Virome genomics: a tool for defining the human virome. Curr. Opin. Microbiol. 16 (4), 479–484.

Xiang, J., et al., 2001. Effect of coinfection with GB virus C on survival among patients with HIV infection. N. Engl. J. Med. 345 (10), 707–714.

Yager, E.J., et al., 2009. γ-Herpesvirus-induced protection against bacterial infection is transient. Viral Immunol. 22 (1), 67–71.

Ye, H., Sugihara, G., 2016. Information leverage in interconnected ecosystems: overcoming the curse of dimensionality. Science 353 (6302), 922–925.

Yen, S., et al., 2015. Metabolomic analysis of human fecal microbiota: a comparison of feces-derived communities and defined mixed communities. J. Proteome Res. 14 (3), 1472–1482.

Yezli, S., Otter, J.A., 2011. Minimum infective dose of the major human respiratory and enteric viruses transmitted through food and the environment. Food Environ. Virol. 3 (1), 1–30.

Zaiss, M.M., Harris, N.L., 2016. Interactions between the intestinal microbiome and helminth parasites. Parasite Immunol. 38 (1), 5–11.

Ze, X., Mougen, F.L., Duncan, S.H., et al., 2013. Some are more equal than others. Gut Microbes 4 (3), 236–240.

Zhang, C., et al., 2016. Ecological robustness of the gut microbiota in response to ingestion of transient food-borne microbes. ISME J., 10, 2235–2245.

Zhou, J., et al., 2010. Functional molecular ecological networks. mBio 1 (4), e00169-10.

Zhou, Y., et al., 2014. Exploration of bacterial community classes in major human habitats. Genome Biol. 15 (5), R66.

ADVANCES IN ECOLOGICAL RESEARCH VOLUME 1-57

 CUMULATIVE LIST OF TITLES

Aerial heavy metal pollution and terrestrial ecosystems, **11**, 218

Age determination and growth of Baikal seals (*Phoca sibirica*), **31**, 449

Age-related decline in forest productivity: pattern and process, **27**, 213

Allometry of body size and abundance in 166 food webs, **41**, 1

Analysis and interpretation of long-term studies investigating responses to climate change, **35**, 111

Analysis of processes involved in the natural control of insects, **2**, 1

Ancient Lake Pennon and its endemic molluscan faun (Central Europe; Mio-Pliocene), **31**, 463

Ant-plant-homopteran interactions, **16**, 53

Anthropogenic impacts on litter decomposition and soil organic matter, **38**, 263

Arctic climate and climate change with a focus on Greenland, **40**, 13

Arrival and departure dates, **35**, 1

Assessing the contribution of micro-organisms and macrofauna to biodiversity-ecosystem functioning relationships in freshwater microcosms, **43**, 151

A belowground perspective on Dutch agroecosystems: how soil organisms interact to support ecosystem services, **44**, 277

The benthic invertebrates of Lake Khubsugul, Mongolia, **31**, 97

Big data and ecosystem research programmes, **51**, 41

Biodiversity, species interactions and ecological networks in a fragmented world **46**, 89

Biogeography and species diversity of diatoms in the northern basin of Lake Tanganyika, **31**, 115

Biological strategies of nutrient cycling in soil systems, **13**, 1

Biomanipulation as a restoration tool to combat eutrophication: recent advances and future challenges, **47**, 411

Biomonitoring of human impacts in freshwater ecosystems: the good, the bad and the ugly, **44**, 1

Bray-Curtis ordination: an effective strategy for analysis of multivariate ecological data, **14**, 1

Species abundance patterns and community structure, **26**, 112

Stochastic demography and conservation of an endangered perennial plant (*Lomatium bradshawii*) in a dynamic fire regime, **32**, 1

Stomatal control of transpiration: scaling up from leaf to regions, **15**, 1

Stream ecosystem functioning in an agricultural landscape: the importance of terrestrial–aquatic linkages, **44**, 211

Structure and function of microphytic soil crusts in wildland ecosystems of arid to semiarid regions, **20**, 180

Studies on the cereal ecosystems, **8**, 108

Studies on grassland leafhoppers (Auchenorrhbyncha, Homoptera) and their natural enemies, **11**, 82

Studies on the insect fauna on Scotch Broom *Sarothamnus scoparius* (L.) Wimmer, **5**, 88

Sustained research on stream communities: a model system and the comparative approach, **41**, 175

Systems biology for ecology: from molecules to ecosystems, **43**, 87

The study area at Zackenberg, **40**, 101

Sunflecks and their importance to forest understorey plants, **18**, 1

A synopsis of the pesticide problem, **4**, 75

The temperature dependence of the carbon cycle in aquatic ecosystems, **43**, 267

Temperature and organism size – a biological law for ecotherms? **25**, 1

Terrestrial plant ecology and ^{15}N natural abundance: the present limits to interpretation for uncultivated systems with original data from a Scottish old field, **27**, 133

Theories dealing with the ecology of landbirds on islands, **11**, 329

A theory of gradient analysis, **18**, 271; **34**, 235

Throughfall and stemflow in the forest nutrient cycle, **13**, 57

Tiddalik's travels: the making and remaking of an aboriginal flood myth, **39**, 139

Towards understanding ecosystems, **5**, 1

Tradeoffs and compatibilities among ecosystem services: biological, physical and economic drivers of multifunctionality, **54**, 207

Trends in the evolution of Baikal amphipods and evolutionary parallels with some marine Malacostracan faunas, **31**, 195

Trophic interactions in population cycles of voles and lemmings: a model-based synthesis **33**, 75

Edwards Brothers Malloy
Ann Arbor MI. USA
March 22, 2017